Pesticide Transformation Products

ACS SYMPOSIUM SERIES 459

Pesticide Transformation Products
Fate and Significance in the Environment

L. Somasundaram, EDITOR
Iowa State University
Joel R. Coats, EDITOR
Iowa State University

Developed from a symposium sponsored
by the Division of Agrochemicals
at the 200th National Meeting
of the American Chemical Society,
Washington, DC,
August 26–31, 1990

American Chemical Society, Washington, DC 1991

Library of Congress Cataloging-in-Publication Data

Pesticide transformation products: fate and significance in the environment/ L. Somasundaram, editor, Joel R. Coats, editor.

    p.   cm.—(ACS symposium series, ISSN 0097–6156; 459)

"Developed from a symposium sponsored by the Division of Agrochemicals at the 200th National Meeting of the American Chemical Society, Washington, D.C., August 26–31, 1990."

Includes bibliographical references and indexes.

ISBN 0–8412–1994–X

1. Pesticides—Environmental aspects. 2. Pesticides—Biodegradation.

I. Somasundaram, L., 1961–   . II. Coats, Joel R. III. American Chemical Society. Division of Agrochemicals. IV. American Chemical Society. Meeting (200th: 1990: Washington, D.C.) V. Series.

TD196.P38P47   1991
628.5′2—dc20   91–2034
                             CIP

The paper used in this publication meets the minimum requirements of American National Standard for Information Sciences—Permanence of Paper for Printed Library Materials, ANSI Z39.48–1984.   ∞

Copyright © 1991

American Chemical Society

ACS Symposium Series

M. Joan Comstock, *Series Editor*

1991 ACS Books Advisory Board

Foreword

THE ACS SYMPOSIUM SERIES was founded in 1974 to provide a medium for publishing symposia quickly in book form. The format of the Series parallels that of the continuing ADVANCES IN CHEMISTRY SERIES except that, in order to save time, the papers are not typeset, but are reproduced as they are submitted by the authors in camera-ready form. Papers are reviewed under the supervision of the editors with the assistance of the Advisory Board and are selected to maintain the integrity of the symposia. Both reviews and reports of research are acceptable, because symposia may embrace both types of presentation. However, verbatim reproductions of previously published papers are not accepted.

Contents

SIGNIFICANCE

INDEXES

Preface

PESTICIDES HAVE BECOME an integral part of intensive agriculture. This has resulted in extensive research on the biological efficacy and environmental fate of pesticides. One area of significant interest is the transformation of pesticides in the environment. The mechanisms of pesticide degradation and the products formed from various physical, chemical and biological processes have been well documented. For most currently used pesticides, however, the fate and significance of their transformation products are yet to be elucidated.

Pesticide transformation, in general, is a detoxification process resulting in inert products. Some transformation products, however, have the potential to control target pests, affect nontarget species, and contaminate environmental resources. An extensive body of literature exists on transformation products such as DDE, heptachlor epoxide, 2,4-dichlorophenol, and ethylenethiourea. Some types of information on transformation products, generated by the chemical industry, are proprietary in nature and are not currently available to the scientific community. Little is known about transformation products of the vast majority of the more than 600 active ingredients used in crop protection, however.

Although numerous books have been published on pesticides, this is the first book on pesticide transformation products. Overviews of the current understanding of the pesticide degradation mechanisms and products are discussed in the first two chapters. The remaining chapters are organized into two sections, the first section focusing on the fate of transformation products in the physical and biological environment, and the second addressing their significance in crop protection, environmental contamination, nontarget effects, and legal implications. Overall, the book presents the benefits and risks associated with pesticide transformation products. Inevitably, some of the viewpoints in the book are controversial, but we consider that to be beneficial for the promotion of concepts and approaches to the study of pesticide transformation products.

A tradition of research on pesticide metabolism was established at Iowa State University by Paul A. Dahm, Distinguished Professor in the Department of Entomology and many of his graduate students. Dr. Dahm was a pioneer in the field of comparative insecticide toxicology, particularly in elucidation of the mechanisms of biotransformation. During his 37 years at this university, he was instrumental in adapting and developing powerful new methods for the study of pesticide persistence, distribution, and breakdown products, including radiotracers and gas chromatography. We acknowledge his many contributions to pesticide degradation and remember him fondly as a scientist, teacher, and friend.

The editors thank the contributors to this volume for their excellent research reviews of current knowledge on pesticide transformation products as well as some of the research and regulatory issues that face individuals and institutions working in this important field. We are grateful for the efforts and expertise of the scientists who served as peer reviewers for the chapters published in this volume. We hope this collective work will serve as a valuable focus of research and opinion on the subject of pesticide transformation products and will stimulate future research and regulatory policies.

We express our appreciation to the Agrochemicals Division of the American Chemical Society for providing a forum for this work. We also thank John Teeple, Amy Marsh, Ellen Kruger, and Maureen Rouhi for their valuable assistance in the preparation of this book.

We dedicate this book to our parents, Nachammai and Lakshmanan Somalay, and William G. and Catherine (Dodds) Coats, for their love, support, and keen interest in our professional achievements.

L. SOMASUNDARAM
Iowa State University
Ames, Iowa 50011

JOEL R. COATS
Iowa State University
Ames, Iowa 50011

December 26, 1990

OVERVIEWS

Chapter 1

Pesticide Transformation Products in the Environment

L. Somasundaram and Joel R. Coats

Pesticide Toxicology Laboratory, Department of Entomology, Iowa State University, Ames, IA 50011

Pesticides applied in the environment are transformed by biological or nonbiological processes into one or more transformation products. For most pesticides, transformation results in detoxification to innocuous products. Major degradation products of some currently used pesticides, however, play an important role in pest control and environmental contamination. Some pesticide degradation products are of significance in crop protection by being effective against the target pests. Some can be responsible for inadequate pest control by inducing rapid degradation of their parent compounds. Degradation products as potential contaminants of environmental and food resources has been reported recently. Although most of the currently used pesticides are biodegradable, their major degradation products should also be considered in evaluating the overall bioactivity and environmental contamination potential of the parent compound.

Role of Pesticides in Agriculture and Public Health

The world population continues to grow at about 2 percent each year (1). This growth rate means that at least 93 million additional people per year must be provided with food. The 43 countries identified by the United Nations as the food-priority countries have the highest birth rates (2), and the current global population of 5.3 billion is expected to increase to 6.3 billion by the year 2000 (3). As global population is incrasing, the total available agricultural land is decreasing (4), mainly because of soil erosion (5). The need to increase the global food supply, and the constraints faced in the population and the land availability fronts reflect the importance of intensive agriculture in providing food to the global population. Pesticides have been an integral part of intensive agriculture, particularly since the Green Revolution in Southeast Asia in the 1970's. Besides their role in crop and animal

0097–6156/91/0459–0002$06.00/0

protection, pesticides also play a vital role in public health. In fact, one of the first such uses of synthetic pesticides was that of DDT for the control of typhus and malaria outbreaks in the 1940's and 1950's (6).

Pesticide Transformation

Pesticide transformation is any process in which a change takes place in the molecular structure of a pesticide. The transformation of a pesticide can occur immediately after, or even before, application, during storage. Most pesticides applied to the environment are ultimately degraded into universally present materials such as carbon dioxide, ammonia, water, mineral salts, and humic substances. Different chemicals, however, are formed before the pesticides are completely degraded. The chemicals formed by the different transformation processes are referred to by several names (Table I). "Transformation products" and "degradation products" are the terms most widely used. If the products are results of biological degradation, then they are referred to as "metabolites." The products formed by sunlight-induced transformations are known as "photoproducts" or "photolysis products." Some products such as phorate sulfoxide are referred to as "pesticidal products" because of their toxicity to target pests. The term "residues" is ambiguously used and can mean either parent compound or the products formed.

Table I. Some of the terms used to refer to products formed from pesticide transformation

---

Transformation products
Biotransformation products
Degradation products
Metabolites
Photoproducts
Photolysis products
Pesticidal products
Byproducts
Breakdown products
Conversion products
Residues
Derivatives
Analytes
Analogs
Intermediates
Degradates
Daughter compounds
Decomposition products
Pesticide equivalent

For the bioactivity of some pesticides to be expressed, they need to first be transformed into biologically active products. These pesticides known as propesticides (Table II), are relatively safer to mammals and crops because selective detoxification can take place in crop or mammalian systems.

Table II. Commonly used propesticides and their biologically active products

Propesticide	Bioactive product	Effect
2,4-DB	2,4-D	herbicide
methazole	DCPMV	herbicide
metham-sodium	methyl isothiocyanate	fumigant
carbosulfan	carbofuran	insecticide
thiodocarb	methomyl	insecticide
tralomethrin	deltamethrin	insecticide

Significance of Transformation Products

Many modern day pesticides are degraded by microbial or chemical processes into innocuous products. Evidence shows, however, that for some pesticides, the pesticidal activity and environmental contamination attributed to parent compounds are partly due to the products formed (7,8). In some situations, pesticides (e.g., aldoxycarb, promecarb) are formed as degradation products of other pesticides (Table III).

Table III. Pesticides formed as degradation products of other pesticides

Parent compound	Degradation product (also a pesticide)
DDT	dicofol
aldicarb	aldoxycarb (aldicarb sulfone)
acephate	methamidaphos
maneb, zineb	etem
promacyl	promecarb
benomyl	carbendazim
chlorthiamid	dichlobenil

The significance of degradation products was illustrated four decades ago for DDT, one of the first synthetic pesticides. Some of the major concerns with DDT were its degradation products: DDE (1,1-dichloro-2-2 bis [chlorophenyl]ethylene) and DDD (1,1 bis[p-chlorophenyl]-2,2-dichloroethane). For more information on the environmental significance of DDT metabolites readers are referred to Metcalf (9). Although the importance of degradation products has long been known, their significance became more evident in the late 1980's, because of their potential to contaminate water and food resources (8,10,11) and to promote rapid degradation of parent compounds by serving as energy or nutrient sources for soil microorganisms (12).

Pesticide degradation results in the formation of various chemicals including substituted phenols, aromatic amines, and chlorobenzenes. Some of these products, such as 2,4-dichlorophenol and p-nitrophenol, are considered priority pollutants. The toxicity of these compounds in the biosphere and their potential to persist in environmental matrices for extended periods illustrates the need to study these compounds (13,14). It is equally important to understand the fate of biologically active products formed from propesticides and pesticides so that adequate pest control is achieved.

Biological Activity. The biological activity of pesticide degradation products is not new to the literature. Two of the metabolites of DDT -DDD and dicofol- were effective against insects and mites, respectively, and were marketed as pesticides. The degradation products of some organophosphorus insecticides are more potent inhibitors of acetylcholinesterase, a major component of the nervous system, than their respective parent compounds (e.g., paraoxon from parathion). Besides cholinesterase inhibition, some metabolites also inhibit the activity of other enzymes such as mammalian ribonucleotide reductase (15). The herbicidal activity of the sulfoxides of carbomothioate herbicides has been previously reported (7,16). In the U.S., in Midwest soils treated with pesticides such as butylate and terbufos, residues of the parent compound were not detected in significant amounts within a few weeks after application. Good control of weeds and corn rootworm larvae, however, was observed in these fields (7, Somasundaram, L., unpublished data.). The good performance of these pesticides, despite their lack of persistence, was probably due in part to subsequently formed pesticidal products (Table IV).

Table IV. Degradation products effective in controlling target pests

Pesticidal Product	Pesticide
DDD, dicofol	DDT
EPTC sulfoxide	EPTC
butylate sulfoxide	butylate
terbufos sulfoxide	terbufos
phorate sulfoxide	phorate

Transformation products as environmental contaminants. The growing concern for preserving the quality of surface-water and groundwater resources stimulated research into the occurrence of pesticides in groundwater in the 1980's. Several pesticides have been frequently detected in groundwater samples in agricultural areas. Until recently, however, monitoring pesticides in groundwater was mainly restricted to the parent compound. Studies on the quality of Long Island's groundwater revealed the primary contaminants to be aldicarb sulfoxide and aldicarb sulfone, which are the degradation products of the insecticide/nematicide aldicarb (17). Since then, degradation products of a number of commonly used pesticides, including atrazine, cyanazine, and carbofuran, have been detected in groundwater (Table V). In the USEPA's recent national survey of drinking water wells, the most frequently detected pesticide product was a metabolite of DCPA (18). Recent studies confirm that many degradation products are more mobile, and some others are more persistent than their respective parent compounds (19,20). In some monitoring studies, degradation products occured in groundwater in which the parent compound was no longer detected (21). These findings reveal that degradation products of certain pesticides could be important contaminants of groundwater. For most pesticides, however, the concentrations of their degradation products in groundwater was well below the USEPA's Maximum Contaminant Level.

Table V. Pesticide degradation products as contaminants of water resources

Pesticide	Degradation product	Reference
DDT	DDE, DDD	(22)
endosulfan	endosulfan sulfate	(23)
carbofuran	3-ketocarbofuran	(24)
	3-hydroxycarbofuran	
aldicarb	aldicarb sulfoxide	(17)
	aldicarb sulfone	
atrazine	deethylatrazine	(24)
	deisopropylatrazine	
cyanazine	cyanazine amide	(25)
simazine	deethylsimazine	(21)
alachlor	hydroxyalachlor	(24)
DCPA	tetrachloroterephthalic acid	(18)
maneb, mancozeb	ethylenethiourea	(22)

In contrast, conversion of some pesticides to their degradation products could prevent the movement of parent compound to groundwater. For example, the greatest potential for groundwater contamination by dicamba would occur immediately after application, before significant degradation to 3,6-dichlorosalicylic acid (DCSA) occurred. This is because of the increased sorption of DCSA by soil (26).

The recent controversy over the presence of the growth regulator Alar in apples in the United States was not caused by the parent compound daminozide (27). A metabolite of daminozide, unsymmetrical dimethylhydrazine (UDMH), was the chemical of concern present in the contaminated apples. Unsymmetrical dimethylhydrazine is a potent carcinogen, and its presence in apples resulted in major public concern that led to the withdrawal of daminozide's use in apples.

Ethylenethiourea (ETU) is a degradation product of ethylenebisdithiocarbamate (EBDC) fungicides such as maneb, zineb, and mancozeb. The source of ETU in plants can be the ETU present in formulations as an impurity produced during EBDC manufacture or formed during storage (11). Soil degradation of dithiocarbamates can result in the formation of ETU, which can be taken up by plants. The presence of ETU in food can be the result of heat treatment of plant products containing residues of EBDC. Ethylenethiourea has been shown to possess carcinogenic, teratogenic, and mutagenic activities and has been detected in groundwater samples (22). Because of these properties of ETU, EBDC fungicides are under special review by the USEPA.

Properties of Pesticide Degradation Products

Characteristics influencing the environmental significance of pesticide degradation products include water solubility, vapor pressure, and carcinogenic and mutagenic potential. Although most degradation products of pesticides are converted into less toxic or nontoxic materials, some degradation products because of the above mentioned characteristics, may be biologically and/or environmentally active. In general, many degradation products are more soluble in water than their parent compounds (19). For example, the sulfoxide and sulfone metabolites of aldicarb are 55 and 1.4 times more soluble respectively, than the parent aldicarb (28). This increased solubility favors their mobility to groundwater.

Vapor pressure is an important factor influencing the evaporation of pesticides from treated surfaces. The vapor pressure of PCCH (r-1,3,4,5,6,pentachlorocyclohexane), a metabolite of lindane, was 14 times greater than that of lindane (29). Thus, much of the lindane volatilized from soil would be in the form of PCCH. The vapor pressure of p-p'-DDE is 8 times greater than that of p-p'-DDT. In well-aerated soils, DDT is converted to DDE, and most of the volatilization occurs in the form of DDE (29). Thus, because of the high vapor pressure, metabolites of DDT and lindane volatilize more readily than do their parent compounds.

Because of the public concern over the spectra of agrochemically-induced carcinogenicity, use patterns are influenced by their potential human-health concerns. A few degradation products, including ETU and UDMH, are carcinogenic. The rates at which these products are formed, their concentrations in food, or their persistence in environmental resources will have a significant impact on the future use of the parent compounds.

Conclusions

Pesticide transformation is mainly a beneficial process resulting in detoxification of the parent compound. For some pesticides, however, the products formed can be of significance in both crop protection and environmental contamination. Thus, as with the case of EBDC fungicides, the transformation products could determine the future use of the parent compound. Yet, for many of the currently used pesticides, the fate and significance of their degradation products is not clearly understood. The chapters in this book present the available information on pesticide transformation products and address the relative significance of these products in the environment.

Acknowledgments

We thank the United States Department of Agriculture's Management
System Evaluation Area Program and North Central Region Pesticide
Impact Assessment Program, and Leopold Center for Sustainable
Agriculture for funding our research on pesticide degradation
products. Journal Paper No. J-14298 of the Iowa Agriculture and
Home Economics Experiment Station, Ames, Iowa, Project No. 2306.

Literature Cited

1. *Population Today* 1990, *18*, (11) 9. Population Resource Bureau
   Inc. Washington, D.C.
2. Knusli, E. In *World Food Production - Environment - Pesticides*;
   Geissbuhler, Ed.; Pergamon Press: New York, NY, 1978.
3. *Statistical Abstract of the United States*, 110th Edition.   The
   National data book, U.S. Dept. of Commerce, Bureau of the
   Census. Washington, D.C., 1990; p 830.
4. Kraus, P. In *Pesticides: Food and Environmental Implications*;
   1987, International Atomic Energy Agency, Vienna, Austria, pp
   1-9.
5. Woolman, M.G. In *Soil Erosion and Crop Productivity*, Follett,
   R.F.; Stewart, B.A., Eds., American Society of Agronomy, Inc.:
   Madison, WI, 1985, pp 9-21.
6. McEwen, F.L.; Stephenson, G.R. *The Use and Significance of
   Pesticides in the Environment*; John Wiley and Sons: New York,
   NY, 1979; pp 1-7.
7. Tuxhorn, G.L.; Roeth, F.W.; Martin, A.R.; Wilson, R.G. *Weed
   Sci.* 1986, *34*, 961-965.
8. Hallberg, G.R. *Agric. Ecosyst. Environ.* 1989, *26*, 299-367.
9. Metcalf, R.L. *J. Agric. Food Chem.* 1973, *21*, 511-519.
10. Leistra, M. Boesten, J.J.T.I.; *Agric. Ecosyst. Environ.* 1989,
    26, 369-389.
11. Lentza-Rizos, C.H. *Rev. Environ. Contam. Toxicol.* 1990, *115*,
    1-37.
12. Somasundaram, L.; Coats, J.R.; Racke, K.D.; *J. Environ. Sci.
    Health*, 1989, *B24*, 457-478.
13. Kaiser, K.L.E.; Dixon, D.G.; Hodson, P.V. In *QSAR in
    Environmental Toxicology;* Kaiser, K.L.E., Ed.; D. Reidel:
    Dordrecht, Holland, 1984, pp 189-206.
14. MacRae, I.C.; Cameron, A.I. *Appl. Environ. Microbiol.* 1985, *49*,
    236-237.
15. Wright, J.A.; Hermonat, M.W.; Hards, R.G. *Bull. Environ.
    Contam. Toxicol.* 1982, *28*, 480-483.
16. Casida, J.E.; Gray, R.A.; Tilles, H. *Science* 1974, *184*, 573-
    574.
17. Cohen, S.Z.; Creeger, S.M.; Carsel, R.F.; Enfield, C.G. In
    Treatment and Disposal of Pesticide Wastes; Krueger, R.F.;
    Seiber, J.N. Eds.; Am. Chem. Soc: Washington, D.C., 1984, pp
    297-325.
18. U.S.E.P.A. National Pesticide Survey, Phase I Report, 1990
    National Technical Information Services, Alexandria, VA.
19. Somasundaram, L.; Coats, J.R.; Shanbhag, V.M.; Racke, K.D.
    Environ. Tox. Chem. 1991, *10*, 185-194.

20. Khan, S.U.; Saidak, W.J. *Weed Research* 1981, *21*, 9-12.
21. Ontario Ministry of the Environment, 1987. *Pesticides in Ontario Drinking Water-1986;* Ont. Min. Environ. Rep. (Water Resources Branch) Toronto, Canada.
22. Holden, P. *Pesticides and Groundwater Quality: Issues and Problems in Four States;* Natl. Res. Counc., Bd. Agric. Natl. Acad. Press: Washington, D.C., 1986, p 124.
23. Frank, R.; Braun, H.E.; Van Hove Holdrinet, M.; Sirons, G.J; Ripley, B.D.   *J. Environ. Qual.* 1982, *11*, 497-505.
24. *Iowa State-Wide Rural Well-Water Survey. Summary of results: Pesticide detections;* Iowa Department of Natural Resources, Des Moines, Iowa, 1990.
25. Muir, D.C.; Baker, B.E. *J. Agric. Food Chem.* 1976, *24*, 122-125.
26. Murray, M.R.; Hall, J.K. *J. Environ. Qual.* 1989, *18*, 51-57.
27. Chemical and Engineering News, March 13, 1989. Am. Chem. Soc.: Washington, D.C.
28. Hornsby, A.G. Rao, P.S.C., Wheeler, W.B., Nkedi-Kizza, P., Jones, R.L. In *Proceedings of the NWWA/USEPA Conference on Characterization and Monitoring of the Vadose Zone*, Nielson, D.M. and Curl, M., Eds.; Natl. Water Well Assoc., Worthington, OH, 1983, pp 936-958.
29. Cliath, M.M.; Spencer, W.F. *Environ. Sci. Technol.* 1972, *6*, 910-914.

RECEIVED December 11, 1990

Chapter 2

Pesticide Degradation Mechanisms and Environmental Activation

Joel R. Coats

Pesticide Toxicology Laboratory, Department of Entomology, Iowa State University, Ames, IA 50011

Pesticides are degraded by many different mechanisms. Physical, chemical, and biological agents play significant roles in the transformation of insecticide, herbicide, and fungicide molecules to various degradation products. Transformation mechanisms include oxidation, hydrolysis, reduction, hydration, conjugation, isomerization, and cyclization. Resultant products are usually less bioactive than the parent psticide molecule, but numerous cases have been documented of metabolites with greater bioactivity. The physical and chemical properties of the degradation products are also different from those of the parent compound, and their fate and significance in the environment also are altered with the structural changes. The concept of "environmental activation" is introduced, to describe the transformation of a pesticide to a degradation product that is of significance in the environment as a result of its environmental toxicology or chemistry.

Degradation of synthetic organic pesticides begins as soon as they are synthesized and purified. Formulation processes can initiate minor amounts of decomposition of active ingredients. During shipping and storage of products, breakdown of the principal components may occur due to harsh environmental conditions (e.g., severe heat) or prolonged periods of storage. Once a pesticide product is prepared as a tank mix, further degradation can occur because of chemical interactions between the

0097–6156/91/0459–0010$06.25/0
© 1991 American Chemical Society

parent molecules and the water used or other pesticides in the tank mixture. Once applied at a site for pest control, the pesticidal compound is exposed to numerous agents capable of transforming it into various other forms. Once inside organisms, both target and nontarget, they are subject to attack by detoxification enzymes. The majority of the pesticide applied does not immediately enter any organism, but remains in the environment. These residues in soil, water, and air are subject to transformation, to transport to a different location, and to uptake by organisms at that site. Once transformation products have been formed in any of the circumstances described, these new compounds are of relevance, as well, and are susceptible to the same forces of movement and/or further degradation.

Agents Involved in Degradation

Processes responsible for the degradation of pesticides can be classified as physical, chemical, and biological. A combination of factors usually influence the breakdown of a pesticidal chemical, over a time period of hours, days, weeks, months, or years, and through many of the situations enumerated above. Factors that influence the relative importance of the various transformation agents depend to a great extent on the chemical's use pattern, physical properties, and chemical structure.

The two primary physical agents involved in the degradation process are light and heat. Photolysis of pesticide residues is extremely significant on vegetation, on the soil surface, in water, and in the atmosphere (1). Thermal decomposition of the chemicals often occurs concomitantly with the photodegradative reactions; solar radiation, therefore, is directly responsible for the decomposition in two ways, through photolysis and thermal decomposition. Cold, especially freezing, temperatures can also contribute occasionally to pesticide degradation if certain formulations are allowed to freeze, and the pesticide is forced out of solution, suspension, or microencapsulation, making it more susceptible to degradative forces before and after application.

Chemical degradation occurs as a result of the various reactive agents in the formulations, tank mixes, and in the environment. Water is responsible for considerable breakdown of pesticides in solution, especially in conjunction with pH extremes. Even slight variance from a neutral pH can elicit rapid decomposition of pH-sensitive compounds. Molecular oxygen and its several more reactive forms (e.g., ozone, superoxide, peroxides) are capable of reacting with many chemicals to generate oxidation products. In all but the most oxygen-poor environments, oxidative transformations are frequently the most common degradation pathways observed. Other chemical oxidations, as well as reductions, can progress in the presence of certain inorganic redox reagents (2). These reactive species of metals function as highly effective catalysts which effect pesticide transformations in soils and in aquatic and marine environments.

Biological agents are also significant degraders of pesticides. Microorganisms (bacteria, fungi, actinomycetes) are the most important group of degraders, based on their prevalence in soil and water and on the tendency of pesticides to collect in those medias (3,4). Three major degradation strategies are exhibited by microbes:

1. Co-metabolism is the biotransformation of a pesticidal molecule coincidental to the normal metabolic functions of the microbial life processes (growth, reproduction, dispersal).
2. Catabolism is the utilization of an organic molecule as a nutrient or energy source. Some pesticides have been observed to be susceptible to enhanced microbial degradation (EMD) by populations of microbes (principally bacteria) that have adapted, following repeated use, to utilize the pesticide molecule as a sole source of carbon or nitrogen (5). Control of pests by these products may be seriously compromised under some circumstances (e.g., a need for weeks of persistence in soil (6). The major factors that affect the susceptibility or resistance of a chemical to EMD are the relative ease of hydrolysis, plus the nutritional value of the hydrolysis products (7), as well as the toxicity of those products to microbes in the soil (8). This specialization of biodegradation has also been exploited for bioremediation processes (9).
3. Microorganisms secrete enzymes into the soil for digestion of substrates. The enzymes, e.g., phosphatases and amidases, may persist in the soil long after the parent microbial cell is dead. The stability and persistence of these extracellular enzymes can provide soil with various important biochemical catalytic capabilities (10).

Other biological degradation agents include invertebrates, vertebrates, and plants. In general, vertebrates possess the most sophisticated enzymatic arsenal for biotransformation of xenobiotics that enter their bodies. In addition to the broader spectrum of reactions possible in vivo, the rates of detoxification and elimination are typically highest in this group, especially mammals and birds.

Transformation Reactions

Three basic types of reactions can occur: degradative, synthetic (e.g., conjugations), and rearrangements.

Although most organic pesticides must undergo many reactions before they are completely degraded, or mineralized, one or two transformations are frequently sufficient to alter the biological activity drastically. The change usually is a detoxification, but numerous examples of intoxication, or activation, have been reported as well (11).

The evolved biochemical strategies of organisms for detoxification of organic foreign substances are four-fold. They transform the xenobiotic molecule to a more water-soluble, more polar molecule to facilitate elimination of it from the organism. Oxygen or oxygen-containing moieties like sugars, amino acids, sulfates, and phosphates serve this purpose well. Secondly, they alter functional groups in an attempt to render the molecule less toxic (e.g., amines and sulfhydryls). The third approach entails breaking of the potentially toxic molecule into two or more fragments to decrease the probability of toxic impacts of the chemical on vital organs or organelles. In addition, the breaking down process also contributes toward possible use of the component parts of the molecule as nutrients, as mentioned regarding microbial catabolism.

Degradative Reactions

Oxidations. A wide spectrum of oxidative reactions occur in organisms and in the greater environment. Nearly all impart some degree of increased water solubility to the pesticide molecule. Most also alter the bactivity of the parent compound. For their impact on the bioactivity and mobility, and, hence, the environmental significance of the chemicals, the oxidation pathways are extremely important transformation reactions.

The mechanisms of oxidation vary: physical and chemical oxidations involve molecular oxygen or more reactive species including various acids, peroxides, or singlet oxygen, and are often enhanced by light, heat, or oxidized metals in soil or water. Biological oxidations are accomplished enzymatically, primarily by the mixed function oxidases (utilizing cytochrome P_{450}) (12,13), the flavin-dependent monooxygenases (14,15), and monoamine oxidases (12). The reactive forms of oxygen that are involved in the various oxidative biotransformations have been reported to be hydroxyl radical, superoxide anion, and hydrogen peroxide. Specific oxidative reactions of significance in pesticide degradation are presented below.

Ring Hydroxylation

carbaryl   4-hydroxy carbaryl   (16)

This reaction occurs on aromatic rings in pesticides of
several types, as well as polycyclic aromatic
hydrocarbons (PAH). It is easily accomplished on single
or multiple-ring molecules that are unsubstituted or
activated. Substitution with bulky or multiple groups
can sterically inhibit the reaction, as can the presence
of deactivating substituents such as halogens.

Side-chain oxidation

diazinon   hydroxy diazinon   (17)

Any methylene group in an aliphatic side-chain is
susceptible to oxidation. Carbofuran undergoes
hydroxylation of a methylene in the furan ring to form 7-
hydroxycarbofuran, followed by further oxidation to 7-
ketocarbofuran (18). Terminal methyl groups are also
susceptible to hydroxylation.

Epoxidation

aldrin   dieldrin   (19)

The conversion of alkenes to epoxides does not radically alter the polarity of the substrate, nor the biological activity in some cases, e.g., aldrin's epoxidation to dieldrin. Epoxide formation may be, in fact, critical to bioactivity, e.g., precocenes require activation to be cytotoxic to the corpora allatum of insects to inhibit the production of juvenile hormone (20). Epoxides can be hydrolyzed to form diols; the net effect of the arene oxide formation followed by hydrolysis, is ring hydroxylation.

O-dealkylation

methoxychlor   desmethyl methoxychlor (21)

Dealkylation of groups larger than methyl also occurs, primarily via an α-hydroxylation, followed by the cleavage of the ether bond. Relative rates of O-dealkylation can be quite species-specific (22).

N-dealkylation

atrazine   desethyl atrazine   (23)

In this example, either or both of the alkyl groups can be oxidatively removed. N-Demethylation is an important step in the activation of chlordimeform (24).

S-dealkylation

prothiophos   despropyl prothiophos   (25)

Di- and tri-alkyl thiophosphate (and thiophosphoramidate) esters also are susceptible to isomerization which entails migration of an alkyl group (12).

Oxidative desulfuration

fonofos   fonofoxon   (26)

This reaction is requisite of nearly all thionophosphates for activation to highly effective inhibitors of acetyl-cholinesterase (AChE). It occurs in vivo primarily, through an oxidative attack on the sulfur atom, as demonstrated with the metabolism of the R and S optical isomers of fonofos, since the oxons (P=O) formed retained the stereochemistry of the parent phosphorothion (27).

Sulfoxidation

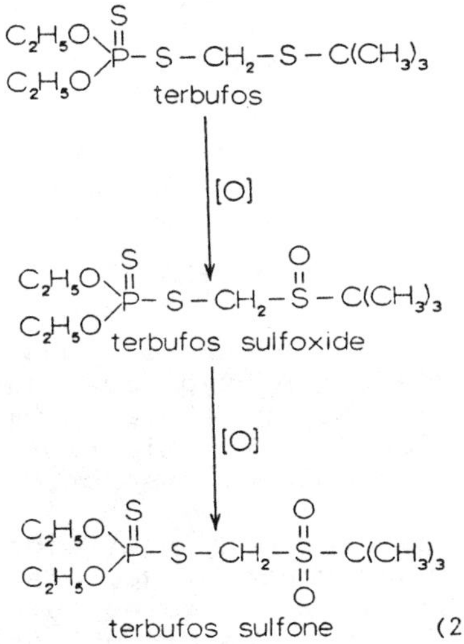

$$C_2H_5O \underset{C_2H_5O}{\overset{S}{\diagdown}} P - S - CH_2 - S - C(CH_3)_3$$
terbufos

[O]

$$C_2H_5O \underset{C_2H_5O}{\overset{S}{\diagdown}} P - S - CH_2 - \overset{O}{\overset{\|}{S}} - C(CH_3)_3$$
terbufos sulfoxide

[O]

$$C_2H_5O \underset{C_2H_5O}{\overset{S}{\diagdown}} P - S - CH_2 - \underset{\underset{O}{\|}}{\overset{\overset{O}{\|}}{S}} - C(CH_3)_3$$

terbufos sulfone   (28)

The oxidation of sulfides to the sulfoxide and sulfone is often rapid. Sulfoxide formation is the result of FAD monooxygenase enzymes in vivo, while sulfones are formed by the cytochrom P_{450} monooxygenase system. The two forms can be generated chemically by peroxides and peracids, respectively. It is often the path of least resistance to degradation, and it results in oxidative metabolites that are often of biological significance. Aromatic alkyl thioethers, e.g., measurol (17) and methiochlor (29), are also susceptible to sulfoxidation. Thiomethyl metabolites of chlorinated aromatic hydrocarbons, formed via a mercapturic acid intermediate, can also be oxidized to the sulfinyl and sulfonyl derivatives (30).

Amine oxidation

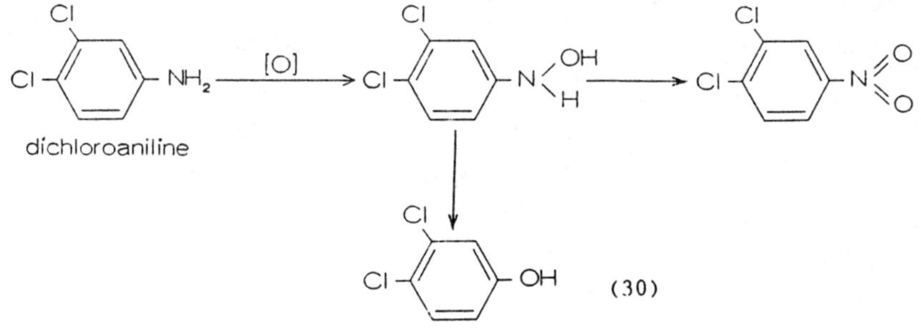

dichloroaniline

(30)

Arylamine N-oxidation can occur via monoamine oxidases or the CytoP-450-dependent monooxygenases (12). The hydroxylamine form is very reactive in vivo (31) and can undergo deamination or further oxidation to form the more stable phenol or nitrobenzene, respectively.

Hydrolysis reactions. For many pesticide molecules, hydrolysis is a primary route of degradation. Many types of esters are hydrolytically cleaved, yielding two fragments with little or no pesticidal activity. The products do possess bioactivity of various other types, however, e.g., microbial toxicity (8) or induction of microbial growth (7). Hydrolysis of esters can occur by chemical means; even mildly alkaline solutions can cause hydrolytic decomposition of some esters (organophosphorus, carbamate, pyrethroid), while acid-catalyzed hydrolysis typically is induced only by strongly acidic solutions (e.g., pH 3-4).

Hydrolysis - of phosphate esters

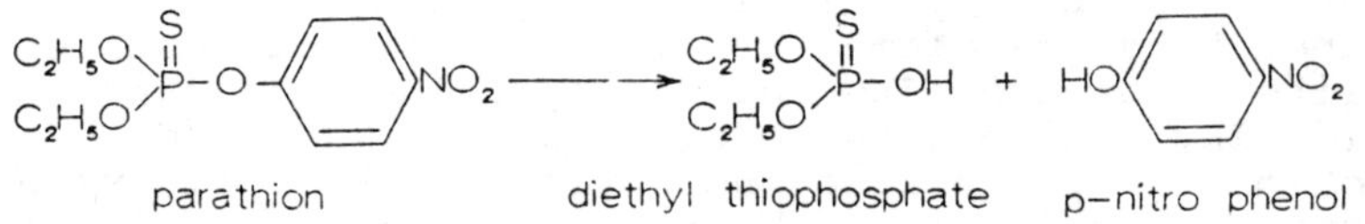

parathion   diethyl thiophosphate   p-nitro phenol   (12)

Biologically, the hydrolysis of parathion has been shown to proceed by two mechanisms, one oxidative, as a result of the action of mixed-function oxidases or by hydrolases (33,34).

Hydrolysis - of carboxyl esters

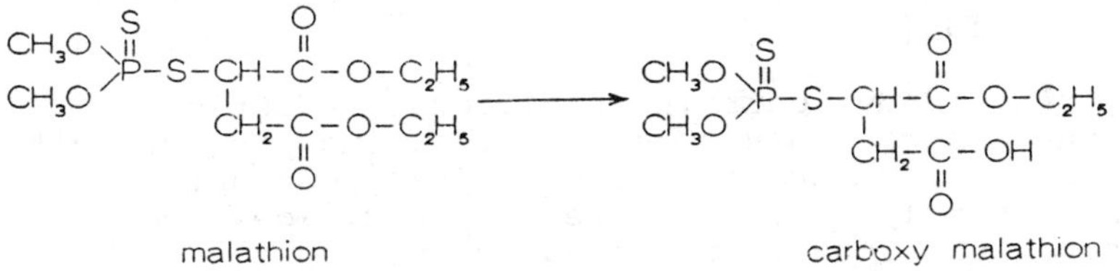

malathion   carboxy malathion   (12)

Carboxyl esterase enzymes of vertebrates rapidly render
malathion non-toxic through this hydrolysis. Malathion-
resistant insects also have been demonstrated to possess
the enzymatic capability (25). Carbamate esters, e.g.,
N-methylcarbamate insecticides, are also degraded by
ester hydrolysis (35).

Hydrolysis - of amides

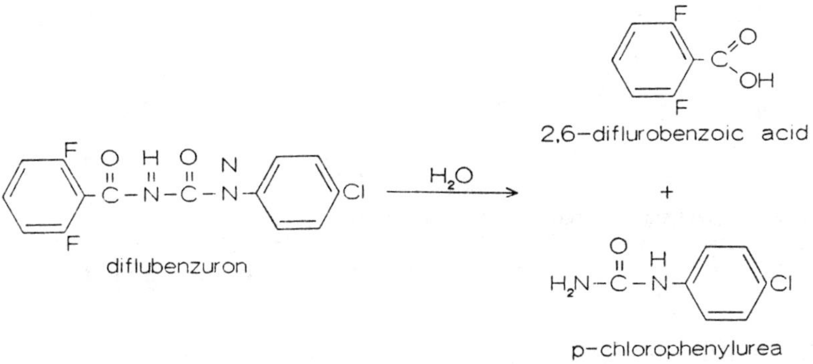

2,6-diflurobenzoic acid

diflubenzuron   +

p-chlorophenylurea   (17)

Hydrolysis - of halogens

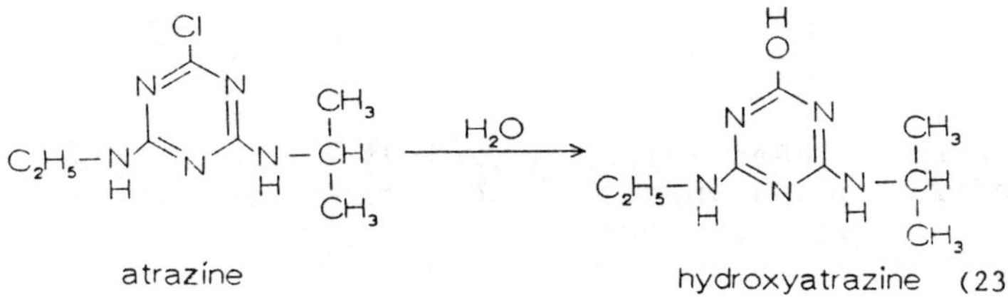

atrazine   hydroxyatrazine   (23)

Chemical hydrolysis of chlorides and bromides can occur,
yielding a hydroxylated product. Hydrolysis of an
epoxide, by an epoxide hydratase, produces a diol (36).
Oximecarbamates can be hydrolyzed to an oxime and a
carbamic acid (35).

Reductions. Under certain conditions, reduction
reactions are commonly observed, yielding products with
lower polarity generally. Their biological activity may
be concomitantly altered as well. Pesticides undergo
reduction reactions in reducing environments,
characterized by low oxygen concentrations, low pH's, and
anaerobic microorganisms. Examples of situations in
which pesticides are reduced include stagnant or
eutrophic ponds and lakes, bogs, flooded rice fields, the
rumens of ruminant livestock, and lower intestines of
some species, including man.

Reductive dehalogenation

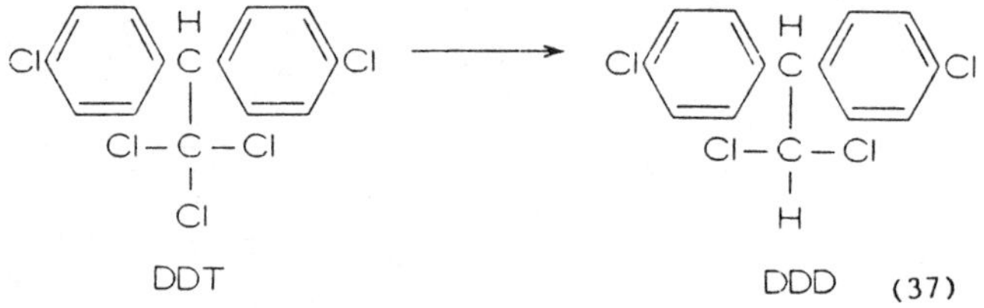

DDT   DDD   (37)

Mammalian systems are more capable of this
biotransformation of DDT than many lower animals (38),
probably due to this conversion by the intestinal flora
(11).

Reduction of a sulfoxide

phorate sulfoxide   phorate   (39)

Phorate soil insecticide is rapidly sulfoxidized in soil,
but under flooded anaerobic conditions can be reduced
back to the sulfide form.

Reduction of a nitro group

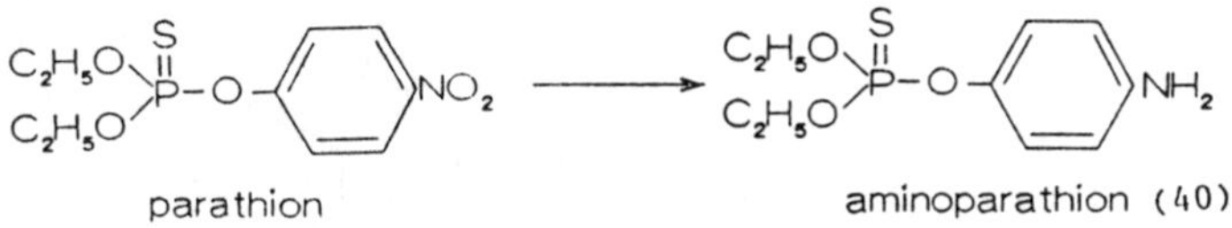

parathion   aminoparathion (40)

This conversion has been observed in flooded soil. The fungicide pentachloronitrobenzene has also been shown to undergo reduction to an amino analog (41).

Synthetic reactions

Conjugations of different types are synthetic in nature, adding a moiety to a xenobiotic, primarily at a hydroxyl, amino, sulfhydryl, or carboxylic acid group. The moiety is usually a relatively polar one, such as glucuronic acid, glucose, sulfate, phosphate, or amino acids (42). Alkylation, usually methylation, of the polar moiety tends to reduce the polarity of that group. Acylation, primarily acetylation, also reduces the hydrophilic nature of the polar group (12). Both methylation and acetylation alter the bioactivity and lipophilicity of primary amines and occur commonly in aquatic systems. Persistence, bioaccumulation, and mobility can also be significantly changed by conjugation reactions.

Rearrangements

Some reactions are neither degradative nor synthetic, but involve changes in the structural arrangement of the molecule. Properties can change considerably.

Isomerizations. The transformation of one isomer to another can occur under various conditions, but is usually effected by physical or chemical agents. The optical activity of some diastereomers can be reversed in some situations. Fenvalerate, a synthetic pyrethroid insecticide, has two chiral carbons, one in the acid portion of the molecule and one in the alcohol moiety. In water, the alcoholic chiral carbon of a pure isomer undergoes racemization, apparently until a stereochemical equilibrium is reached. No change in the stereochemistry of the acid-moiety chiral carbon was noted. Other protic solvents also facilitated the racemization of the S,S isomer of fenvalerate to the S,R isomer, but aprotic ones did not (43). Deltamethrin isomers also can undergo racemization in aqueous solutions (44). Optical isomers of EPN and its oxon exert different types of effects in chickens. Both sterioisomers inhibited the neuropathy target esterase (NTE), but aging of the phosphorylated

NTE occurred primarily with the L(-)-isomers and
correlated closely the potency of those isomers to cause
organophosphate induced delayed polyneuropathy (45).

(acid moiety)   (alcohol moiety)

fenvalerate

Dimerization

$$2BrCH_2 - CH_2Br \longrightarrow BrCH_2 - CH_2 - S - S - CH_2 - CH_2Br$$

EDB   dialkyl disulfide   (46)

Under extremes of heat and light many pesticides can form
dimers, trimers, or polymers, but under conditions of
less severity, dimers can be formed, sometimes after an
initial degradative or
conjugation step.

Cyclization

dieldrin   photodieldrin   (19)

Closure of acyclic systems to form a ring has been shown
to occur in several cases, usually catalyzed by light,
pH, or enzymes. Some organophosphates undergo ring
closure to form cyclic phosphates or phosphates with
cyclic leaving groups (34).

Migration

diclofop methyl

An example of migration of a halogen from one position to another, more favored, position on the molecule is the NIH shift. The para-hydroxyl metabolite of diclofop methyl is produced through the formation of the arene oxide intermediate of the para-chloro to a meta position in wheat plants. The hydroxylation creates a more significant change in the biological and physical properties than the migration of the chlorine does (47).

Activation Reactions

The biological significance of degradation products is sometimes greater than that of the parent pesticide molecule. Several types of activations have been reported, some programmed into the molecule and others inadvertent.

Organophosphorus Insecticides

Most organophosphorus insecticides are activated *in vivo* to potent inhibitors of acetylcholinesterase (AChE). The oxidative desulfuration of the P=S to P=O is the most critical step of activation (48), but sulfoxidation of side-chain sulfides (above) is also important. Both enhance the $\delta+$ charge on the phosphorus atom and, subsequently, the affinity of the P for the serine hydroxyl at the active site of the enzyme. However, the oxidatively activated forms (oxons, sulfoxides, and sulfones) are also more susceptible to hydrolysis of the phosphate ester and also do not penetrate insect cuticle as readily. Therefore, the thion and sulfide forms, which are more stable during storage and application and penetrate quickly into the insect pest, are preferred for OP insecticide products. The bioactivity of oxidative

degradation products against insect pests has been
previously investigated for some OP's.
     The insecticidal activity of phorate oxidation
products is equivalent to that of the parent materials
when applied topically to corn rootworm larvae (49), but
in soil bioassays on crickets, they were not as
efficacious as the parent compound (50). Oral LD_{50}
determinations on chickens revealed essentially
equivalent acute toxicity of phorate and several
metabolites, all in the range of 0.3 to 1.7 mg/kg (51).
Isofenphos and three of its oxidative metabolites have
been compared for their bioactivity. Their
anticholinesterase activity and injection toxicity to
American cockroaches showed the metabolites to be
superior inhibitors of AChE. The N-dealkylated oxon
metabolite was the most effective inhibitor, with
evidence indicating further oxidation of that product was
necessary for optimal inhibitory activity (52).
Mortality of corn rootworm larvae and adults by topical
application of isofenphos and three of its oxidative
metabolites revealed very similar toxicity of all four
compounds, at 1.7-2.7 µg/g for third-instar larvae (53).
However, in soil bioassays, the degradation products were
not as effective as isofenphos, probably due to less
efficient uptake out of the soil.

Proinsecticides

carbosulfan   carbofuran

Compounds that are designed for activation as a result of
degradation reactions are called proinsecticides (54).
Carbofuran derivatives have been designed to be
hydrolyzed to carbofuran in insects, but not in mammalian
systems (27). These derivatives have a lower toxicity to
mammals than the carbofuran toxaphore.

The synthetic pyrethroid tralomethrin decomposes, via the elimination of Br_2, to deltamethrin a more potent insecticidal molecule (55).

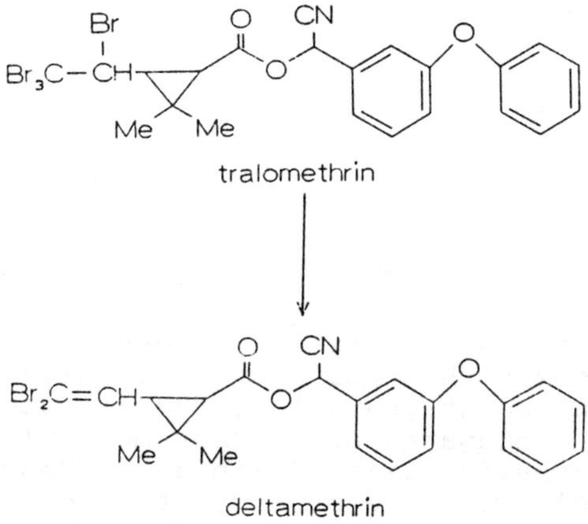

tralomethrin

deltamethrin

Promutagens and Procarcinogens

Pesticides that can be transformed into mutagens or carcinogens are called promutagens or procarcinogens. Ethylene bis-dithiocarbamate fungicides degrade under certain conditions, including cooking and processing, to ethylene thiourea (ETU), a carcinogen (56).

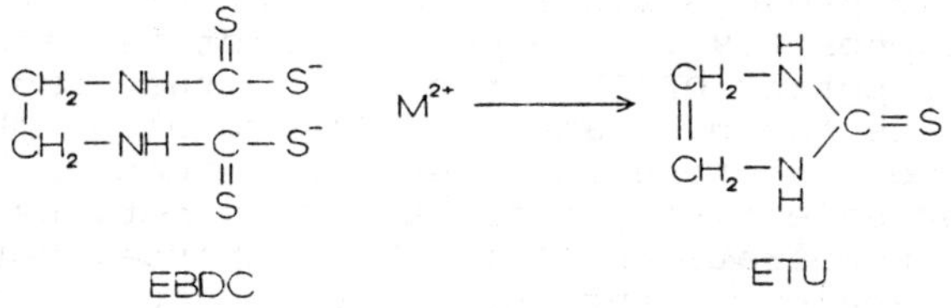

EBDC   ETU

The plant growth regulator/ripening agent daminozide (Alar) degrades via a hydrolysis to unsymetrical dimethylhydrazine (UDMH) which is carcinogenic (57).

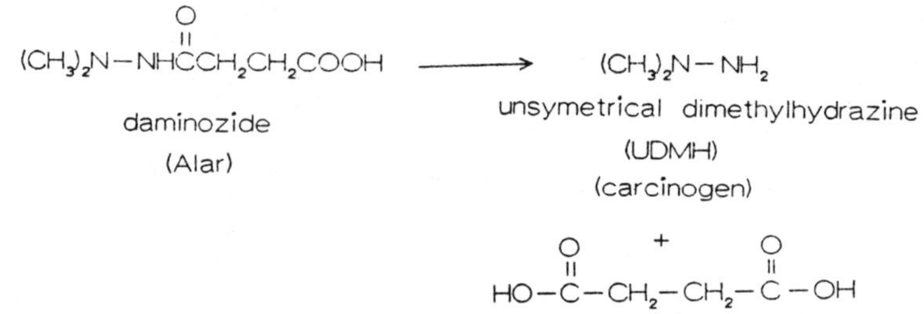

$$(CH_3)_2N-NH\overset{\overset{O}{\|}}{C}CH_2CH_2COOH \longrightarrow (CH_3)_2N-NH_2$$

daminozide

(Alar)

unsymetrical dimethylhydrazine

(UDMH)

(carcinogen)

$$HO-\overset{\overset{O}{\|}}{C}-CH_2-CH_2-\overset{\overset{O}{\|}}{C}-OH$$

Although conjugation reactions generally result in products less toxic than the pesticide molecule, examples have emerged in recent years that demonstrate deleterious bioactivity of conjugates or further metabolized conjugates. Aromatic primary amines can undergo acetylation, followed by a hydroxylation of the acetyl group and sulfate conjugation of the hydroxy acetyl group. The resultant sulfates can be carcinogenic (42). A cholesterol conjugate formed from the acid portion of the R,S-isomer of fenvalerate has been implicated in the formation of a liver granuloma (58).

Environmental Activation

Environmental activation is any transformation that enhances the environmental significance of a chemical as a result of changes in its environmental toxicology or chemistry. An environmental activation may occur due to changes in a pesticide's (1) bioactivity, (2) persistence, or (3) mobility.

The organophosphorus insecticide leptophos has been shown to undergo an aromatic debromination, in sunlight especially, to produce significant quantities of desbromoleptophos (59). The photoproduct is capable of inducing more potent OPIDN effects than the parent molecule, as is the oxon analog. Desbromoleptophos is not metabolized or excreted more slowly than the parent molecule, nor selectively distributed to nervous tissue in greater concentrations (60). The enhanced potency of the desbromo analog is thought to be due to a greater absorption rate from the gut (61) and/or greater affinity in vivo for the neuropathy target esterase (62).

leptophos   desbromoleptophos

Degradation products of some insecticides and herbicides have been shown to be more mobile than the parent molecules (63), and often move to locations where degradation may be slower, e.g., groundwater (64). Mobility of pesticides and their transformation products must be better understood, regarding mechanisms, prevention, and significance of those residues in water

Conclusions

1. Degradation of pesticides begins at formulation and continues for hours, days, weeks, months, years, or decades, depending on the chemical and the conditions.
2. Pesticide degradation products occur in air, water, soil, and organisms.
3. Activation reactions are of special significance and are demanding greater attention in the future.

Acknowledgment

The author acknowledges grants from the U. S. Department of Agriculture (North Central Region Pesticide Impact Assessment Program and the Management Systems Evaluation Area) and the Leopold Center for Sustainable Agriculture for supporting our research on pesticide degradation products. This chapter is journal paper No. J-14304 of the Iowa Agriculture and Home Economics Experiment Station, Project 2306.

Literature Cited

1. Matsumura, F. Toxicology of Insecticides; Plenum Press: New York, 1985; Chapter 5, 9.
2. Manahan, S. E. Environmental Chemistry; Lewis Publishers: Ann Arbor, MI, 1990; Chapter 3, 10, 11.
3. Alexander, M. Science 1981, 211, 132-138.
4. Bollag, J-M. In Advances in Applied Microbiology; Perlman, D., Ed.; Academic: New York, 1974, Vol. 18; pp 75-151.

5.   Racke, K. D.; Coats, J. R. J. Agric. Food Chem. 1987, 35, 94-99.

6.   Racke, K. D.; Coats, J. R., Eds. In Enhanced Biodegradation of Pesticides in the Environment; American Chemical Society: Washington, DC, 1990; Chapter 6.

7.   Somasundaram, L.; Coats, J. R.; Racke, K. D. J. Environ. Sci. Health 1990, B24, 457-478.

8.   Somasundaram, L.; Coats, J. R.; Racke, K. D. Bull. Environ. Contam. Toxicol. 1990, 44, 254-259.

9.   Felsot, A. S.; Dzantor, E. K. In Enhanced Biodegradation of Pesticides in the Environment; Racke, K. D.; Coats, J. R., Eds; American Chemical Society: Washington, DC, 1990; Chapter 19.

10. Dick, W. A.; Ankumah, R. O.; McClung, G.; Abou-Assaf, N. In Enhanced Biodegradation of Pesticides in the Environment; Racke, K. D.; Coats, J. R., Eds; American Chemical Society: Washington, DC, 1990; Chapter 8.

11. Miyamoto, J.; Kaneko, H.; Hutson, D. H.; Esser, H. O.; Gorbach, S.; Dorn, E. Pesticide Metabolism: Extrapolation from Animals to Man; Blackwell Scientific Publications: Palo Alto, CA, 1988; Chapter 2, 3.

12. Hodgson, E.; Dauterman, W. C. In Introduction to Biochemical Toxicology; Hodgson, E.; Guthrie, F. E., Eds.; Elsevier: New York, 1980; Chapter 4.

13. Wilkinson, C. F. In The Enzymatic Oxidation of Toxicants; Hodgson, E., Ed.; North Carolina State University: Raleigh, NC, 1968; pp 113-149.

14. Smyser, B. P.; Sabourin, P. J.; Hodgson, E. Pestic. Biochem. Physiol. 1985, 24, 368-374.

15. Tynes, R. E.; Hodgson, E. J. Agric. Food Chem. 1985, 33, 471-479.

16. Menzie, C. M. Metabolism of Pesticides; U.S. Dept. Interior: Washington, DC, 1969; pp 72-76.

17. Menzie, C. M. Metabolism of Pesticides - Update II; U. S. Dept. Interior: Washington, DC, 1978; pp 109-111.

18. Metcalf, R. L.; Fukuto, T. R.; Collins, C.; Borck, K.; El-Azia, S. A.; Munoz, R.; Cassil, C. C. J. Agric Food Chem. 1968, 16, 300-303.

19. Brooks, G. T. Chlorinated Insecticides; CRC Press: Boca Raton, FL, 1974; Vol I, Chapter 3.

20. Bowers, W. S. In Insecticide Mode of Action; Coats, J. R., Ed.; Academic: New York, 1982; Chapter 12.

21. Coats, J. R.; Metcalf, R. L.; Kapoor, I. P.; Chio, L. C.; Boyle, P. A. J. Agric. Food Chem. 1979, 27, 1016-1022.

22. Hansen, L. G.; Metcalf, R. L.; Kapoor, I. P. Comp. Gen. Pharmacol, 1974, 5, 157-163.

23. Kahn, S. U.; Marriage, P. B. J. Agric. Food Chem. 1977, 25, 1408-1413.

24. Hollingworth, R. M.; Murdock, L. L. Science 1980, 208, 74-76.
25. Aizawa, H. Metabolic Maps of Pesticides, Vol. II Academic: New York, 1989; pp 173, 178.
26. Dahm, P. A.; Kopecky, B. E.; Walker, C. B. J. Toxicol. Appl. Pharmacol. 1962, 4, 683-696.
27. Fukuto, T. R. Environ. Hlth Perspect. 1990, 87, 245-254.
28. Fukuto, T. R.; Wolf, J. P., III; Metcalf, R. L.; March, R. B. J. Econ. Entomol. 1956, 49, 147-151.
29. Kapoor, I. P.; Metcalf, R. L.; Hirwe, A. S.; Coats, J. R.; Khalsa, M. S. J. Agric. Food Chem. 1973, 21, 310-315.
30. Dauterman, W. C. In Intermediary Xenobiotic Metabolism in Animals; Hutson, D. H.; Caldwell, J.; Paulson, G. D., Eds.; Taylor and Francis: New York, 1989; pp 139-150.
31. Kato, Y.; Kogure, T.; Sato, M.; Kimura, R. Toxicol. Appl. Pharmacol. 1988, 96, 550-559.
32. Kadlubar, F. F.; Fu, P. F.; Hyewook, J.; Ali, U. S.; Beland, F. A. Environ. Hlth Perspect. 1990, 87, 233-236.
33. Nakatsugawa, T.; Dahm, P. A. Biochem. Pharmacol. 1967, 16, 25-38.
34. Eto, M. Organophosphorus Pesticides: Organic and Biological Chemistry; CRC Press: Cleveland, Ohio, 1974; Chapter 4.
35. Kuhr, R. J. Carbamate Insecticides; CRC Press: Cleveland, Ohio, 1975; Chapter 5.
36. Sipes, I. G.; Gandolfi, A. J. In Toxicology; Klaassen, C. D.; Amdur, M. O.; Doull, J., Eds.; MacMillan: New York, 1986; Chapter 4.
37. Pfaender, F. K.; Alexander, M. J. Agric. Food Chem. 1972, 20, 842-846.
38. Kapoor, I. P.; Metcalf, R.L.; Hirwe, A. S.; Lu, P. Y.; Coats, J. R.; Nystrom, R. F. J. Agric, Food Chem. 1972, 20, 1-6.
39. Walter-Echols, G.; Lichtenstein, E. P. J. Econ. Entomol. 1977, 70, 505-509.
40. Katan, J.; Lichtenstein, E. P.; J. Agric. Food Chem. 1977, 25, 1404-1408.
41. Chacko, C. I.; Lockwood, J. L.; Zabik, M. Science 1966, 154, 893-895.
42. Hutson, D. H. In Intermediary Xenobiotic Metabolism in Animals; Hutson, D. H.; Caldwell, J.; Paulson, G. D., Eds.; Taylor and Francis: New York, 1989; pp 179-204.
43. Bradbury, S. P.; Symonik, D. M.; Coats, J. R.; Atchison, G. J. Bull. Environ. Contam. Toxicol. 1987, 38, 727-735.
44. Day, K. E.; Maguire, R. J. Environ. Toxicol. Chem. 1990, 9, 1297-1300.
45. Johnson, M. K.; Read, D. J. Toxicol. Appl. Pharmacol. 1987, 90, 103-115.

46. Schwarzenbach, R. P.; Geiger, W.; Schaffner, C.;
     Wanner, O. Environ. Sci. Technol. 1985, 19, 322-
     327.
47. McFadden, J. J.; Frear, D. S.; Mansager, E. R.
     Pestic. Biochem. Physiol. 1989, 34, 92-100.
48. Chambers, J. E.; Forsyth, C. S.; Chambers, H. S.
     In Intermediary Xenobiotic Metabolism in Animals;
     Hutson, D. H.; Caldwell, J.; Paulson, G. D.,
     Eds.; Taylor and Francis: New York, 1989; pp
     99-115.
49. Waller, J. B., Ph.D. Dissertation, Iowa State
     University, Ames, Iowa, 1972.
50. Chapman, R. A.; Harris, C. R. J. Econ. Entomol.
     1980, 73, 536-543.
51. Kim, J. H. J. Korean Agric. Chem. Soc. 1988, 31,
     92-99.
52. Heppner, T. J.; Coats, J. R., Drewes, C. D.
     Pestic. Biochem. Physiol. 1987, 27, 76-85.
53. Coats, J. R.; Hsin, C. Y. Abstracts of the
     American Chemical Society, Pesticide Chemistry
     Div.; 186th National Meeting, Washington, DC,
     Aug. 28-Sept. 2, 1983.
54. Prestwich, G. D. In Safer Insecticides:
     Development and Use; Hodgson, E.; Kuhr, R. J.,
     Eds; Marcell Dekker: New York, 1990; Chapter 8.
55. Drabek, J.; Neumann, R. In Insecticides; Hutson,
     D. H.; Roberts, T. R., Eds.; Wiley: New York,
     1985, Chapter 2.
56. Hawkins, D. R. In Intermediary Xenobiotic
     Metabolism in Animals; Hutson, D. H.; Caldwell,
     J.; Paulson, G. D., Eds.; Taylor and Francis: New
     York, 1989; pp 225-242.
57. Brown, M. A.; Casida, J. E. J. Agric. Food Chem.
     1988, 36, 819-822.
58. Okuna, Y.; Seki, T.; Ito, S., Kaneko, H.;
     Watanabe, T.; Yamada, T.; Miyamoto, J. Toxicol.
     Appl. Pharmacol. 1986, 83, 157-169.
59. Sanborn, J. R.; Metcalf, R. L.; Hansen, L. G.
     Pestic. Biochem. Physiol. 1977, 7, 142-145.
60. Cozzi, E. M., Bermudez, A. J.; Hansen, L. G. J.
     Agric. Food Chem. 1988, 36, 108-113.
61. Hansen, L. G., Cozzi, E. M., Metcalf, R. L., Hansen,
     T. K. Pestic. Biochem. Physiol. 1985, 24, 136-148.
62. Reinders, J. H.; Hansen L. G.; Metcalf, R. L.;
     Metcalf, R. A. Pestic. Biochem. Physiol. 1983, 20,
     67-75.
63. Somasundaram, L.; Coats, J. R.; Racke, K. D.
     Environ. Toxicol. Chem. 1991, 10, 185-194.
64. Leistra, M.; Boesten, J. J. T. I. Agric.Ecosyst.
     Environ. 1989, 26, 369-389.

RECEIVED December 12, 1990

FATE

Chapter 3

Biotransformation of Organophosphorus Insecticides in Mammals

Relationship to Acute Toxicity

Janice E. Chambers[1] and Howard W. Chambers[2]

Departments of Biological Sciences[1] and Entomology[2],
Mississippi State University, Mississippi State, MS 39762

The organophosphorus insecticides or their activated
metabolites are potent anticholinesterases which display
a wide range of acute toxicity levels in mammals. While
target site sensitivity does not effectively explain the
acute toxicity levels, the magnitude of various aspects
of the parent insecticide's biotransformation, including
both bioactivation and detoxication pathways, can predict
an insecticide's overall toxicity level. Thus,
biotransformation is a critical factor in determining
mammalian sensitivity to the acute lethality of
organophosphorus insecticides.

The organophosphorus (OP) insecticides are a very effective and
widely used group of pesticides of both historical and current
significance. Their toxicology is important because of the likely
accidental exposures to humans and other mammals during this
widespread use. This paper describes the common biotransformation
routes for OP insecticides and assesses their significance in
determining the acute toxicity level displayed by the insecticide.

Organophosphorus Insecticide Toxicity

The organophosphorus insecticides are potent neurotoxicants in
vertebrates and invertebrates. The mechanism of acute toxicity is
currently accepted to be the inhibition of the enzyme
acetylcholinesterase (AChE) in nervous tissue (1). AChE is
responsible for the hydrolysis of the important neurotransmitter
acetylcholine which transmits information across cholinergic
synapses. Cholinergic synapses in vertebrates are widely
distributed, occurring within the somatic nervous system which
activates skeletal muscles, within the autonomic nervous system
(both parasympathetic and sympathetic divisions) which innervates

0097–6156/91/0459–0032$06.00/0

smooth muscles, cardiac muscle and numerous glands, and also at various locations in the central nervous system. Therefore alterations in the cholinergic pathways can cause widespread disturbances and diverse symptoms of intoxication. This hydrolysis by AChE rapidly inactivates the acetylcholine, leading to the transient effect of the neurotransmitter during normal nervous system function. When AChE is inhibited, the neurotransmitter accumulates, which results in hyperactivity within cholinergic pathways and serious imbalances within the nervous system. If doses are high, the main signs of poisoning are parasympathomimetic effects such as dilated pupils, headache and blurred vision, and nicotinic effects such as tremors and, in some cases, convulsions. Respiratory system failure is usually the cause of death in lethal intoxications in mammals, resulting from four factors: bronchoconstriction, an increase in bronchiolar secretion of mucus, paralysis of the respiratory muscles and an inhibition of the respiratory control centers in the medulla oblongata/pons regions of the brain. Long term effects on memory and other cognitive skills have been reported in humans as a result of single high-dose poisonings with an OP insecticide (2).

Thus, numerous target areas exist throughout the central and peripheral nervous systems of vertebrate non-target organisms, including humans. The threat of lethal or sub-lethal but severe poisonings is of great concern for both humans in occupational settings and for other non-target organisms which might receive inadvertent exposures. It has been estimated that 500,000 humans are poisoned annually by pesticides throughout the world and that over 20,000 of these are fatal. Some 6,000-10,000 poisonings occur in the United States each year which result in 3,000 farm worker hospitalizations and 50-60 deaths. OP insecticides comprise a large fraction of the agents involved in these poisonings (3). In addition to the deaths, the long-term sequelae in survivors of these intoxications are also of continuing concern.

Thus, the OP insecticides continue to be a health threat from the standpoint of their acute toxicity. Over half of the registered OP insecticides have rat oral LD_{50}'s less than 50 mg/kg placing them in the most toxic category in the U.S. Environmental Protection Agency's classification scheme. A number of the OP insecticides are considerably less toxic, however, and thus would constitute a lesser threat of lethality to non-target populations. The acute oral LD_{50}'s to rats range from 3-15 mg/kg for such highly toxic insecticides as disulfoton, parathion and azinphos-methyl to 1,000-13,000 mg/kg for such insecticides as malathion and temephos (4,5).

This wide range of toxicities suggests that there are important differences in the OP insecticides with respect to metabolism and disposition and/or target sensitivity which can explain this range. Such differences could be exploited in the development of safer insecticides which would pose less of a threat to non-target species.

Xenobiotic Metabolism

The reactions of xenobiotic metabolism have been classified into
two general categories, called Phase I and Phase II (6). The Phase
I reactions are involved in placing a more polar and reactive group
into the xenobiotic to make the compound less lipophilic and more
likely to be excreted. The enzymes involved are primarily
monooxygenases (both cytochrome P-450-dependent and flavin
monooxygenases), reductases and hydrolases, and they catalyze a
variety of oxidations (for example, hydroxylations, epoxidations or
dealkylations) reductions and hydrolyses (for example, ester and
amide cleavages). The oxidations and reductions are energy-
expensive requiring reducing equivalents, usually in the form of
NADPH, to proceed. Although these reactions are frequently
detoxications, they are also the reactions most likely to form
reactive intermediates (for example, epoxides or oxons) and
therefore can be activation reactions. These bioactivations, then,
are largely responsible for the toxicity displayed following
exposure to the parent xenobiotic. However, it should be borne in
mind that these reactive metabolites will be chemically labile and
would be expected to undergo chemical reactions with numerous non-
target molecules. Therefore, it should be expected that a
relatively small proportion of the activated metabolite formed
would react with the molecular target and induce toxicity.
    The Phase II reactions are conjugation reactions involved with
placing a highly polar or charged group on the
xenobiotic/metabolite which renders it readily excretable in either
the urine or the bile of vertebrates. The group added is typically
a sugar (glucuronic acid), sulfate or an amino acid/amino acid
derivative. With few known exceptions, the Phase II reactions are
detoxications. Phase II reactions are also energy-expensive,
requiring energy rich cofactors such as uridine diphosphoglucuronic
acid (UDPGA), phosphoadenosinephosphosulfate (PAPS) or the
tripeptide glutathione (GSH; glutamic acid-cysteine-glycine) to
proceed.
    The animal can rid itself of the conjugates resulting from
Phase II reactions, especially small ones, quite readily by
excretion in the urine. Excretion of the conjugates from the liver
into the bile is also a likely route, especially for larger
conjugates. However, these conjugates can also be hydrolyzed by
the intestinal microflora and the aglycone can be reabsorbed via
the hepatic portal circulation back into the liver. This
enterohepatic circulation allows some potential recycling of the
compound. If the compound being recycled is a toxic one, then the
animal remains at risk.
    A number of elements within the Phase I and Phase II reactions
bear striking similarity if not identity to reactions involving the
endogenous steroid hormones. For example, cytochrome P-450-
mediated hydroxylations and conjugations to glucuronic acid are
common reactions for many of the steroids. Also, the enterohepatic
circulation routinely reabsorbs the vast majority of bile salts so
that daily synthesis need be minimal. Therefore, many of the
xenobiotic metabolizing reactions share a commonality with
metabolism of endogenous compounds. By this token, xenobiotic

metabolism could compete with normal biochemistry, and has the potential of interfering with the metabolism and disposition of endogenous biochemicals.

Generally, the highest specific activities of the xenobiotic metabolizing enzymes occur in the liver with considerably lower activities in other tissues. In the laboratory rat, the enzymes sometimes display sex differences and developmental differences, with higher activities of some enzymes in male livers than female livers, and higher hepatic activities in adults than immatures. Also, several of the enzymes can be induced to higher specific activities following in vivo exposures to selected chemicals such as the barbiturate drug phenobarbital, male sex hormones, polycyclic aromatic hydrocarbons (for example, 3-methylcholanthrene or β-naphthoflavone), or ethanol.

Numerous reactions may be available for any given xenobiotic, and a variety of exogenous and endogenous factors may influence the likelihood of those reactions taking place. Understanding the roles of metabolism and disposition in the toxicity of any particular xenobiotic will require concurrent consideration of numerous factors.

Organophosphorus Insecticide Metabolism

Many of the OP insecticides are phosphorothionates, which are characterized by one =S and three -OR groups bonded to the central phosphorus atom (Figure 1). The phosphorothionate molecule, such as parathion, is inherently a weak anticholinesterase, with an I_{50} in the range of 10^{-5}M (7). Thus the phosphorothionate molecule would not be expected to be particularly toxic itself.

However, the phosphorothionates can be bioactivated very effectively by the cytochrome P-450-dependent monooxygenases to highly reactive metabolites, the phosphates or oxons. This bioactivation, called a desulfuration reaction, is hypothesized to occur is a result of an attack on the phosphorus by oxygen and the formation of an unstable phosphooxathiiran intermediate, which then undergoes a release of sulfur and the formation of the oxon (Figure 1, pathway 1a) (8,9). The sulfur released by the desulfuration reaction is highly reactive and can destroy nearby molecules, such as the cytochrome P-450 which produced it (8,10).

The oxon is about 3 orders of magnitude more potent as an anticholinesterase than its corresponding phosphorothionate (7). It inhibits the AChE by phosphorylating the serine hydroxyl present at the active site; this phosphorylation is extremely persistent, with inhibition lasting hours to days (11). For some organophosphates the persistence is partially a result of a process known as "aging", a poorly understood dealkylation which leaves the phosphorylated AChE charged at physiological pH. This phosphorylated and aged AChE cannot be reactivated by either spontaneous hydrolysis or the activity of therapeutic oximes (such as 2-PAM), and, therefore, is permanently deactivated. We have found rat brain AChE I_{50}'s ranging from 1.8 nM for chlorpyrifos-methyl-oxon to 89.3 nM for methyl paraoxon (12). The oxons are also potent inhibitors of other serine esterases which are less critical to survival such as aliesterases (carboxylesterases) (12),

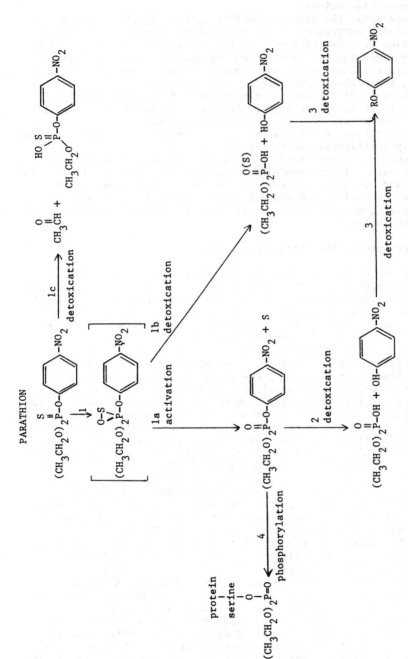

Figure 1. The metabolism of the phosphorothionate parathion. Reactions labelled 1 are mediated by the cytochrome P-450-dependent monooxygenases requiring NADPH and O_2 (Phase I), 2 A-esterases (Phase I) and 3 conjugation enzymes (Phase II). Paraoxon can persistently phosphorylate the hydroxyl at the active site of serine esterases and serine proteases (4). In the case of acetylcholinesterase, this phosphorylation results in neurotoxicity.

and butyrylcholinesterase, and serine proteases such as trypsin or chymotrypsin (Figure 1, pathway 4). As we and others have documented for the aliesterases (*12-14*), phosphorylation of these enzymes provides some detoxication and protection from intoxication through the stoichiometric destruction of some of the molecules of oxon; this point will be discussed further below. However, this phosphorylation could also be involved with some of the long term effects of OP insecticide exposure.

Because of the great difference in anticholinesterase potency between the phosphorothionate and its corresponding oxon, this bioactivation reaction is critical for the phosphorothionate insecticide to display appreciable toxicity. However, the potency of the oxons as anticholinesterases do not predict the overall acute toxicity levels of the parent insecticides. In fact, of the six phosphorothionates/oxons we have studied in detail (i.e., parathion, methyl parathion, EPN, leptophos, chlorpyrifos and chlorpyrifos-methyl, and their corresponding oxons), there is no correlation. As examples, the least potent of the oxons was methyl paraoxon (I_{50}: 89.3 nM) and the most potent were chlorpyrifos-oxon and chlorpyrifos-methyl-oxon (I_{50}: 4.0 and 1.8 nM, respectively) (*12*) while methyl parathion was highly toxic (rat oral LD_{50} 14-24 mg/kg) and chlorpyrifos and chlorpyrifos-methyl were least toxic (LD_{50} 82-163 and 1630-2140 mg/kg, respectively) (*4,5,15*). This indicates that metabolic and/or dispositional factors must be of greater significance than target site sensitivity in determining the acute toxicity level.

The cytochrome P-450-mediated desulfuration reaction mentioned above which forms the toxic oxon metabolite is clearly required for toxicity. The mammalian liver, with its large size and high monooxygenase specific activities, obviously has the greatest potential for forming the reactive metabolite. Yet the hepatic specific activities for desulfuration of the compounds do not correlate with acute toxicity levels either (*16*). It has been shown that much of the phosphorothionate entering the liver fails to exit the liver either in its unchanged form or as the oxon metabolite (*17,18*), indicating that the liver serves as an effective filter for the phosphorothionate either by trapping the insecticide and/or the oxon, or by detoxifying them. Serving as a trap is not surprising, at least from the perspective of the aliesterases, as mentioned above, since most of these have such a high affinity for the oxons (*12*). Thus, phosphorylation of the aliesterases will remove a large amount of the oxon formed until the aliesterases are saturated; this constitutes an important but non-catalytic mechanism of detoxication. Partitioning into the lipid of the liver could also be a mechanism of trapping lipophilic xenobiotics.

To expand further on the importance of the aliesterases, we have recently demonstrated that rat liver aliesterases were about an order of magnitude more sensitive to inhibition by five of the six oxons tested than was rat brain AChE (*12*). A reversal of this trend was observed with methyl paraoxon. These data suggested that the hepatic aliesterases should be able to offer appreciable protection in poisonings with five of the six insecticides, but little protection against methyl parathion. And, indeed, in

animals treated with the six insecticides and sampled at 90
minutes, liver aliesterase activity was inhibited to a greater
extent than brain AChE activity in all cases except methyl
parathion. Liver homogenates added to a brain homogenate offered
protection from the oxons while liver homogenates from parathion
treated rats offered less protection. Finally, the ability of
liver homogenates from phosphorothionate-treated rats to detoxify
paraoxon was directly proportional to the residual aliesterase
activity level. Thus, the aliesterases appear to offer in most
instances substantial detoxication ability. However, it must be
kept in mind that this detoxication results from a phosphorylation
of the serine hydroxyl moiety at the aliesterase's active site, so
the resultant detoxication is non-catalytic and stoichiometric, and
therefore saturable. Treatments of rats with parathion have
resulted in a more rapid maximal inhibition of the hepatic and
serum aliesterases than brain AChE, indicating that this protection
is occurring in vivo (19).

The other mechanism for removing the compounds would be by
active (catalytic) detoxication. The cytochrome P-450-dependent
monooxygenases are also capable of two detoxication reactions on
phosphorothionates. The phosphorothionate can also undergo a
dearylation reaction in which the phosphooxathiiran intermediate
dissociates in a different manner to yield detoxified products,
such as 4-nitrophenol plus diethyl phosphoric acid or diethyl
phosphorothioic acid, in the case of parathion (Figure 1, reaction
1b) (8). If the unstable intermediate breaks down in this way,
then the phosphorothionate has been detoxified and will pose no
further threat to the organism. The ratio of desulfuration to
dearylation is a characteristic of each cytochrome P-450 isoform
(20). Therefore, the overall ratio of desulfuration to dearylation
in any given tissue should reflect the cytochrome P-450 isoform
composition of that tissue.

Another route of detoxication of the phosphorothionate also
mediated by cytochrome P-450 is a dealkylation reaction in which
one of the ethyl or methyl groups may be removed and oxidized to
acetaldehyde or formaldehyde, respectively (21,22). This reaction
is illustrated as pathway 1c in Figure 1, with reaction products of
acetaldehyde and desethylparathion resulting from parathion. The
significance of this reaction has not been as well researched as
the other two cytochrome P-450-dependent reactions.

Several detoxication reactions which are not mediated by
cytochrome P-450 can also further modify the OP insecticide or its
metabolites. The oxon can be hydrolyzed by calcium-dependent A-
esterases, which produce the detoxified products 4-nitrophenol and
diethyl phosphate from paraoxon (Figure 1, reaction 2). Although
they typically display considerably higher in vitro specific
activities than the cytochrome P-450-dependent reactions, they also
usually possess high Km values for their substrates (23). These
Km's for the A-esterases, in the range of about 150 μM, are much
higher than would be expected to occur within the cell, and are
high enough to require the presence of extremely toxic levels of
the oxons (I_{50}'s for AChE of the oxons are the nM range, 12).
Thus, at realistic levels of the oxon, little hydrolysis would be

expected to occur in vivo, and oxon is expected to accumulate, as suggested by Wallace and Dargan (23).

The 4-nitrophenol produced by either dearylation or hydrolysis can be conjugated and made more excretable by such enzymes as uridine diphosphoglucuronosyl transferases, which form conjugates of the ligand with glucuronic acid and require UDPGA, or by sulfotransferases, which form conjugates with sulfate and require PAPS (Figure 1, pathway 3). These conjugates are more water soluble and therefore can be readily excreted in the urine or the bile. However, as mentioned above, the intestinal microflora can hydrolyze these conjugates, allowing the absorption of the ligand into the enterohepatic circulation allowing some potential recycling. Since the metabolites being recirculated here are the detoxified ones, the recycling should not constitute an appreciable hazard to the organism.

Another possible route of detoxication is by the glutathione transferases which can dealkylate the insecticide. These appear to be more effective with dimethyl than diethyl compounds (24,25). Although it appears that this pathway is occurring in animals treated with OP insecticides since hepatic glutathione depletion occurs, the overall significance of this pathway in the disposition of the insecticide is unclear. In some cases, prior depletion of glutathione failed to alter the toxicity of an insecticide administered subsequently (26), and in other cases glutathione depletion enhanced the toxicity of the insecticides. At this point, considering evidence from several laboratories, it appears that glutathione conjugation represents a minor pathway in the overall disposition of an insecticide in vivo.

As is true for a variety of drugs and toxicants, a prior exposure to a chemical capable of affecting the activity levels of the xenobiotic metabolizing enzymes can alter the toxicity of phosphorothionate insecticides. Exposures to phenobarbital, and other agents which can induce the xenobiotic metabolizing enzymes, attenuate the toxicity of phosphorothionate insecticides (19,27,28). This attenuation of toxicity in our laboratory was observed as a slowing of the inhibition of AChE in the cerebral cortex and medulla oblongata of the brain following exposure of phenobarbital-pretreated rats to parathion compared to untreated controls (19). Paradoxically, in vitro hepatic parathion desulfuration activity is enhanced in the liver following phenobarbital exposure (29). Thus, despite a heightened ability to bioactivate the phosphorothionate parathion, the toxicity was reduced, indicating that simultaneous induction of detoxication mechanisms must be ultimately more important in determining a phosphorothionate's toxicity level. This same study indicated that there was a slight degree of slowing of brain AChE inhibition in phenobarbital-pretreated rats who were subsequently administered paraoxon (19), suggesting that the detoxication pathway(s) induced included one or more which could also act upon the oxon. Therefore, in predicting the effect of an inducer on a xenobiotic's toxicity, a variety of metabolic routes possible for the xenobiotic need to be considered simultaneously.

Target Site Activation

As indicated above, the phase I reactions can produce highly
reactive metabolites. The relevant bioactivation for the
phosphorothionate insecticides is the cytochrome P-450-mediated
desulfuration which produces the oxon. The oxons, being highly
reactive, are capable of phosphorylating serine hydroxyl moieties
on a variety of esterases and proteases. A number of these
esterases, such as aliesterases and butyrylcholinesterase, are
present in high levels in the liver and the plasma, therefore
serving an important role in protection against hepatically-
generated oxon. Coupled with the active detoxication mechanisms
present in the liver, these serine esterases provide for much
detoxication of oxon before it has an opportunity to reach the
target nervous tissue. We have hypothesized that, except for
extremely high doses of phosphorothionate which would overwhelm all
of the available protective mechanisms, little of the hepatically-
generated oxon will be able to escape both the liver and the blood
intact to reach the target, and that target site activation is an
important factor in toxicity. Indeed, while liver desulfuration
activities among six phosphorothionates of widely varying acute
toxicities (4-2140 mg/kg) are very similar (3.6-27.5 nmol/g/min)
and bear no relationship to acute toxicity level, the brain
specific activities (composite values of the microsomal plus
mitochondrial activities) correlate very well with toxicity. For
parathion, methyl parathion and EPN which are highly toxic (rat ip
90 min lethal doses of 8 to 60 mg/kg; 12), brain activities are
relatively high (0.2-1.8 nmol/g/min; 7,30), while for leptophos,
chlorpyrifos and chlorpyrifos-methyl which were non-lethal in 90
min at 500 mg/kg, brain activities were considerably lower. Within
this group of less toxic insecticides, the brain desulfuration
activities also correlate with toxicity, i.e., 0.0456-0.0511,
0.0127-0.0130, and 0.0072-0.0076 nmol/g/min, for leptophos,
chlorpyrifos and chlorpyrifos-methyl, respectively, which are
listed in decreasing order of toxicity (16). In order to determine
whether such low desulfuration activities as occur in the brain are
actually capable of occurring in vivo, rat brains were surgically
isolated from circulation to the liver by ligating the aorta
immediately posterior to the diaphragm such that the posterior part
(about 60%) of the circulation was not functioning (31). These
rats were then injected with parathion and the brain AChE
activities were monitored 15 min after injection. Although the
lower dose (2.4 mg/kg) tested (which yielded about 95% brain AChE
inhibition in the intact animal) failed to yield significant
inhibition of brain AChE, the higher dose (48 mg/kg) yielded 70%
inhibition after 15 min, strongly suggesting that the brain is
capable of generating its own activated metabolite in vivo.
Therefore, it appears that despite the low desulfuration activity
present in the brain compared to the liver, the close proximity of
this bioactivation to the target AChE may be an extremely important
factor in setting the overall toxicity class of a phosphorothionate
insecticide.

Summary

Our study of several metabolic routes on six phosphorothionates of widely ranging acute toxicity levels has indicated that the biochemical disposition of a phosphorothionate and its metabolites varies greatly with the compound. No one single metabolic route in mammals is solely responsible for the acute toxicity displayed, although both the potential for the brain to bioactivate the phosphorothionate and the ability of the liver aliesterases to sequester the oxon to greater or lesser degrees appear to be important factors in setting the acute toxicity levels. As examples, the relatively high bioactivation of parathion in rat brain allows its high toxicity despite relatively high affinity of the hepatic aliesterases for paraoxon, while the very low bioactivation of chlorpyrifos-methyl coupled with the very high affinity of the hepatic aliesterases for chlorpyrifos-methyl-oxon leads to its low toxicity. A knowledge of structure-activity relationships for target site sensitivity and bioactivation potential, plus sensitivity of alternate phosphorylation sites as well as other routes of non-target site detoxication will allow a better understanding of the wide range of acute toxicities displayed by organophosphorus insecticides. When better characterized, the biochemical disposition of phosphorothionate insecticides will be useful information in the development of insecticides which will pose less of a danger of severe accidental poisonings in man and other mammals.

Acknowledgments

The author gratefully acknowledges research support from NIH grant ES04394, as well as the support of Research Career Development Award ES00190. The author also appreciates the technical assistance and animal care rendered by Marilynn Alldread, Michael Bassett, Scott Boone, Russell Carr, Shawn Clemmer, Amanda Holland, Jana Munson, John Snawder, Jeffery Stokes, and Sherrill Wiygul.

Literature Cited

1. Murphy, S.D. In *Cassarett and Doull's Toxicology*; Klaassen, C.D.; Amdur, M.; Doull, J., Ed. MacMillan Publishers, New York, NY, 1988; pp. 519-581.
2. Savage, E.P.; Keefe, T.J.; Mounce, L.M.; Heaton, R.K.; Lewis, J.A.; Burcar, P.J. *Arch. Environm. Hlth.* 1988, 43,38-45.
3. *Silent Spring Revisited*; Marco, G.J.; Hollingworth, R.M.; Durham, W. Eds; American Chemical Society: Washington, DC, 1987.
4. Gaines, T.B. *Toxicol. Appl. Pharmacol.* 1960, 2,88-99.
5. Gaines, T.B. *Toxicol. Appl. Pharmacol.* 1969, 14,515-534.
6. Sipes, I.G.; Gandolfi, A.J. In *Cassarett and Doull's Toxicology*; Klaassen, C.D.; Amdur, M.; Doull, J., Ed. MacMillan Publishers, New York, NY, 1988; pp. 64-98.
7. Forsyth, C.S.; Chambers, J.E. *Biochem. Pharmacol.* 1989, 10,1597-1603.

8. Neal, R.A. In *Reviews in Biochemical Toxicology*; Hodgson, E.;
    Bend, J.R.; Philpot, R.M., Ed.; Elsevier-North Holland: New
    York, NY, 1980, Vol. 2; pp. 131-171.
9. Kulkarni, A.P.; Hodgson, E. *Ann. Rev. Pharmacol. Toxicol.*
    1984, *24*,19-42.
10. Norman, B.J.; Poore, R.E.; Neal, R.A. *Biochem. Pharmacol.*
    1974, *23*,1733-1744.
11. Chambers, H.W.; Chambers, J.E. *Pestic. Biochem. Physiol.*
    1989, *33*,125-131.
12. Chambers, H.; Brown, B.; Chambers, J.E. *Pestic. Biochem.
    Physiol.* 1990, *36*,308-315.
13. Clement, J.G. *Fundam. Appl. Toxicol.* 1984, *4*,596.
14. Maxwell, D.M.; Brecht, K.M.; Lenz, D.E.; O'Neill, B.L. *J.
    Pharmacol. Exp. Ther.* 1988, *246*,986-991.
15. *The Pesticide Manual*, 8th edition; Worthing, C.R.; Walker,
    S.B., Eds.; British Crop Protection Council: Thornton Heath,
    United Kingdom, 1987.
16. Chambers, J.E.; Chambers, H.W. *J. Biochem. Toxicol.* 1989,
    4,201-203.
17. Tsuda, S.; Sherman, W.; Rosenburg, A.; Timoszyk, J.; Becker,
    J.M.; Keadtisuke, S.; Nakatsugawa, T. *Pestic. Biochem.
    Physiol.* 1987, *28*,201-215.
18. Sultatos, L.G.; Minor, L.D. *Drug Metab. Disp.* 1986, *14*,214-
    220.
19. Chambers, J.E.; Chambers, H.W. *Toxicol. Appl. Pharmacol.*
    1990, *103*,420-429.
20. Levi, P.E.; Hodgson, E. *Toxicol. Lett.* 1985, *24*,221-228.
21. Donninger, C.; Hutson, D.H.; Pickering, B.A. *Biochem. J.*
    1972, *126*,701-707.
22. Appleton, H.T.; Nakatsugawa, T. *Pestic. Biochem. Physiol.*
    1977, *7*,451-465.
23. Wallace, K.B.; Dargan, J.E. *Toxicol. Appl. Pharmacol.* 1987,
    90,235-242.
24. Motoyama, N.; Dauterman, W.C. In *Reviews in Biochemical
    Toxicology*; Hodgson, E.; Bend, J.R.; Philpot, R.M., Ed.;
    Elsevier-North Holland: New York, NY, 1980, Vol. 2; pp. 49-
    69.
25. Fukami, J. *Pharmac. Ther.* 1980, *10*,473-514.
26. Sultatos, L.G.; Woods, L. *Toxicol. Appl. Pharmacol.* 1988,
    96,168-174.
27. Alary, J.-G., Brodeur, J. *J. Pharmacol. Exp. Ther.* 1969,
    169,159-167.
28. Sultatos, L.G. *Toxicol. Appl. Pharmacol.* 1986, *86*,105-111.
29. Chambers, J.E.; Forsyth C.S. *J. Biochem. Toxicol.* 1989,
    4,65-70.
30. Chambers, J.E.; Forsyth, C.S.; Chambers, H.W. In *Intermediary
    Xenobiotic Metabolism: Methodology, Mechanisms and
    Significance*; Caldwell, J.; Hutson, D.H.; Paulson, G.D., Ed.;
    Taylor and Frances: Basingstoke, United Kingdom, 1988, pp.
    99-115.
31 Chambers, J.E.; Munson, J.R.; Chambers, H.W. *Biochem.
    Biophys. Res. Comm.* 1989, *165*,327-333.

RECEIVED December 11, 1990

Chapter 4

Degradation Products of Commonly Used Insecticides in Indian Rice Soils

Degradation and Significance

N. Sethunathan[1], T. K. Adhya[1], S. Barik[2], and M. Sharmila[1]

[1]Division of Soil Sciences and Microbiology, Central Rice Research Institute, Cuttack—753 006, Orissa, India
[2]ARCTECH, Inc., 5390 Cherokee Avenue, Alexandria, VA 22312

The study of pesticide metabolism in soil, sediment and water systems is of great importance from an environmental, ecological and economic standpoint. The degradation and the significance of the degradation products from commonly used pesticides are discussed. Following treatment of flooded rice field with γ-hexachlorocyclohexane, γ-tetrachlorocyclohexene was detected as a major but transitory metabolite in soils up to 60 cm depth. The route and the rate of degradation of pesticides such as parathion, which can be degraded by more than one metabolic pathway, are governed by moisture regime, temperature, organic matter content and frequency of pesticide application. Degradation of parathion and methyl parathion by a *Bacillus* sp. in mineral salts medium proceeded by hydrolysis in the presence of 0.05% yeast extract and by nitro group reduction when 0.5% yeast extract was present. A shift from nitro group reduction to hydrolysis occurred in the parathion degradative pathway after repeated addition of the pesticide to a flooded soil. Accelerated degradation of pesticides, such as parathion, occurs after repeated applications of the pesticide or its hydrolysis product, i.e. *p*-nitrophenol. In anaerobic flooded soils, slow ring cleavage of many pesticides could lead to the formation of persistent soil-bound residues.

Researchers over the last four decades have generated vast amounts of literature on the significance and fate of many complex organic pesticides in soil, sediment and aqueous environments. A majority of these studies focused on the kinetics of degradation of parent molecules, although major degradation products were characterized in many instances. Reports of almost complete mineralization of

0097–6156/91/0459–0043$06.00/0

complex organic molecules to inorganic end products, such as carbon dioxide, in microbially active ecosystems under favorable conditions, are not uncommon in the literature. However, for most pesticides currently used, the ultimate fate of their degradation products has not been determined. Pesticide degradation products are important because they may 1) persist for longer periods than the parent compounds; 2) be equally or more toxic than the parent compounds; and 3) interact with other substances to form more toxic products. Formation and eventual mineralization of the pesticide metabolites in the environment are influenced by moisture, temperature, pH, redox potential and the availability of active microflora. Degradation patterns can be very complex for pesticides which undergo degradation via more than one primary pathway.

Rice soils are inundated with water at some point during the growth of a rice plant. Such flooded soils differ from non-flooded soils in physical, chemical and biological characteristics (1). Under submerged conditions, soils become predominantly reduced and anaerobic microorganisms become dominant (2). The degradation of a pesticide and extent to which degradation products accumulate in a flooded soil may not necessarily be the same as in non-flooded soils. This review highlights the progress made in studies on the formation and significance of pesticide degradation products in rice soils at our laboratory.

Factors Affecting the Formation of Degradation Products

Metabolism of pesticides in a complex and dynamic flooded soil ecosystem is influenced by many factors such as soil type, moisture regime, organic matter and temperature.

Soil Type. It is well known that the soil type has a profound influence on the persistence of pesticides and their degradation products. Soil pH is an important parameter affecting the persistence of chemically unstable pesticides, especially those belonging to the organophosphorus and carbamate groups, although the role of soil pH on the persistence of pesticides and their metabolites may be less pronounced in flooded soils than in non-flooded soils (1). Generally, pesticides and their degradation products persist longer in sandy soils than in organic rich soils, because of low microbial activity in the sandy soils. Surface-catalyzed hydrolysis of organophosphorus and carbamate pesticides can be common in the soil environment. In a study, Adhya (3) demonstrated positive correlation between the persistence of three organophosphorus insecticides, parathion, methyl parathion and fenitrothion, and silt content in five soils under flooded conditions; whereas no correlation was found in non-flooded conditions.

Soil type can also affect the nature and amount of degradation products formed from soil-applied pesticides such as parathion, methyl parathion and fenitrothion. These phosphorothioate insecticides were degraded by hydrolysis with concomitant accumulation of respective nitrophenols in selected soils, particularly under non-flooded conditions (4). Under flooded conditions, the degradation products from these pesticides differ with regard to soil type: i) only amino-analog in organic-rich pokkali soil; ii) only nitrophenol in Sukinda and

Canning soils; iii) both amino-analog and nitrophenol in alluvial soil (Figure 1). In a majority of soils, the amino analog was formed as the major metabolite under flooded conditions. Differences in redox potential when soils are flooded and the impact of flooding on pesticide-degrading microorganisms might be responsible for such variations in the degradation pattern of these pesticides in different soils.

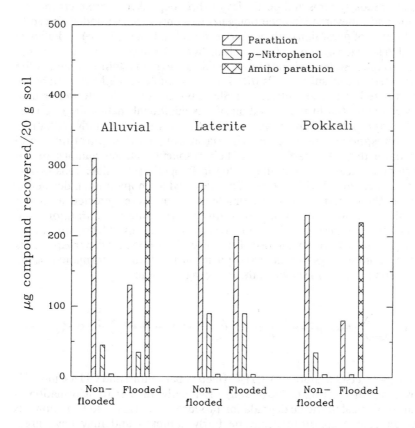

Figure 1. Degradation of parathion and formation of degradation products in flooded and non-flooded soils, 6 days incubation.

Nitro group reduction was the major route of parathion degradation in five coastal saline soils of varying electrical conductivity under flooded conditions. The accumulation of aminoparathion was less pronounced in saline soils than in non-saline soils concomitant with slow degradation of parathion occurring in saline soils (5). However, the effect of salts on nitro group reduction and accumulation of the amino analog appears to be more indirect. The degree of nitro group reduction in saline soils of varying salt content appeared to be related to microbial activity. In laboratory studies, however, the salinity had no effect on the formation and accumulation of the hydrolysis product of carbofuran in contrast to the inhibition of parathion hydrolysis which occurred as the salinity levels increased (6).

Sulfur is an important redox system in environments such as acid sulfate soils. There is increasing evidence that chemically catalyzed interactions between the dominant redox systems and pesticides or their degradation products may be common and widespread in anaerobic environments such as flooded rice soils. Equilibration of parathion with pre-reduced acid sulfate (pokkali) and alluvial soils led to its extensive degradation only in flooded acid sulfate soil (7); but not in alluvial soil. Besides their respective amino analogs an additional metabolite was detected in acid sulfate soils (8). Formation of this additional metabolite was also noted in ferrous sulfate-amended flooded soils. Sulfate rather than Fe^{2+} was implicated in the formation of this additional metabolite (9), later identified as desethyl aminoparathion (8). In a more detailed study (10), three organophosphorus insecticides viz. parathion, methyl parathion, and fenitrothion degraded rapidly past their respective amino analogs to the corresponding dealkylated amino analog in flooded acid sulfate soils or low sulfate soils amended with sulfate. Thus desethyl aminoparathion, desmethyl aminoparathion, and desmethyl aminofenitrothion were detected as major metabolites in addition to aminoparathion, aminomethyl parathion, and aminofenitrothion respectively in flooded acid sulfate soils. Hydrogen sulfide, the end product of sulfur metabolism in an anaerobic flooded soil, perhaps reacts with the amino analogs of organophosphates to result in the formation of the dealkylation products as shown in the following reaction:

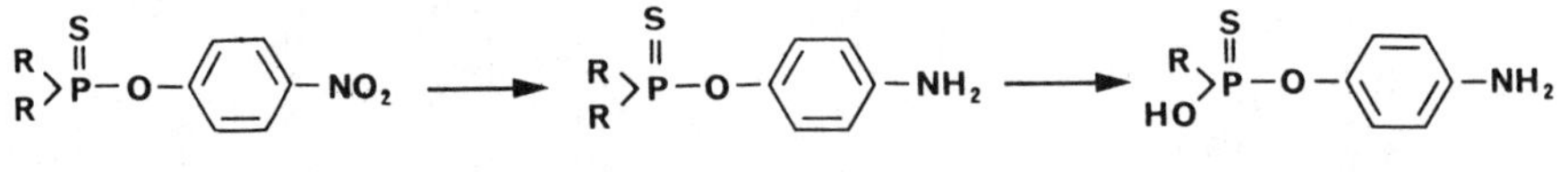

where R = C_2H_5O (parathion), CH_3O (methyl parathion) and CH_3O and CH_3 at position 3 of the ring (fenitrothion). Evidently, further transformation of primary biological pesticide degradation products by inorganic soil components under anaerobic conditions may be fairly common and may have great environmental significance.

Water Regime. The water regime also affects the extent of degradation, the type of products formed and product persistence in soils. For example, parathion and related organophosphorus insecticides, methyl parathion and fenitrothion undergo hydrolysis forming p-nitrophenol under non-flooded conditions and essentially by nitro group reduction under flooded conditions (4, 11). Likewise, aerobic decomposition of DDT in non-flooded soils is slow, resulting in the formation of DDE as the major metabolite (12). DDD, another major intermediate of DDT degradation, is seldom detected. In contrast under flooded soil conditions DDT was rapidly converted to TDE, a more stable and equally toxic compound via reductive dehalogenation (*13, 14*). DDE, DBP, DDOH, DBH, dicofol and certain unknown metabolites of DDT were also detected in flooded soils besides DDD (*15*). Endrin, another organochlorine insecticide, was converted to six metabolites in a flooded soil as compared to only four metabolites in a non-flooded soil (*16*). Under continued flooding these metabolites persisted for a longer period than endrin (*17*). Carbofuran (*18, 19*) and carbaryl (*20, 21*) were hydrolyzed in both flooded and non-flooded soils, but at a slightly faster rate under flooded conditions.

Degradation of pesticides in soils is closely related to the redox potential following flooding. The redox status of most soils rich in native organic matter decreases within a few days after flooding. Thus, in an organic matter rich pokkali soil where the redox potential drops from + 200 mV at flooding to - 210 mV after 10 days of flooding, organophosphorus insecticides such as parathion, methyl parathion and fenitrothion undergo rapid nitro group reduction to their respective amino analogs. On the other hand, in a laterite soil containing low native organic matter, the redox potential drops to - 60 mV after 10 days of flooding when the organophosphorus insecticides undergo hydrolysis (*4*).

Organic Matter. Organic matter, native or added, is known to enhance the degradation of several pesticides by increasing the biomass of active microbial populations. Addition of rice straw or green manure to a flooded soil accelerated the degradation of several pesticides. The application of rice straw (*22*) or green manure (*23*) also enhanced the degradation of different isomers of hexachlorocyclohexane. Similarly, addition of green manure to a flooded soil enhanced the selective pathway of reductive dechlorination of DDT to DDD (Table I) without the enhanced formation of DDE (*15*). The addition of rice straw also accelerated the hydrolysis of carbofuran to 7-phenol which

Table I. Persistence of DDT and Formation of its Degradation Products in a Soil (Black Clay) Amended with Green Manure

Soil	Treatments	% Recovery as[1]		
		DDT	DDD	DDE
Non-flooded	- green manure	74.0	4.0	4.0
	+ green manure	77.0	7.0	4.5
Flooded	- green manure	25.0	26.0	3.0
	+ green manure	10.0	30.0	4.5

SOURCE: Reprinted with permission from ref. 15. Copyright 1988 Selper Ltd., Publications Division.

[1] 40 days incubation.

accumulated in anaerobic flooded soil (24). However, the effect of organic matter is not always stimulatory and hydrolysis of parathion was inhibited when rice straw was added to a flooded soil inoculated with a parathion hydrolyzing enrichment culture (25).

Temperature. Very little is known about the effect of temperature on the formation and persistence of pesticide degradation products in tropical environment. The rate of the formation of γ-pentachlorocyclohexene from γ-hexachlorocyclohexane in a flooded soil increased as the temperature increased from 25 to 35° C (26), probably due to increased microbial activity and a rapid decrease in the redox potential at high temperature. Methyl parathion was degraded in a flooded alluvial soil incubated at 25° C, primarily by nitro group reduction with concomitant formation of aminomethyl parathion. Despite more reduced conditions of the flooded soil held at 35° C, hydrolysis was more pronounced than nitro group reduction leading to the accumulation of p-nitrophenol (Figure 2). Evidently, an increase in the temperature effected a shift in the pathway of methyl parathion degradation from nitro group reduction to hydrolysis (27). However, such temperature-dependent shift in the degradation pathway of methyl parathion and accumulation of the respective degradation products are governed by soil type (28). In another study, carbofuran and its analog carbosulfan, were degraded faster at 35° C than at 25° C concomitant with the formation of carbofuran from carbosulfan (29).

Biological Degradation of Pesticides after Repeated Pesticide Application or Metabolite Addition

Accelerated biodegradation of many pesticides after their intensive use for pest control in agriculture has been reported (30, 31). Rapid biodegradation of pesticides can occur due to selective build-up of pesticide-degrading microorganisms when a pesticide is used by the microorganisms as a source of carbon and/or energy for growth. Cometabolism, a process in which a complex organic molecule is degraded by a microorganism but not utilized for microbial growth, has also been implicated in the transformations of many pesticides in the environment.

The major degradation products of parathion are aminoparathion and p-nitrophenol. Aminoparathion is formed by a nitro group reduction and p-nitrophenol is formed via hydrolysis. The rate of parathion degradation and the accumulation of aminoparathion and p-nitrophenol were monitored in a flooded soil (32). The rate of parathion degradation distinctly increased after each successive application of parathion. Moreover, the frequency of application of parathion influenced the nature of the degradation product(s). After the first pesticide application, the major parathion metabolite was aminoparathion. Both aminoparathion and p-nitrophenol were detected after the second pesticide application. p-Nitrophenol was the only metabolite detected after the third addition (Table II). This shift in the parathion degradation pathway from nitro group reduction to hydrolysis occurred due to

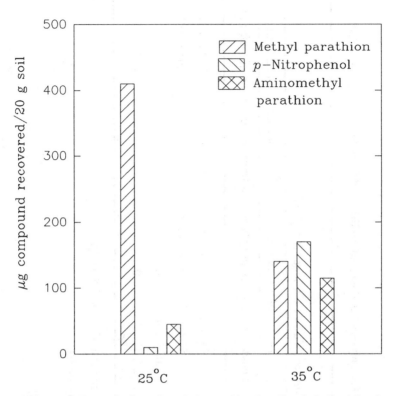

Figure 2. Degradation of methyl parathion in a flooded alluvial soil.

rapid proliferation of parathion-hydrolyzing microorganisms that utilized *p*-nitrophenol as the carbon source (*32*).

Accelerated biodegradation of a pesticide after its repeated additions was noticed primarily with organophosphorus and carbamate pesticides (*31, 33*). Interestingly, hydrolysis, the primary pathway in the degradation of many organophosphorus and carbamate pesticides is not an energy yielding reaction and no proliferation of microorganisms is expected. Evidence suggests that pesticide metabolites derived from the parent pesticide or added, may serve as the energy source for proliferation of microorganisms which can readily hydrolyze the parent molecule. For example, repeated applications of parathion to a flooded soil resulted in accelerated hydrolysis of the pesticide (*32, 34*). The population of parathion-hydrolyzing microorganisms increased from undetectable levels after one addition to 14×10^6 and 4300×10^6/g of soil after two and three successive applications of parathion. Cometabolism is implicated in the primary hydrolysis of parathion. Rapid hydrolysis of parathion also

Table II. Shift from Nitro Group Reduction to Hydrolysis after Repeated Additions of Parathion to a Flooded Soil[a]

Incubation after each parathion addition (days)	Compounds Recovered, $\mu mol/20g$ of soil					
	1st parathion addition		2nd parathion addition		3rd parathion addition	
	aminoparathion	p-nitrophenol	aminoparathion	p-nitrophenol	aminoparathion	p-nitrophenol
1	0	0	0.48 ± 0.03	0.30 ± 0.04	0	0.92 ± 0.02
3	0.11 ± 0.03	0	0.51 ± 0.03	0.27 ± 0.02	0	0.07 ± 0.01
7	0.30 ± 0.02	0	0.74 ± 0.02	0	0	0
15	0.39 ± 0.02	0	0.37 ± 0.02	0	0	0

SOURCE: Reprinted from ref. 32. Copyright 1979 American Chemical Society.

[a] Parathion (1.72 μmol) was added to the soil at 15-day intervals.

occurred in a flooded soil pretreated with its hydrolysis product, p-nitrophenol (Table III) (32). Ferris and Lichtenstein (34) also reported rapid biological mineralization of parathion to carbon dioxide in a flooded soil pretreated with p-nitrophenol.

A *Pseudomonas* sp. isolated from parathion-treated flooded soil, readily hydrolyzed parathion and metabolized p-nitrophenol to nitrite (35) and carbon dioxide (36). Although parathion-hydrolyzing microorganisms proliferated during the metabolism of p-nitrophenol, rapid hydrolysis of parathion was not demonstrated in a flooded soil pretreated with aminoparathion. Interestingly, aminoparathion was not hydrolyzed by the parathion-hydrolyzing enrichment culture.

Decreased efficacy of diazinon due to rapid biodegradation following repeated pesticide application was first reported by Sethunathan (30) in rice fields. Diazinon, like parathion, undergoes hydrolysis at P-O-C linkage in soils as well as in mineral salts medium under laboratory conditions. Although 2-isopropyl-4-methyl-6-hydroxypyrimidine (IMHP), the diazinon metabolite, appears to be stable in predominantly anaerobic flooded soil (37), IMHP may be mineralized by adapted microorganisms in soils pretreated with diazinon. These microorganisms can also hydrolyze the parent diazinon molecule. Enrichment cultures from diazinon-treated soils readily converted diazinon to IMHP (38), which in turn, was subsequently mineralized to carbon dioxide (39). A *Flavobacterium* sp. (ATCC 27551) isolated from a diazinon-treated rice field, readily hydrolyzed diazinon and mineralized IMHP to carbon dioxide (40). Evidently, p-nitrophenol and IMHP can be degraded by microorganisms capable of hydrolyzing the parent molecules, parathion and diazinon.

Decreased efficacy of carbofuran, other methylcarbamate and oxime carbamate pesticides after repeated application has also been reported (31,33). Build-up of carbofuran-hydrolyzing microorganisms in carbofuran-treated soil probably occurs during the metabolism of its hydrolysis product, 7-phenol. Pretreatment of a flooded soil with 7-phenol or 1-naphthol (hydrolysis product of carbaryl) accelerated the hydrolysis of subsequently added carbofuran and carbaryl (41). Moreover, an enrichment culture (42) from flooded soil treated with carbofuran and a pure culture of *Arthrobacter* sp. (43) readily mineralized carbofuran to carbon dioxide at 35° C. 7-Phenol accumulated only in small concentrations in these cultures probably because of its further metabolism. In contrast, an *Achromobacter* sp., isolated from a problem soil in which rapid degradation of carbofuran occurred, readily hydrolyzed carbofuran to 7-phenol which resisted further degradation (44). The mechanism for enrichment of *Achromobacter* sp. in carbofuran-treated soil, is however, not known.

Persistence and Degradation of Pesticide Metabolites

Degradation Products. Nitrophenols, the major metabolites of parathion degradation and its analogs, and other nitro-substituted herbicides, are further metabolized in soils, soil enrichment cultures and by pure microbial cultures. The position and number of nitro groups on the benzene ring determine the

Table III. Shift in the Pathway of Parathion Degradation in Flooded Soils Pretreated with *p*-Nitrophenol[a]

Incubation after each parathion addition (days)	Compounds Recovered, μmol/20g of soil					
	Soil not pretreated with *p*-nitrophenol			Soil pretreated with *p*-nitrophenol		
	parathion	*p*-nitrophenol	aminoparathion	parathion	*p*-nitrophenol	aminoparathion
0	1.67 ± 0.02	0	0	1.63 ± 0.01	0	0
0.5	1.65 ± 0.02	0	0	1.38 ± 0.03	0.29 ± 0.03	0
1	1.65 ± 0.03	0	0	0.46 ± 0.02	1.19 ± 0.05	0
3	0.69 ± 0.01	0	0.86 ± 0.01	0.30 ± 0.01	0	0
12	0.26 ± 0.03	0	0.81 ± 0.01	0	0	0

SOURCE: Reprinted from ref. 32. Copyright 1979 American Chemical Society.

[a] Parathion (1.72 μmol) was added to flooded soil that was never treated with *p*-nitrophenol or was previously treated with two applications of *p*-nitrophenol.

degree of susceptibility of nitrophenols to biodegradation. When nitrophenols (*p*-, *o*-, and *m*-isomers and 2,4-dinitrophenol) were incubated with alluvial (CRRI) and acid sulfate (pokkali) soils under flooded conditions, the concentration of each of four nitrophenols declined rapidly, irrespective of nitro-group substitution and reached low levels in 10 days (*45*). No nitrite was detected in either soil. These nitrophenols disappeared from both soils when inoculated with a soil enrichment culture developed from parathion-treated flooded soils. Degradation of nitrophenols was less rapid in uninoculated soils. The rate of degradation of these nitrophenols depended on soil type and on position and number of nitro groups. Moreover, during the metabolism of nitrophenols in inoculated soils, nitrite accumulated in alluvial soil irrespective of the position and number of nitro groups. But, in inoculated pokkali soil, no nitrite was detected despite rapid disappearance of all the four nitrophenols. Possibly, the nitrophenols undergo oxidation to yield nitrite in alluvial soils whereas nitro group reduction to respective amino analogs takes place in pokkali soil. Alternatively, nitrite may be formed in pokkali soil, but is immediately denitrified due to intense reducing conditions in this organic-rich soil.

In a mineral salts medium inoculated with parathion-enrichment culture from an alluvial soil, *p*-, *o*-, and *m*-nitrophenols were readily degraded with concomitant release of nitrite (*45*); however, 2,4-dinitrophenol was not degraded. An enrichment culture derived from pokkali soil degraded only *p*-nitrophenol yielding nitrite.

Ring Cleavage. Biological ring cleavage is generally mediated by the enzymes known as oxygenases. Reactions mediated by such enzymes are inhibited under predominantly anaerobic conditions found in flooded soils (*2*). Although some pesticides undergo very rapid degradation in flooded soils, their degradation products contain intact ring moieties which may persist due to slow ring cleavage under anaerobic conditions. For example, diazinon and parathion are easily hydrolyzed in flooded soils and their hydrolysis products can accumulate in substantial amounts (*37*). However, these hydrolysis products are further degraded in soils which have previously been treated with the parent pesticide (*38, 45*).

Ring cleavage reactions can, to some extent, occur in oxygen-rich standing water over flooded soils and in oxidized surface soil layers. Moreover, in a flooded soil planted with rice, aerobic conditions would prevail in the microbially active rhizosphere region due to transport of oxygen from foliage to the root region. Parathion degradation was monitored in an unplanted flooded soil and in a flooded soil planted with rice. Approximately 23% of the ^{14}C present in the uniformly labeled ring-^{14}C-parathion was released as $^{14}CO_2$ from the flooded soil planted with rice whereas only 3.3% of the $^{14}CO_2$ was evolved from the unplanted flooded soil (*46*). A similar experiment monitored the degradation of ^{14}C-γ-HCH. $^{14}CO_2$ evolution was negligible (1-2%) under both planted and unplanted conditions (*47*). Evidence suggests that dehydrochlorination of γ-HCH takes place in flooded rice soils and in anaerobic cultures with the eventual formation of chlorine-free volatile metabolites (*48, 49*). The addition of green manure to a flooded soil seems to stimulate the mineralization of γ- and β-isomers of HCH as measured by $^{14}CO_2$ evolution from ring-^{14}C-HCH (Raghu, K. 1990, personal communication).

7-Phenol and 1-naphthol, hydrolysis products of carbofuran and carbaryl, persist in flooded soils possibly due to slow ring-cleavage. Alternate flooded and dry conditions may assist in accelerating the primary degradation pathway which takes place under flooded conditions with subsequent mineralization of the aromatic metabolites under succeeding non-flooded conditions.

Microbial Transformation of Pesticide Degradation Products

Degradation of pesticides in flooded rice soils is far from complete and usually results in the accumulation of intermediary metabolites and/or bound residues. Flooded soils harbor consortia of microorganisms, bacteria in particular, that can mediate, either singly or in combination, degradation of complex pesticide molecules using both aerobic and anaerobic reactions. Nutrients can affect the mineralization of intermediary metabolites to inorganic end products. Some pesticides and their metabolites do persist for long periods of time depending upon soil conditions. IMHP, formed by rapid hydrolysis of diazinon in flooded soil, is stable under anaerobic conditions, but is completely mineralized in diazinon-treated soil under aerobic conditions (38). A *Flavobacterium* sp. (ATCC 27551) isolated from diazinon-treated soil readily hydrolyzed diazinon and then metabolized IMHP as sole sources of carbon and energy (40) under laboratory conditions.

Degradation of nitrophenols in soils, by enrichment cultures and by pure cultures of bacteria has been demonstrated (36). A *Pseudomonas* sp. (ATCC 29353) and a *Bacillus* sp. (ATCC 29355) isolated from parathion-treated flooded alluvial soil (35), metabolized *p*-nitrophenol to nitrite and CO_2. The *Pseudomonas* sp. (ATCC 29353) also produced nitrite from 2,4-dinitrophenol, after a lag period; whereas *o*- and *m*-nitrophenols resisted degradation by this bacterium. The *Bacillus* sp. (ATCC 29355) effected rapid transformation of *m*-nitrophenol to nitrite and phenol (36). Generally meta-substituted aromatic compounds are known for their recalcitrance to biodegradation. Yet another bacterium, *Pseudomonas* sp. (ATCC 29354) isolated from parathion-treated flooded soil, converted *p*-nitrophenol to 4-nitrocatechol (50) which was not further degraded. In a study on the metabolism of the methyl carbamate pesticides, carbofuran and carbaryl, by soil enrichment and a *Bacillus* sp. isolated from flooded soils, carbaryl was hydrolyzed to 1-naphthol which were then converted to 1,4-naphthoquinone, a more persistent metabolite (51). Enrichment cultures from flooded soil and pure cultures isolated from these enrichments transformed carbofuran to 7-phenol as the major metabolite. Other metabolites, 3-hydroxycarbofuran and 3-ketocarbofuran, were further metabolized while 7-phenol resisted further degradation. However, an *Arthrobacter* sp. isolated from carbofuran-treated flooded soil, mineralized carbofuran to CO_2 possibly via 7-phenol which was detected only in small amounts (43).

γ-Hexachlorocyclohexane undergoes rapid degradation in flooded rice soils (*52*). Anaerobic degradation of the γ-isomer of HCH in flooded rice soils (*53*) and in microbial culture (*54*) leads to the formation of γ-tetrachlorocyclohexene, a major transitory metabolite. Following treatment of flooded rice field with γ-hexachlorocyclohexane, γ-tetrachlorocyclohexene was detected as a major but transitory metabolite in soils up to 60 cm depth (Adhya, T.K. 1990, unpublished data).

The rate and pathway by which a pesticide is degraded by pure microbial cultures is affected by nutrients, such as yeast extract. A *Bacillus* sp. isolated from a flooded soil pretreated with methyl parathion, degraded methyl parathion, parathion and fenitrothion in a mineral salts medium supplemented with yeast extract (*55*). An increase in the concentration of yeast extract not only increased the rate of degradation of these phosphorothioates but also altered the metabolic pathways by which methyl parathion and parathion were degraded. Bacterial degradation of methyl parathion and parathion led to the formation of *p*-nitrophenol at 0.05% yeast extract concentration. Both *p*-nitrophenol and amino-analogs of the pesticides were formed in the presence of 0.1% concentration of yeast extract, whereas only amino-analogs were detected when 0.5% yeast extract was added to the medium (Figure 3). This phenomenon was quite interesting because the same bacterium produced both the amino analogs and *p*-nitrophenol during degradation of parathion and methyl parathion depending on the concentration of yeast extract used. Moreover, the degradation of fenitrothion by the bacterium proceeded only by hydrolysis to 3-methyl- 4-nitrophenol and then to nitrite at all concentrations of yeast extract. This is analogous to the inhibition of hydrolysis in a flooded soil amended with organic matter such as rice straw (*25, 56*).

Bound Residues. Bound or unextractable residues may represent the undegraded parent pesticides and/or their degradation products. These residues are tightly bound to the soil and are not recovered by conventional extraction procedures used in pesticide residue analysis (*57*). The formation and accumulation of soil-bound pesticide residues is of increasing environmental concern. For most pesticides, the exact mechanism of soil-binding and the ultimate fate of bound residues as well as their toxicological significance are not understood.

Accumulation of soil bound residues in flooded soil was first demonstrated with parathion (*58, 59*). More than 66% of the ^{14}C in the ring-^{14}C-parathion applied to a flooded loam soil was accounted for as soil-bound residues after 2 weeks. In the flooded soil, parathion was first converted to aminoparathion which appeared to bind to the soil rapidly. In non-flooded soil, formation of bound residues was negligible. When ^{14}C-labeled nitro compounds (parathion, paraoxon and p-nitrophenol) and their respective amino analogs were incubated with moist soil for 2 hours, amino compounds were bound to the soil to a greater extent (14-26 x) than the parent nitro compounds (*60*). Conditions (flooding, high organic matter content, high temperature and increased

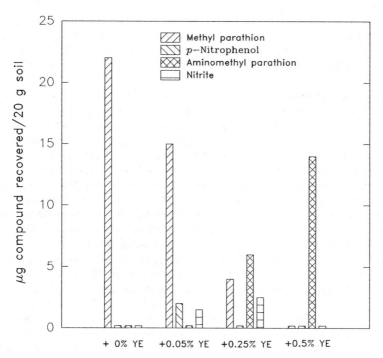

Figure 3. Degradation of methyl parathion by a *Bacillus* sp. in a mineral salts medium supplemented with yeast extract (YE), 2 days incubation. Reproduced with permission from ref. 55. Copyright 1989 National Research Council of Canada.

microbial activity) which catalyze the formation of aminoparathion enhanced eventual accumulation of soil-bound residues (*57*).

The extent of soil-bound residues from [14]C-*p*-nitrophenol was related to soil type and aerobic status of the flooded soil. Aeration of the flooded soil by stirring decreased the amount of soil-bound pesticide residues. Soil-binding was more pronounced in organic matter rich pokkali soil (8.2% organic matter) than in alluvial soil (1.6% organic matter). More than 55% of the ^{14}C in [14]C-*p*-nitrophenol was transformed to the soil-bound residues in pokkali soil as compared to 28 to 44% in alluvial soil under anaerobic flooded soil conditions (*61*). Metabolism of *p*-nitrophenol was negligible under these conditions and nitrite was not detected.

Inoculation of flooded soils with a parathion-hydrolyzing enrichment culture resulted in the accumulation of soil-bound residues from [14]C-*p*-nitrophenol, similar to the soil bound residues in the uninoculated soil. The amount of soil-bound residues in inoculated and uninoculated soils aerated by stirring was almost the same although differences in CO_2 evolution were observed. Carbon dioxide evolution from inoculated pokkali soil was approximately 13% and from inoculated alluvial soil was 83%. Insignificant amounts of CO_2 were evolved from both uninoculated soils. It appears that the degradation of p-nitrophenol under anaerobic conditions can lead to significant formation of soil-bound residues. Evidence suggests that both hydrolysis and nitro group reduction of parathion can result in the eventual accumulation of soil-bound residues.

Hydrolysis products of carbofuran and carbaryl, 7-phenol and 1-naphthol respectively, end up as soil-bound residues in flooded soils. Application of ring-labeled and carbonyl-labeled ^{14}C-carbofuran to a flooded soil, resulted in accumulation of more soil bound residues from ring-^{14}C (26%) than from the carbonyl (7.3%) label (24). Conversely, more $^{14}CO_2$ was released from carbonyl label (27%) than from ring label (0.3%). The aryl group may contribute significantly to the bound residues from carbofuran in flooded soil.

Interestingly, unlike parathion, flooding retarded the accumulation of bound residues from ^{14}C-γ-HCH in some of the selected Indian soils and appeared to accelerate the formation of volatile metabolites (62). The addition of green manure to the flooded soil resulted in further decrease in the soil-bound residues which probably explains the low concentration of bound residues from γ-HCH found under flooded conditions. Non-flooded and flooded soils containing bound residues of γ-HCH were incubated under moist conditions. During the incubation, extractable residues were formed from bound residues in both flooded and non-flooded soils, but to a less extent in the flooded soil. More $^{14}CO_2$ was evolved from the bound residues in flooded soil than from non-flooded soil.

With DDT, an organochlorine insecticide, formation of bound residues in soil was greater in green manure-amended soil (30%) than in unamended soil (16%) at 80 days under flooded conditions (15).

In a classical study, Bartha and Pramer (63) demonstrated formation of persistent soil-bound residues from propanil, a herbicide, after its hydrolytic conversion to 3,4-dichloroaniline in flooded rice soils (64).

Discussion

Little attention has been paid to the environmental significance and fate of degradation products of commonly used pesticides in tropical rice growing soils. An aerobic-anaerobic interface, prevalent in flooded rice fields, may facilitate the complete mineralization of a pesticide to innocuous inorganic end products. There is evidence that anaerobic flooded conditions may favor the formation and accumulation of soil-bound residues from certain pesticides. Accumulation of pesticide residues in soils may result from binding of metabolites and inhibition of ring cleavage under anaerobic conditions. Pesticides, such as ethyl parathion or methyl parathion, undergo degradation via more than one primary pathway, therefore, the type and amount of metabolite(s) formed are dependent on many factors, i.e. soil type, organic matter, moisture regime, temperature, redox status, sulfate content and number of pesticide applications. Generally, no energy for growth is derived during hydrolytic reactions, a common route for biodegradation of many pesticides. However, accelerated biodegradation after repeated pesticide application is common with many organophosphorus and carbamate pesticides which undergo primary degradation by hydrolysis. Degradation products from these pesticides appear to serve as an energy source for growth and proliferation of microorganisms capable of hydrolyzing the parent molecule.

Acknowledgments. We thank Dr. S. Patnaik for permission to publish this manuscript. M. Sharmila was supported by a fellowship grant from the Council of Scientific and Industrial Research, New Delhi. This research was supported in part by US-India Fund (Research Grant No. FG-IN-693).

Literature Cited

(1) Ponnamperuma, F.N. *Adv. Agron.* **1972,** *24,* 29.

(2) Sethunathan, N.; Rao, V.R.; Adhya, T.K.; Raghu, K. *CRC Crit. Rev. Microbiol.* **1983,** *10,* 125.

(3) Adhya, T.K. *Ph.D. Thesis, Calcutta University, West Bengal, India,* **1986,** pp. 140.

(4) Adhya, T.K.; Wahid, P.A.; Sethunathan, N. *Biol. Fert. Soils.* **1986,** *4,* 36.

(5) Reddy, B.R.; Sethunathan, N. *Soil Biol. Biochem.* **1985,** *17,* 235.

(6) Panda, S.; Sethunathan, N. *Pestic. Res. J.* **1989,** *1,* 1.

(7) Wahid, P.A.; Ramakrishna, C.; Sethunathan, N. *J. Environ. Qual.* **1980,** *9,* 127.

(8) Wahid, P.A.; Sethunathan, N. *Nature (London)* **1979,** *282,* 401.

(9) Rao, Y.R.; Sethunathan, N. *J. Environ. Sci. Health* **1979,** *14B,* 335.

(10) Adhya, T.K.; Barik, S.; Sethunathan, N. *Pestic. Biochem. Physiol.* **1981,** *16,* 14.

(11) Sethunathan, N.; Siddaramappa, R.; Rajaram, K.P.; Barik, S.; Wahid, P.A. *Residue Rev.* **1977,** *68,* 91.

(12) Lichtenstein, E.P.; Fuhremann, T.W.; Schulz, K.R. *J. Agric. Food Chem.* **1971,** *19,* 718.

(13) Guenzi, W.D.; Beard, W.E. *Science* **1967,** *156,* 1116.

(14) Castro, T.F.; Yoshida, T. *J. Agric. Food Chem.* **1971,** *19,* 375.

(15) Mitra, J.; Raghu, K. *Environ. Tech. Lett.* **1988,** *9,* 847.

(16) Gowda, T.K.S.; Sethunathan, N. *Soil Sci.* **1977,** *124,* 5.

(17) Gowda, T.K.S.; Sethunathan, N. *J. Agric. Food Chem.* **1976,** *24,* 750.

(18) Venkateswarlu, K.; Gowda, T.K.S.; Sethunathan, N. *J. Agric. Food Chem.* **1977**, *25*, 533.

(19) Siddaramappa, R.; Seiber, J.M. *Prog. Water Technol.* **1979**, *11*, 103.

(20) Venkateswarlu, K.; Chendrayan, K.; Sethunathan, N. *J. Environ. Sci. Health* **1980**, *15B*, 421.

(21) Rajagopal, B.S.; Chendrayan, K.; Reddy, B.R.; Sethunathan, N. *Plant Soil* **1983**, *73*, 35.

(22) Siddaramappa, R.; Sethunathan, N. *Pestic. Sci.* **1975**, *6*, 395.

(23) Ferreira, J.; Raghu, K. *Environ. Tech. Lett.* **1981**, *2*, 357.

(24) Venkateswarlu, K.; Sethunathan, N. *J. Environ. Qual.* **1979**, *8*, 365.

(25) Sethunathan, N. *Soil Biol. Biochem.* **1973**, *5*, 641.

(26) Yoshida, T.; Castro, T.F. *Soil Sci. Soc. Amer. Proc.* **1970**, *34*, 440.

(27) Sharmila, M.; Ramanand, K.; Adhya, T.K.; Sethunathan, N. *Soil Biol. Biochem.* **1988**, *20*, 399.

(28) Sharmila, M.; Ramanand, K.; Sethunathan, N. *Bull. Environ. Contam. Toxicol.* **1989**, *43*, 45.

(29) Sahoo, A.; Sahu, S.K.; Sharmila, M.; Sethunathan, N. *Bull. Environ. Contam. Toxicol.* **1990**, *44*, 948.

(30) Sethunathan, N. *PANS.* **1971**, *17*, 18.

(31) Felsot, A. *Ann. Rev. Entomol.* **1989**, *34*, 453.

(32) Barik, S.; Wahid, P.A.; Ramakrishna, C.; Sethunathan, N. *J. Agric. Food Chem.* **1979**, *27*, 1391.

(33) Rajagopal, B.S.; Brahmaprakash, G.P.; Reddy, B.R.; Singh, U.D.; Sethunathan, N. *Residue Rev.* **1984**, *93*, 1.

(34) Ferris, I.G.; Lichtenstein, E.P. *J. Agric. Food Chem.* **1980**, *28*, 1011.

(35) Siddaramappa, R.; Rajaram, K.P.; Sethunathan, N. *Appl. Microbiol.* **1973**, *26*, 846.

(36) Barik, S.; Siddaramappa, R.; Sethunathan, N. *Antonie von Leeuwenhoek, J. Microbiol. Serol.* **1976**, *42*, 461.

(37) Sethunathan, N.; Yoshida, T. *J. Agric. Food Chem.* **1969**, *17*, 1192.

(38) Sethunathan, N.; Pathak, M.D. *J. Agric. Food Chem.* **1972**, *20*, 586.

(39) Sethunathan, N.; Pathak, M.D. *Can. J. Microbiol.* **1971**, *17*, 669.

(40) Sethunathan, N.; Yoshida, T. *Can. J. Microbiol.* **1973**, *19*, 873.

(41) Rajagopal, B.S.; Panda, S.; Sethunathan, N. *Bull. Environ. Contam. Toxicol.* **1986**, *36*, 827.

(42) Ramanand, K.; Panda, S.; Sharmila, M.; Adhya, T.K.; Sethunathan, N. *J. Agric. Food Chem.* **1988**, *36*, 200.

(43) Ramanand, K.; Sharmila, M.; Sethunathan, N. *Appl. Environ. Microbiol.* **1988**, *54*, 2129.

(44) Karns, J.S.; Mulbry, W.W.; Nelson, J.O.; Kearney, P.C. *Pestic. Biochem. Physiol.* **1986**, *25*, 211.

(45) Barik, S.; Sethunathan, N. *J. Environ. Qual.* **1978**, *7*, 349.

(46) Reddy, B.R.; Sethunathan, N. *Appl. Environ. Microbiol.* **1983**, *45*, 826.

(47) Brahmaprakash, G.P.; Reddy, B.R.; Sethunathan, N. *Biol. Ferti. Soils* **1985**, *1*, 103.

(48) Sethunathan, N.; Yoshida, T. *Plant Soil* **1973**, *38*, 663.

(49) Haider, K.; Jagnow, G. *Arch. Microbiol.* **1975**, *104*, 113.

(50) Barik, S.; Siddaramappa, R.; Wahid, P.A.; Sethunathan, N. *Antonie von Leeuwenhoek, J. Microbiol. Serol.* **1978**, *44*, 171.

(51) Rajagopal, B.S.; Rao, V.R.; Nagendrappa, G.; Sethunathan, N. *Can. J. Microbiol.* **1984**, *30*, 1458.

(52) Raghu, K.; MacRae, I.C. *Science* **1966**, *154*, 263.

(53) Tsukano, Y.; Kobayashi, A. *J. Agric. Biol. Chem.* **1972**, *36*, 166.

(54) Heritage, A.O.; MacRae, I.C. *Appl. Environ. Microbiol.* **1977**, *33*, 1295.

(55) Sharmila, M.; Ramanand, K.; Sethunathan, N. *Can. J. Microbiol.* **1989**, *35*, 1105.

(56) Rajaram, K.P.; Sethunathan, N. *Soil Sci.* **1975**, *119*, 296.

(57) Lichtenstein, E.P. *Residue Rev.* **1980**, *76*, 147.

(58) Katan, Y.; Fuhremann, T.W.; Lichtenstein, E.P. *Science* **1976**, *193*, 891.

(59) Lichtenstein, E.P.; Katan, Y.; Anderegg, B.A. *J. Agric. Food Chem.* **1977**, *25*, 43.

(60) Katan, Y.; Lichtenstein, E.P. *J. Agric. Food Chem.* **1977**, *25*, 1401.

(61) Barik, S. *Ph.D. Thesis, Utkal University, Orissa, India* **1978**, pp. 143.

(62) Raghu, K.; Drego, J. *Quantification, Nature and Bioavailability of Bound ^{14}C-Pesticide Residues in Soil, Plants and Food; International Atomic Energy Agency, Vienna, Austria* **1986**, 41-50.

(63) Bartha, R.; Pramer, D. *Adv. Appl. Microbiol.* **1970**, *13*, 317.

(64) Bartha, R. *J. Agric. Food Chem.* **1971**, *19*, 385.

RECEIVED December 6, 1990

Chapter 5

Degradation Products of Sulfur-Containing Pesticides in Soil and Water

C. J. Miles

Department of Agricultural Biochemistry, University of Hawaii, Honolulu, HI 96822

Pesticide degradation products, especially those containing sulfur, are important groundwater contaminants. Sulfoxidation is one of the most important pesticide degradation pathways since the sulfoxide product has much higher groundwater contamination potential. Reactions involving sulfur chemistry of aldicarb, ethylene dibromide, ethylenethiourea, and malathion in soil and water are discussed.

Occurrence of pesticides in groundwater is well documented. Although most of this contamination is from non-point agricultural sources, point sources such as pesticide mixing and loading sites (1), agrochemical dealerships (2), and pesticide applicator cleaning sites (3) can contribute to this problem significantly. Most groundwater contamination occurs by percolation of chemicals through overlying soils and other geological formations. Thus, physical, chemical, and biological reactions within these soils will determine the form and concentration of the chemical if and when it appears in groundwater.

Many pesticide degradation products have appeared in groundwater. The U.S. Environmental Protection Agency (USEPA) is conducting a national pesticide survey of groundwater and many of the 123 analytes are pesticide degradation products. Hydrolysis, oxidation, and other chemical and biological reactions often convert the parent pesticides into products that are more water soluble and thus more mobile in soil. Because of the reactivity of sulfur, several of these degradation products contain oxidized sulfur moieties. This chapter will focus on sulfur-containing pesticide degradation products in soil and water. Formation and reaction of these compounds in soil and water and by general synthetic routes is discussed along with movement in soils to groundwater. Cases studies of aldicarb, ethylene dibromide (EDB), ethylenethiourea (ETU), and malathion are presented.

0097–6156/91/0459–0061$06.00/0

Pesticide Degradation Products Found in Groundwater

The USEPA has assembled a data base of pesticides in groundwater and the 1988 interim report lists a total of 54 pesticides (14 S-containing) with confirmed detections (*4*). This report also lists several pesticide metabolites that are potential groundwater contaminants and most of these compounds are sulfur-containing products. Many are oxidation products since sulfoxidation increases water solubility and subsequent groundwater contamination potential. The final report of the U.S. Environmental Protection Agency (EPA) National Pesticide Survey (NPS) of groundwater is due soon and should contain a representative account of pesticides in groundwater. Table I lists sulfur-containing pesticides and degradation products that are confirmed or suspected groundwater contaminants including those on the NPS list. Health Advisory Levels (HAL) developed by the USEPA are also listed and help identify toxicologically significant compounds.

Formation of Degradation Products of Sulfur-Containing Pesticides in Soil, Water and Model Systems

Chemical and biologically mediated reactions can transform S-containing pesticides in soils and water (*5*)(Figure 1). Usually microbial activity is the major degradative pathway, but abiotic (acellular) reactions can dominate for some compounds and/or environmental conditions. Pesticide selectivity is often achieved by using relatively stable, nonpolar compounds (propesticides) that are metabolized to the active toxicant. Most S-containing pesticides are propesticides that are metabolically activated by reactions involving or initiated by sulfoxidation (*6*). These oxidations alter the reactivity, solubility, and ease of translocation of systemic pesticides.
    One of the most important degradative pathways for sulfide pesticides is oxidation to the sulfoxide and the sulfone (reaction 1; Figure 1)(*7*). In the laboratory, sulfoxides are prepared by addition of one equivalent of oxidant to the sulfide although careful selection and addition of oxidant is important. Sulfones are easily prepared by addition of excess oxidant although oxidation of the sulfoxide is usually slower than oxidation of the sulfide (*8*).
    In soils, sulfide pesticides are often rapidly oxidized to sulfoxides while sulfones are formed more slowly (*7*). Sulfoxidation has been observed on compounds with an alkylthio group attached to another alkyl group (aldicarb; Figure 2) and to an aromatic group (fenamiphos). This oxidation is so rapid and complete that sulfoxides are often the dominant species found in soil shortly after application of the parent sulfide. In most cases, the sulfoxide and sulfones also have pesticidal activity. Aldicarb oxidation is slowed considerably in sterilized soils and with increasing depth in the soil profile suggesting that microbes mediate this oxidation (*9*). Sulfoxidation of alkylthiocarbamates (molinate) and alkylthiotriazines (ametryn) has been observed in biological systems (*6*) but these sulfoxides have not been reported as major soil or water degradation products.
    Sulfoxides can be reduced to sulfides by $LiAlH_4$ and other reagents but many sulfones are stable to reducing agents (*8*) (reaction 2). Both sulfoxides and sulfones can be reduced by heating with sulfur (which is oxidized to SO_2), and the reaction

Table I.   **Sulfur-containing pesticides and degradation products that are confirmed or potential groundwater contaminants**

PESTICIDE	DEGRADATION PRODUCT(S)	STATUS[a]	HAL (μg/L)[b]
aldicarb	sulfoxide and sulfone	C	10, 10, 40[c]
ametryn	des-ethyl	C	60
bentazon		NPS	20
butylate		NPS	700
carboxin	sulfoxide (and sulfone)	NPS	700
cycloate		NPS	
diazinon	diazoxon	C	0.6
disulfoton	sulfoxide and sulfone	NPS	0.3
EDB, etc.	(alkylmercaptans, dialkylsulfides, and dialkyldisulfides)	C	0.0004
EPTC		NPS	
endosulfan	sulfate	C	
ethoprop		C	
EBDC	ethylenethiourea	C	0.2
ethyl parathion	ethyl paraoxon	C	
fenamiphos	sulfoxide and sulfone	NPS	2
malathion	maloxon	C	
merphos		NPS	
methamidophos		C	
methiocarb		C	
methomyl		C	200
methyl parathion	methyl paraoxon	C	
metribuzin	(metribuzin DA)	C	200
molinate	sulfoxide	C	
oxamyl		C	200
pebulate		NPS	
prometryn		NPS	
simetryn		NPS	
sulprofos		C	
tebuthiuron		NPS	500
terbufos		NPS	1
terbutryn		NPS	
triallate		C	
tricylazole		NPS	
vernolate		NPS	

[a] C = confirmed, NPS = from the USEPA National Pesticide Survey list.
Note: most of those confirmed are also on NPS list. Parentheses denote a known degradation product but not previously listed.
[b] Health Advisory Levels for parent compounds except aldicarb products and ETU.
[c] aldicarb, aldicarb sulfoxide, and aldicarb sulfone.

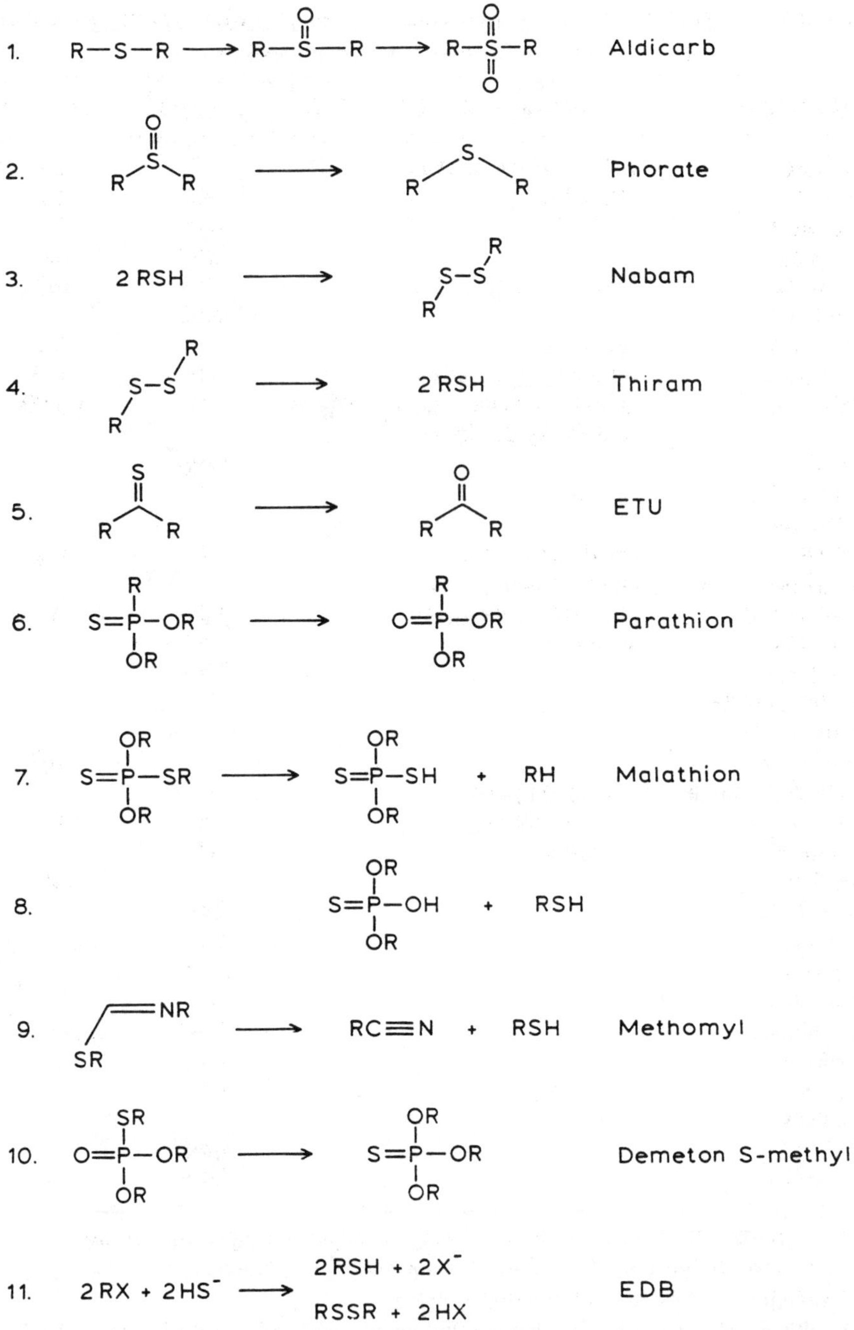

Figure 1.   Reactions of sulfur-containing pesticides in soil and water and examples of each.

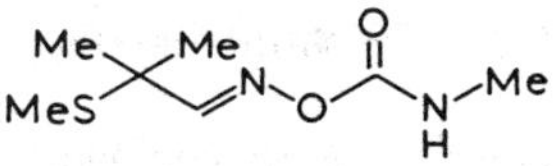

ALDICARB

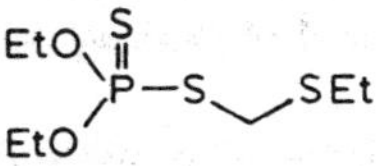

PHORATE

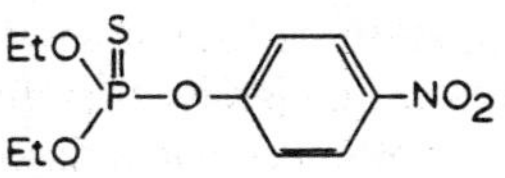

PARATHION

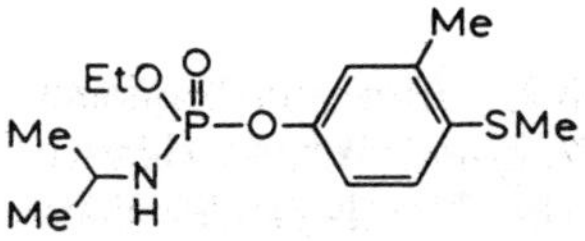

FENAMIPHOS

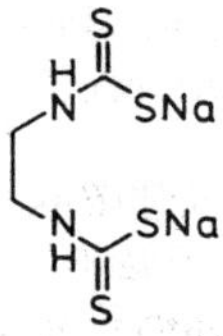

NABAM

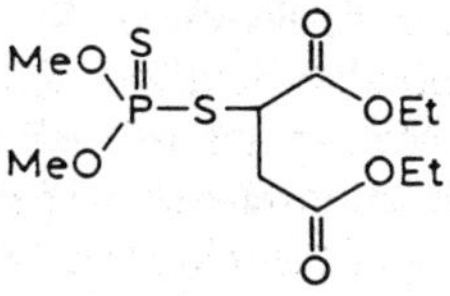

MALATHION

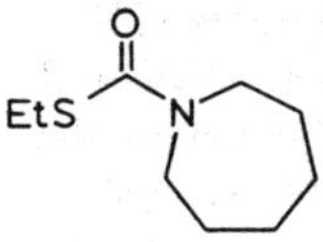

MOLINATE

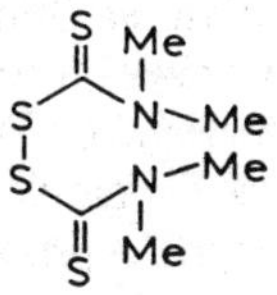

THIRAM

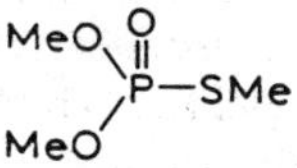

O,O,S-TRIMETHYL
PHOSPHOROTHIOATE

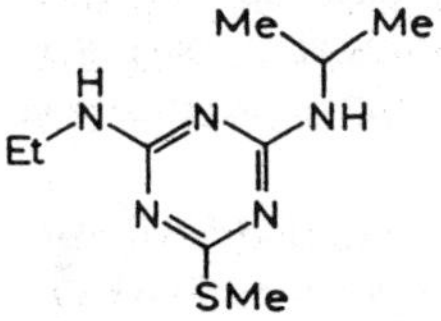

AMETRYN

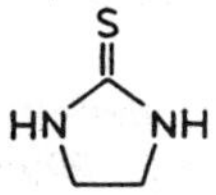

ETU

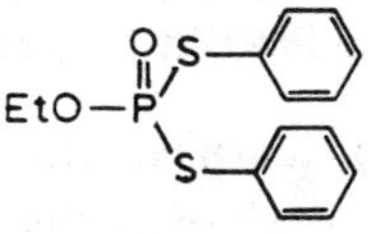

HINOSAN

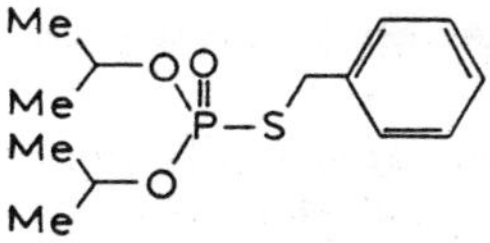

KITAZIN P

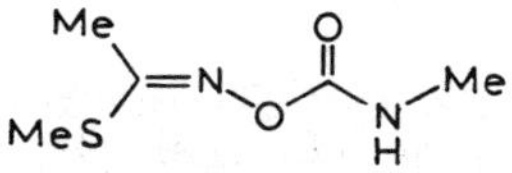

METHOMYL

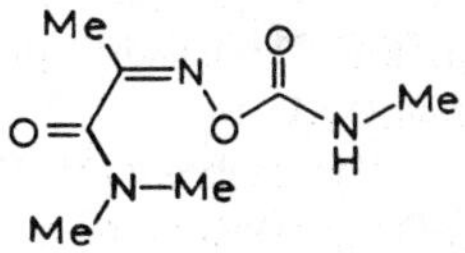

OXAMYL

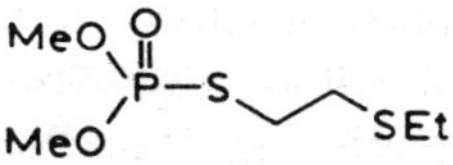

DEMETON S-METHYL

Figure 2. Structures of the pesticides discussed.

with sulfoxides proceeds at lower temperatures. Experiments with ^{35}S-labelled substrates show that the sulfoxides transfer oxygen to elemental sulfur. On the other hand, sulfones lose most of the label indicating that the sulfur in the sulfide product comes from elemental sulfur (8).

Reduction of aldicarb sulfoxide to aldicarb was observed in groundwater treated with limestone (10, 11). Enhancement of reduction upon addition of glucose and the observation of turbid solutions suggested a microbially-mediated reaction. Phorate sulfoxide was reduced to phorate in lake sediment microcosms (12). These reactions demonstrate the importance of redox conditions on the fate of sulfur-containing pesticides in soils and water.

Mercaptans are easily oxidized to disulfides (reaction 3). With a small amount of base, atmospheric oxygen can be the oxidant. The mechanism involves oxidative coupling of the thiolate anion via a free radical mechanism (8). The fungicide nabam (sodium ethylenebisdithiocarbamate) oxidizes to the disulfide in the presence of atmospheric oxygen (13) and this disulfide is an intermediate in EBDC degradation to ethylenethiourea (ETU) (14). Conversely, disulfides can be reduced to mercaptans by mild reducing agents such as zinc in dilute acid (reaction 4) (8). The fungicide thiram is reduced to dimethyldithiocarbamate in soils and microbial cultures (15).

ETU is a manufacturing byproduct and environmental degradation product of the EBDC fungicides. These widely used fungicides are under Special Review by the USEPA largely because of the carcinogenic properties of ETU. ETU is rapidly oxidized to ethyleneurea (EU) in some soils by an acellular mechanism (reaction 5)(16).

The thione moiety, in organophosphates, can be oxidized to the oxon analog which is often accompanied by an increase in toxicity, water solubility, and susceptibility to hydrolysis (reaction 6). Parathion is oxidized to paraoxon in plants and animals and also by some soil microbes (17, 18). On the other hand, malathion was not oxidized in oxygen-saturated, acidic water microcosms for several weeks (19).

With malathion, the sulfur which is bound to an alkyl chain and phosphorous, can cleave at either position. Under basic conditions, elimination produces O,O-dimethyl phosphorodithioic acid and diethyl fumarate (reaction 7) while hydrolysis produces O,O-dimethyl phosphorothioic acid and diethyl thiosuccinate (reaction #8) (19). The organophosphates Hinosan (S,S,O-diphenylethyl phosphorodithioate) and Kitazin P (O,O,S-diisopropylphenyl phosphorodithioate) follow similar degradation pathways (20).

The carbamoyloximes, methomyl and oxamyl, can undergo a ferrous iron-catalyzed reductive degradation in anaerobic soils (21). A good leaving group on the imino nitrogen facilitates formation of an imino radical which is further reduced to the anion. The anion ultimately forms methanethiol and a nitrile (reaction 9). Aldicarb undergoes a similar reductive reaction to form the corresponding nitrile and aldehyde. Aldicarb sulfoxide and sulfone degraded faster in these reducing soils systems but degradation products were not identified (21). Aldicarb nitrile and aldehyde were the major products observed in aqueous, Cu^{2+}-catalyzed decomposition of aldicarb (22). Aldicarb nitrile has been identified in anaerobic groundwater microcosms enriched with microorganisms and limestone (23).

Demeton (ethyl esters) is a 2:1 mixture of the isomeric thiono and thiolo esters. Isomerization of the thiono form into the thiolo form occurs in aqueous solution at increased temperatures through an ionic mechanism (reaction 10) (7, 24). Lack of rearrangement with the sulfoxide or sulfone supports the formation of a sulfonium cation. The thiolo ester has greater mammalian toxicity and insecticidal activity than the thiono isomer (7). The thiolo ester is also much more water soluble (2000 mg/L) than the thiono ester (60 mg/L) and is more susceptible to hydrolysis (7, 24). Demeton has methyl ester analogs with similar chemistry (7).

Although halogenated alkanes are not S-containing, their reaction with reduced sulfur species (reaction 11) forms important S-containing pesticide degradation products. Because alkyl halides are common groundwater contaminants, this reaction is discussed in a subsequent section.

Movement of Pesticide Metabolites to Groundwater

Considerable effort has been expended to predict movement of pesticides through the unsaturated soil zone into groundwater. Several transport models have been developed and have had varying success predicting pesticide concentrations in soil and groundwater. Two of the most important physical-chemical parameters of pesticides and metabolites used for modeling efforts are the soil sorption coefficient (K_{oc}: organic carbon normalized sorption coefficient) and the degradation rate expressed as the half-life ($t_{1/2}$). In general, the stronger a pesticide is bound to the soil and the faster it degrades, the less likely it will percolate into groundwater. Literature values for these two parameters can deviate widely and this variability significantly affects the validity of transport model predictions.

A problem for transport modelers is lack of physical-chemical data for important pesticide metabolites. Aldicarb and fenamiphos are good examples. Rapid oxidation of the parent sulfide to the sulfoxide in soils makes transport modelling with parent input data irrelevant. In fact, when one considers that water solubilities increase and sorption and degradation rates decrease for these sulfoxide metabolites, the potential to contaminate groundwater increases significantly.

When physical-chemical parameters are not available in the literature for important pesticide metabolites, transport modelers may employ certain techniques to derive these values. Among the possibilities are equations that correlate other known parameters to the desired parameter (25), structure-activity models (26, 27), and indirect chemical methods such as reversed-phase liquid chromatography (LC) (28, 29).

Partition coefficients of pesticides into soil organic matter (K_{oc}) can be related to the octanol/water partition coefficient (P or K_{ow}). Values for K_{ow} derived by the techniques described above have varying success when compared with empirically derived (literature) values (Table II). One must assume that the experimental values are correct but review of the methods used in the original literature is wise. Validation of estimation techniques is also a prudent practice and the two computer models discussed here have been validated for several types of simple organic compounds (26, 27).

Table II. Water solubilities and octanol/water partition coefficients of selected sulfur-containing pesticides

pesticide	water solubility (mg/L)	octanol/water partition coefficients				
		lit.[a]	corr.[b]	model[c]	model[d]	rpLC[e]
aldicarb	6.0×10^3	2.1×10^1	2.8	1.9×10^1	1.7×10^1	1.6×10^1
aldicarb SO	3.3×10^5	0.3	4.0×10^{-2}	NA	0.5	0.7
aldicarb SO2	8.0×10^3	0.3	2.0	NA	0.5	1.2
fenamiphos	5.2×10^2	1.6×10^3	4.0×10^1	2.5×10^3	2.3×10^2	1.5×10^2
fenamiphos SO	NA	NA	NA	NA	3.1	1.9×10^1
fenamiphos SO2	NA	NA	NA	NA	2.8	2.4×10^1
disulfoton	4.6×10^1	1.0×10^4	5.4×10^2	NA	2.8×10^3	2.4×10^2
disulfoton SO	$>4.0 \times 10^3$	6.5×10^1	<4.3	NA	5.4×10^1	4.7×10^1
disulfoton SO2	8.8×10^2	6.0×10^1	2.2×10^1	NA	7.4×10^1	6.0×10^1
demeton S-methyl	3.3×10^3	NA	5.3	NA	1.7×10^2	2.6×10^1
demeton SO	$>3.3 \times 10^3$	NA	$<1.3 \times 10^{-1}$	NA	6.6×10^{-2}	1.3
demeton SO2	3.3×10^3	NA	5.3	NA	3.6×10^{-2}	2.4

SO = sulfoxide; SO2 = sulfone.
a from literature (47, 48).
b from correlation with water solubility [log WS = -0.922 (log K_{ow}) + 4.184](25).
c from Bodar model (27).
d from CLOGP model (26).
e from reversed-phase liquid chromatography (this study).

It is apparent from the selected compounds that oxidation from the sulfide to the sulfoxide significantly increases water solubility and lowers hydrophobic partitioning. Further oxidation to the sulfone has a smaller effect on partitioning behavior. Despite the importance of using reliable water solubility, partitioning data and degradation rates on pesticidal sulfoxides, Table II shows several deficiencies. A major obstacle to obtaining this data is lack of reliable analytical techniques for measuring sulfoxides. Lack of standard materials due to synthetic difficulties, poor extraction efficiency and gas chromatographic tailing caused by the polar nature of the sulfoxide, make acquisition of these data difficult. Problems with standard materials can be overcome with proper attention to details. Extraction and analysis problems can be alleviated by using liquid chromatography with selective detection methods.

Case Studies of S-Containing Pesticide Degradation Products in Soil and Water

Aldicarb. Aldicarb is the active ingredient in Temik insecticide/nematicide and is one of the most acutely toxic pesticides registered for use (LD_{50} = 1 mg/kg rat, oral)(*24*). Wide use of this relatively water-soluble chemical in sandy soils of New York and Florida led to groundwater contamination, extensive monitoring, and Special Review by the USEPA. Monitoring has included over 46,000 analyses in over 28,000 wells of 34 states (*9*). Aldicarb and EDB residues in groundwater represent over half the pesticide contamination episodes in the 1980's and are partially responsible for increased efforts to monitor groundwater for other pesticides.

The primary mode of aldicarb degradation in the root zone is oxidative metabolism by microorganisms, although some chemical hydrolysis may occur (*30-34*). Oxidation to the sulfoxide is rapid while further oxidation to the sulfone is much slower (*31, 34*). In two New York soils, degradation (hydrolysis + oxidation) of aldicarb was about eight times faster than that seen with aldicarb sulfoxide (*34*). Aldicarb sulfone degradation (hydrolysis only) was slightly slower than aldicarb sulfoxide. In these same soils, sorption coefficients for aldicarb were six-fold higher than aldicarb sulfoxide and three-fold higher than aldicarb sulfone. The much higher water solubility (Table II), lower sorption to soil, and slower degradation rate give aldicarb sulfoxide, and to some extent the sulfone, a higher potential for groundwater contamination. By the time a given pulse or application event reaches deep groundwater, the parent pesticide is oxidized completely and not detected. This helps account for the approximate 1:1 ratio observed for aldicarb sulfoxide and aldicarb sulfone in groundwater (*35*).

In water, the primary degradation and detoxification mechanism is acid or base hydrolysis in the order (sulfone > sulfoxide > > sulfide) and the reaction is mostly chemical (*9, 10*). Bank and Tyrrell (*36*) studied degradation of aldicarb over a pH range 3-8.6 and observed aldicarb oxime, nitrile, aldehyde, carbinolamine, 1,3-dimethylurea, and methylamine as products. In base, the oxime was the major product produced by an E1cB hydrolysis mechanism while under acidic conditions, protonation of the ester oxygen and elimination produced primarily the nitrile (36). Metal-ion catalyzed degradation of aldicarb was reported with Cu^{2+} and Fe^{2+} and the nitrile and aldehyde were the major products (*21, 22*). The observation of these products in groundwater microcosms supports the involvement of this mechanism under environmental conditions (*10, 11, 23*). No oxidation of aldicarb was observed in these groundwater microcosms probably because of low concentrations of oxidizing microbes.

None of these nontoxic degradation products has been reported in field water samples largely because of the increased analysis cost and little interest in these low toxicity compounds. Speciation of total aldicarb residues into aldicarb, aldicarb sulfoxide, and aldicarb sulfone for GC analysis required great effort until the LC-postcolumn fluorogenic labeling method made trace water analysis of all three compounds feasible (*9*). This analytical method greatly improved environmental fate studies of aldicarb in soil and water.

EDB and Other Halogenated Alkanes. Ethylene dibromide (1,2-dibromoethane; EDB) has many uses including as a soil fumigant and fuel additive. Since its first reported incidence in Hawaii groundwater in 1980, EDB has been detected in about 13% of over 15,000 wells surveyed (*37*). Although EDB degrades rapidly in surface soils ($t_{1/2} \approx$ days) and has moderate sorption ($K_{oc} \approx 65$), the high water solubility (4250 mg/L) and the significant drop in microbe populations and organic carbon content below the upper soil layer contribute to the high potential for groundwater contamination (*37*). Also, an extremely low method detection limit (0.01 μg/L) and the fact that most surveys were conducted in vulnerable areas contribute to the high detection percentage. The forthcoming EPA National Pesticide Survey should give a more representative account of EDB contamination in groundwater.

Halogenated alkanes, such as EDB, can react with reducing sulfur species, such as bisulfide (HS⁻), in anaerobic groundwater to form several products. Depending upon environmental conditions and substrate structure, these alkyl halides may undergo nucleophilic substitutions, dehydrohalogenation, or reductive dehalogenation reactions (*38*).

Primary alkyl halides react with the strong nucleophile HS⁻ by an S_N2 mechanism (*8*). Tertiary halides react by an S_N1 mechanism while secondary alkyl halides probably react by a mixture of both mechanisms (*39*). Alkyl bromides react faster than the corresponding chlorides because bromide is a better leaving group. Aqueous solutions of HS⁻ and alkyl halide produced alkylmercaptans, dialkyldisulfides, and dialkylsulfides in groundwater and laboratory microcosms (*40*). EDB, an alkyl dihalide, reacted by both inter- and intra-molecular pathways producing mostly cylic alkylsulfides and cyclic disulfides in groundwater microcosms (*41*).

Bisulfide can also promote dehydrohalogenation of alkyl halides (E2 mechanism) by functioning as a Bronsted base in general base catalysis. Reducing sulfur species (i.e HS⁻, SO_3^{2-}, and $S_2O_3^{2-}$) can also transform polyhalogenated alkanes by reductive dehalogenation. These reducing agents may function either as free radicals (in one-electron transfer reactions) or as nucleophiles (in two-electron transfer reactions)(*38*). Substrate structure and environmental conditions will determine the relative rates of the above reactions and other competing reactions such as hydrolysis.

ETU. Ethylenethiourea (ETU), a carcinogenic byproduct and environmental degradation product of the EBDC fungicides, has been found in groundwater (*42*). ETU is relatively mobile in most soils except those high in organic matter. Half-lives in soil range from 1 day to 4 weeks (*43*). Groundwater contamination potential measured by simple transport models (i.e. Attenuation Factor) range from very high contamination probability to very low contamination probability depending upon which half-life is used. This suggests that specific environmental conditions and agricultural practices may have significant effects on groundwater contamination.

ETU was relatively stable in groundwater microcosms ($t_{1/2}$ ca. 3 yrs)(*16*). However, addition of basalt to simulate groundwater conditions accelerated degradation about 10-fold suggesting either microbes introduced with the basalt or the surface effect discussed below were responsible.

ETU degraded rapidly in three Hawaiian soils ($t_{1/2}$ = 2.1 d) to form ethyleneurea (EU) and sulfate ion (*16*). EU subsequently degraded to ethylenediamine and carbon dioxide. Sterilization of soils with gamma irradiation had little effect on degradation rate while azide decreased the rate about 10-fold. The rate of ETU degradation in water was decreased by either mannitol addition or low oxygen concentrations. Since the rate increased upon addition of soil surfaces, ferric iron and metal chelators, an enzymatic or chemical degradation mechanism involving an acellular oxidizing species was suggested. A mechanism of ETU oxidation in these soils consistent with the experimental observations requires the chelation of ferric iron by soil surface components, reduction of the complex to the ferrous state by unidentified electron donors and subsequent reaction with molecular oxygen to generate hydroxyl radicals, the ultimate oxidant. Although ETU is very water soluble (20,000 mg/L) and weakly sorbed to these Hawaiian soils (avg. K_{oc} = 3.7), the short half-life in soils indicates that groundwater contamination potential is low.

Malathion and O,O,S-Trimethyl phosphorothioate. Malathion is a widely used organophosphate insecticide. The minor contaminants, O,O,S-trimethyl phosphorothioate and O,S,S-trimethyl phosphorodithioate occur by side reactions during manufacture of several organophosphates and cause delayed neurotoxicity in mammals (*44*).

Contamination of groundwater by malathion is unlikely since sorption to soils is high (K_{oc}=1797) and degradation is very fast ($t_{1/2}$= 1d)(*45*). Surface water contamination is likely, however, since runoff may release sorbed malathion and aerial applications are often made to control fruit fly infestation. In addition to elimination (reaction #7) and hydrolysis (reaction #8), malathion also undergoes ester hydrolysis. At low temperatures hydrolysis is predominant while elimination is favored at higher temperatures (*19*). Degradation of malathion in river, ground, and seawater ($t_{1/2}$ = 4.7 ± 0.7 d) occurred predominantly through the elimination reaction while photolysis and biodegradation played minor roles (*46*).

O,O,S-trimethyl phosphorothioate degraded more slowly in water than malathion ($t_{1/2}$ = 32 ± 21 d) and biodegradation was an important pathway (*46*). This compound was weakly bound to soils (K_{oc} = 9.8) and had relatively slow soil degradation rates ($t_{1/2}$ = 6 d) which were also controlled by microbes. Although low sorption and slow degradation rates indicate a high groundwater contamination potential, extremely low concentrations found in dilute formulations applied in fruit-fly control should not present a groundwater problem. On the other hand, point source contamination by O,O,S-trimethyl phosphorothioate would be a greater problem than malathion.

Summary

Pesticide degradation products are important groundwater contaminants. This paper discusses formation of sulfur-containing pesticide degradation products in soil and water and transport to groundwater. Sulfoxidation is one of the most important pesticide degradation pathways. The sulfoxide product has a much higher water solubility, lower sorption to soil, and slower degradation rate than the parent sulfide thus a higher groundwater contamination potential. The reaction mechanisms and cases studies discussed demonstrate the importance of detailed environmental fate investigations to understand and control groundwater contamination.

Acknowledgments

I thank Marc Brewster and Sondra Hollister for log P calculations, Daniel Doerge for helpful discussions, and Wendy Oshiro for help preparing the manuscript. This is Journal Series Number 3490 from the Hawaii Institute of Tropical Agriculture and Human Resources.

Literature Cited

1.   Miles, C.J.; Yanagihara, K.; Ogata, S.; Van De Verg, G.; Boesch, R. *Bull. Environ. Contam. Toxicol.* **1990**, *44*, 955-962.
2.   Mueller, W. *Agrichem. Age* **1989**, *33*, 10-12.
3.   *Treatment and Disposal of Pesticide Wastes;* Krueger, R.F.; Seiber, J.S., Eds.; ACS Symposium Series 259; American Chemical Society: Washington, D.C., 1984.
4.   Williams, W.M.; Holden, P.W.; Parsons, D.W.; Lorber, M.N. *Pesticides in Ground Water Data Base 1988 Interim Report;* U.S. Environmental Protection Agency: Washington, D.C., December, 1988.
5.   Alexander, M. *Science* **1981**, *211*, 132-138.
6.   *Sulfur in Pesticide Action and Metabolism;* Rosen, J.D.; Magee P.S.; Casida, J.E., Eds.; ACS Symposium Series 158; American Chemical Society: Washington, D.C., 1981.
7.   *Chemistry of Pesticides;* Buchel, K.H., Ed., John Wiley and Sons: New York, N.Y., 1983.
8.   March, J. *Advanced Organic Chemistry;* John Wiley and Sons: New York, N.Y., 1985; Third Edition.
9.   Moye, H.A.; Miles, C.J. *Rev. Environ. Contam. Toxicol.* **1988**, *105*, 99-146.
10.   Miles, C.J.; Delfino, J.J. *J. Agric. Food Chem.* **1985**, *33*, 455-460.

11. Lightfoot, E.N.; Thorne, P.S.; Jones, R.L.; Hansen, J.L.; Romaine, R.R. *Environ. Toxicol. Chem.* **1987,** *6,* 377-394.
12. Walter-Echols, G.; Lichtenstein, E.P. *J. Econ. Entomol.* **1977,** *70,* 505-509.
13. Thorn, G.D.; Ludwig, R.A. *The Dithiocarbamates and Related Compounds;* Elsevier: Amsterdam, 1962.
14. Hylin, J.W. *Bull. Environ. Contam. Toxicol.* **1973,** *10,* 227-233.
15. Odeyemi, O.; Alexander, M. *Appl. Environ. Microbiol.* **1977,** *33,* 784-790.
16. Miles, C.J.; Doerge, D.R. *J. Agric. Food Chem.,* in press.
17. Yu., C.C.; Sanborn, J.R. *Bull. Environ. Contam. Toxicol.* **1975,** *13,* 543-550.
18. Munnecke, D.M.; Hsieh, D.P.H. *Appl. Environ. Microbiol.* **1976,** *31,* 63-69.
19. Wolfe, N.L.; Zepp, R.G.; Gordon, J.A.; Baughman, G.L.; Cline, D.M. *Environ. Sci. Technol.* **1977,** *11,* 88-93.
20. Uesugi, C.; Tomizawa, T.; Murai, T. In *Environmental Toxicology of Pesticides;* Matsumura, F; Boush, G.M.; Misato, T, Eds.; Academic Press: New York, N.Y., 1972, pp 327-339.
21. Bromilow, R.H.; Briggs, G.G.; Williams, M.R.; Smelt, J.H.; Tunistra, L.G.M. Th.; Traag, W.A. *Pestic. Sci.* **1986,** *17,* 535-547.
22. Bank, S.; Tyrrell, R.J. *J. Org. Chem.* **1985,** *50,* 4938-4943.
23. Trehy, M.L.; Yost, R.A.; McCreary, J.J. *Anal. Chem.* **1984,** *56,* 1281-1285.
24. *The Agrochemicals Handbook;* The Royal Society of Chemistry: Nottingham, England, 1987; Second Edition.
24. *Handbook of Chemical Property Estimation Methods;* Lyman, W.J.; Reehl, W.F.; Rosenblatt, D.H., Eds.; McGraw-Hill: New York, N.Y., 1990; Chapter 1.
26. Leo, A.; Jow, P.Y.C.; Silipo, C.; Hansch, C. *J. Med. Chem.* **1975,** *18,* 865-868.
27. Bodar, N.; Gabanyi, Z.; Wong, C.K. *J. Am. Chem. Soc.* **1989,** *18,* 3703-3706.
28. Unger, S.H.; Cook, J.; Hollenberg, J. *J. Pharm. Sci.* **1978,** *67,* 1364-1367.
29. Veith, G.D.; Morris, R.T. *Water Res.* **1979,** *13,* 43-47.
30. Richey, F.A., Jr.; Bartley, W.J.; Sheets, K.P. *J. Agric. Food Chem.* **1977,** *25,* 47-50.
31. Smelt, J.H.; Leistra, M.; Houx, N.W.H.; Dekker, A. *Pestic. Sci.* **1978,** *9,* 279-285.
32. Bromilow, R.H.; Baker, R.J.; Freeman, M.A.H.; Gorog, K. *Pestic. Sci.* **1980,** *11,* 371-378.
33. Ou, L.T.; Edvardsson, K.S.V.; Thomas, J.E.; Rao, P.S.C. *J. Agric. Food Chem.* **1985,** *33,* 72-78.
34. Zhong, W.Z.; Lemley, A.T.; Wagenet, R.J. In *Evaluation of Pesticides in Ground Water;* Garner, W.J.; Honeycutt, R.C.; Nigg, H.N., Eds.; ACS Symposium Series 315, American Chemical Society: Washington, D.C., 1986, pp 61-77.
35. Cohen, S.Z.; Eiden, C.; Lorber, M.N. In *Evaluation of Pesticides in Ground Water;* Garner, W.J.; Honeycutt, R.C.; Nigg, H.N., Eds.; ACS Symposium

Series 315, American Chemical Society: Washington, D.C., 1986, pp 170-196.

36. Bank, A,; Tyrrell, R.J. *J. Agric. Food Chem.* **1984**, *32*, 1223-1232.
37. Pignatello, J.J.; Cohen, S.Z. *Rev. Environ. Contam. Toxicol.* **1990**, *112*, 1-47.
38. Roberts, A.L.; Sanborn, P.N.; Gschwend, P.M. *Abstracts of Papers,* 199th National Meeting of the American Chemical Society, Boston, MA; American Chemical Society: Washington, D.C., 1990, Division of Environ. Chem., paper #43.
39. Schwarzenbach, R.P.; Geiger, W. *In Ground Water Quality;* Ward, C.H.; Geiger, W.; McCarty, P.L., Eds.; John Wiley and Sons: New York, N.Y., 1985, pp 446-471.
40. Schwarzenbach, R.P.; Geiger, W.; Schaffner, C.; Wanner, O. *Environ. Sci. Technol.* **1985**, *19*, 322-327.
41. Moye, H.A.; Weintraub, R.A.; Jex, G. *Proceedings of the First Annual Environmental Toxicology Research Conference;* Florida Department of Environmental Regulation, Tallahassee, FL, 1988, 195-235.
42. Frakes, R.A. *Regul. Toxicol. Pharmacol.* **1988**, *8*, 207-218.
43. IUPAC. *Pure Appl. Chem.* **1974**, *49*, 675-689.
44. Mallipudi, N.M.; Umetsu, N.; Toia, R.F.; Talcott, R.F.; Fukuto, T.R. *J. Agric. Food Chem.* **1979**, *27*, 463-466.
45. *U.S. Department of Agriculture/Agricultural Research Service Interim Pesticide Properties Database;* Wauchope, R.D., Ed.; U.S. Department of Agriculture: Washington, D.C., 1988.
46. Miles, C.J.; Takashima, S. University of Hawaii, *Arch. Environ. Contam. Toxicol.*, in press.
47. Briggs, G.G. *J. Agric. Food Chem.* **1981**, *29*, 1050-1059.
48. *MedChem Software Database;* Department of Chemistry, Pomona College, Claremont, CA, 1984.

RECEIVED November 8, 1990

Chapter 6

Atrazine Metabolite Behavior in Soil-Core Microcosms

Formation, Disappearance, and Bound Residues

D. A. Winkelmann and S. J. Klaine

Environmental Health and Toxicology Research Institute, Department of Biology, Memphis State University, Memphis, TN 38152

Atrazine (2-chloro-4[ethylamino]-6[isopropylamino]-1,3,5-triazine) is a moderately persistent herbicide used extensively to control annual grasses in cornfields. Intact soil-core microcosms were developed and utilized to follow the degradation of atrazine and the formation and subsequent fate of its major metabolites, deethylatrazine (DEA, 2-chloro-4-[amino]-6-[isopropylamino]-1,3,5-triazine), deisopropylatrazine (DIA, 2-chloro-4-[ethylamino]-6-[amino]-1,3,5-triazine), dealkylatrazine (DAA, 2-chloro-4,6-[diamino]-1,3,5-triazine) and hydroxyatrazine (HYA, 2-hydroxy-4-[ethylamino]-6-[isopropylamino]-1,3,5-triazine). Field studies in west Tennessee were also conducted to determine the predictive potential of the microcosms. Microcosm studies indicated an atrazine half-life of 21 days while field data resulted in a 14-day half-life. Metabolite formation and subsequent disappearance were similar in both microcosm and field studies. Bound residue formation (chemical residue not recovered from standard soxhlet extraction) was significant and may result in the underestimation of chemical half-life. This is particularly true for the parent compound and the chlorinated monoalkylatrazines. Soil bound residues for these compounds after 180 days were as high as 60 percent of the initial radioactivity applied to the microcosms.

Atrazine (2-chloro-4[ethylamino]-6[isopropylamino]-1,3,5-triazine) is a moderately persistent herbicide used extensively to control annual grasses in cornfields. The processes that dictate the fate of atrazine include degradation as determined by dissipation of extractable parent molecule (1-3), mineralization as measured by $^{14}CO_2$ evolution from radiolabeled parent (4-7), migration to subsurface levels (8-11); and sorption by soil organic matter, activated carbon and plants (12-16).

Atrazine can be degraded to several metabolites in soil. Among those are deethylatrazine (DEA, 2-chloro-4-[amino]-6-[isopropylamino]-1,3,5-triazine), deisopropylatrazine (DIA, 2-chloro-4-[ethylamino]-6-[amino]-1,3,5-triazine), dealkylatrazine (DAA, 2-chloro-4,6-[diamino]-1,3,5-triazine) and hydroxyatrazine (HYA, 2-hydroxy-4-[ethylamino]-6-[isopropylamino]-1,3,5-triazine). Degradation of the parent herbicide may result in production of both phytotoxic and non-phytotoxic metabolites. DEA and DIA are phytotoxic degradation products. DEA

0097–6156/91/0459–0075$06.00/0

is almost as phytotoxic as the parent atrazine, while DIA is five times less phytotoxic (*17*). DAA has been shown to be nonphytotoxic in oat bioassays (*18*). HYA is also a nonphytotoxic degradation product. Dechlorination and subsequent hydroxylation of atrazine to HYA is considered a major step in the deactivation of the herbicide (*19*).

N-dealkylation, dechlorination and hydroxylation at the 2- position, deamination and ring cleavage are degradative processes that contribute to the mineralization of atrazine. Degradation of the herbicide can be accomplished by biotic and abiotic processes. Bacteria and fungi isolated from soil enrichment cultures have been shown to convert atrazine to various metabolites. Giardina *et al.* (*20*) isolated a *Nocardia* species that was capable of N-dealkylation. Kaufman and Blake (*21*) showed that two fungi, *Aspergillus fumigatus* and *Rhizopus stolonifer*, produced DEA and DIA when incubated with atrazine. Behki and Khan (*22*) isolated three species of *Pseudomonas* which dealkylated the herbicide. They further showed that the two monoalkylated metabolites were hydroxylated at the 2-position by two of the bacteria. Atrazine can also be utilized as a sole carbon and/or nitrogen source by bacteria (*20,22*).

The mineralization and persistence of the herbicide in soil, determined by using ^{14}C ring-labeled atrazine, has been examined (*1,23-26*). Mineralization rates, determined by measuring $^{14}CO_2$ evolution, ranged from 0.005% of the radioactivity after 12 weeks incubation (*11*) to 28% after 24 weeks (*1*). However, the majority of mineralization rates were in the lower half of this range. Few studies have looked at the mineralization of dealkylated metabolites. These metabolites are difficult to recover from soils because they become intimately associated with the soil and are not recoverable by standard solvent extraction techniques.

Persistence, determined by measuring extractable residues of herbicides, does not consider soil bound residues which have been shown to be bioavailable to plants in some cases (*27-30*). Bound residues have been defined as those molecules which remain associated with various fractions of soil after exhaustive extraction with polar and nonpolar solvents (*31*). Recently, Schiavon (*32,33*) examined the persistence of several radiolabeled metabolites of atrazine, based on extractable and bound ^{14}C residue concentrations, and movement of radioactivity through a soil column. He found that organic matter and loss of alkyl chain were factors that enhanced bound residue formation, while hydroxylation diminished this process. Capriel *et al.* (*34*) showed that a portion of the bound residues extracted from soil still contained the parent atrazine 9 years after application. Since it has been shown that soil bound pesticides are bioavailable (*35-37*), this would suggest that the reported concentrations of the parent compound and metabolites, as determined by conventional extraction procedures, may underestimate true total soil residues.

This paper describes the fate of atrazine metabolites, both in laboratory microcosms and in a west Tennessee field. Further, we will consider the degradation of these metabolites, including bound residue formation.

Materials and Methods

Chemicals. Uniformly ^{14}C ring-labeled atrazine and metabolites of atrazine were supplied by the Ciba-Geigy Corporation. Specific activities of the radiolabels were:

[U-ring-^{14}C]atrazine (ATZ, 97.6% pure), 22.0 μCi/mg;
[U-ring-^{14}C]deethylatrazine (DEA, 99.0% pure), 21.3 μCi/mg;
[U-ring-^{14}C]deisopropylatrazine (DIA, 99.0% pure), 13.7 μCi/mg;
[U-ring-^{14}C]dealkylatrazine (DAA, 98.0% pure), 14.1 μCi/mg; and
[U-ring-^{14}C]hydroxyatrazine (HYA, 99.0% pure), 23.5 μCi/mg.

Soil. Soil for microcosm studies and field studies was obtained from two field locations, A (18 ha) and B (27 ha) at the Agricenter International in Shelby County, Tennessee which is located in the southwestern corner of the state. Soil in both fields was classified as Falaya silt loam. Surface layers were an acidic, friable silt loam, approximately 15 cm in depth, with low organic carbon content (Table I).

Soil microcosms. Intact soil cores for microcosms were taken using sterile bottomless 125 ml serum bottles (Fisher Scientific). Each bottle was pressed 3 cm into the soil and the cores cut at the base of the bottle using a sterile spatula. This provided a coring device which also served as a microcosm when sealed at the bottom with a sterile 8 cm^2 pane of glass and a bead of silicone sealer. This procedure minimized handling of the soil core and maintained its integrity. Soil cores were transported to the lab and aseptic microcosms constructed the same day. Microcosms were treated with radiolabeled test substances.

Sterilized soil microcosms were prepared by irradiation through the courtesies of Oak Ridge National Laboratory and the Research Reactor Facility, University of Missouri. Soil was added to sterile intact 125 ml serum bottles (Fisher Scientific) which were sealed with sterile slotted gray butyl stoppers (13 by 20 mm, Wheaton Scientific) and aluminum seals (20 mm, Wheaton Scientific), and exposed to 5.02×10^6 RAD.

Three nonirradiated and three irradiated microcosms, one from each site, were prepared for each mineralization study. Separate microcosms were prepared for residue analysis on each scheduled sampling day. Solutions of [^{14}C]atrazine or metabolite were filter sterilized and aseptically added to microcosms. Soil microcosms were treated equivalent to a field application rate of 2.20 kg/ha for atrazine or 0.5 kg/ha for a metabolite, brought to 80% field moisture holding capacity and incubated at 25°C. Filter sterilized air flowed through microcosms, 20 ml/min, to sweep evolved $^{14}CO_2$ into a trapping solution (Figure 1), which consisted of scintillation grade phenethylamine (Eastman Kodak Co.) and Scinti Verse I (Fisher Scientific) (1:9,v/v). Trapping efficiency of the solution was 96.3%.

Radioactivity analysis. Aliquots of trapping solution were analyzed by liquid scintillation counting using a Beckman Model LS-7000 scintillation counter and a quenched ^{14}C standard set (Nuclear-Chicago).

Studies using radiolabeled compounds were conducted for a period of 180 d. Triplicate microcosms were sacrificed for analysis on each sampling day (4, 14, 28, 63 and 180).

Field study. Atrazine was applied to field B on May 22, 1987 at a rate of 3.2 kg/ha using an emulsifiable concentrate. Surface soil cores, 3 cm in depth, were taken for residue analysis. Cores were transported on ice and analyzed immediately upon arrival at the lab. The field study was conducted for a period of 180 d. Triplicate samples from each site were taken for analysis on days 0, 4, 14, 28, 63 and 180.

Irradiated microcosm sterility and soil microbial enumeration. Soil from selected irradiation-sterilized and nonirradiated soil microcosms and from the field was tested for microbial growth prior to the start of each assay. Bacteria and fungi were enumerated using plate count agar (Difco) and rose bengal agar base (Difco) respectively. Soil dilutions were prepared using sterile 0.01% (w/v) peptone water (Difco). Soil to diluent ratio was 1:10 (w/v). A spread plate

Table I. Properties of soils used in this study

Field	Site	Sand (%)	Silt (%)	Clay (%)	Carbon (%)[a]	pH[b] Microcosm	Field Study
A	1	5.5	80.7	13.8	0.50	5.3	
A	2	8.2	77.5	14.3	0.29	5.6	
A	3	8.1	76.2	15.7	0.36	5.5	
B	1	3.7	80.7	15.6	0.47		5.3
B	2	5.9	79.3	14.8	0.43		5.5
B	3	6.5	78.4	15.1	0.50		5.4

[a]Percentage organic carbon.
[b]Means of values of microcosm soil for all sampling days.

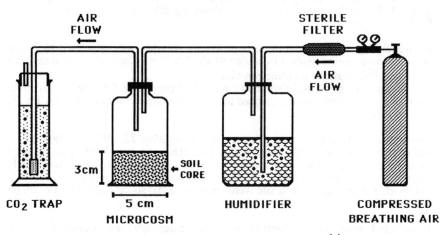

Figure 1. Microcosm design used to examine [U-ring-^{14}C]atrazine degradation in soil cores.

technique was used to inoculate the media. After incubation for 2 weeks at room temperature, no growth was detected in irradiated soil samples.

Chemical analyses. Background atrazine and metabolite concentrations were measured in soil samples taken from the three sampling sites in both experimental fields prior to the start of each study. Control soil cores taken at the start of each radiolabeled study were also analyzed for background residue concentrations remaining at each sampling date. Moist soil samples were measured into 500 ml teflon stoppered Erlenmeyer flasks and extracted with acetonitrile:methanol:distilled water (7:1.5:1.5,v/v, pH 9.0 with ammonium hydroxide). Moist soil to solvent ratio was 1:2.5 (w/v). Samples were shaken on a rotary shaker for 0.5 hr, allowed to stand for 18.0 hr and then shaken again for 0.5 hr. The supernatant was centrifuged at 2000 rpm for 15 min and an aliquot transferred to a round bottom flask and evaporated to just about dryness at 40°C with a rotary evaporator. Residues were resuspended in 100 ml of methylene chloride:distilled water (1:1,v/v), transferred to a separatory funnel and the organic layer separated from the aqueous layer. The round bottom flask was rinsed with another 50 ml of methylene chloride:distilled water (1:1,v/v) and transferred to the separatory funnel to reextract the aqueous phase. Combined methylene chloride extracts were evaporated to dryness on the rotary evaporator at 40°C, the residue dissolved in an aliquot of isooctane, filtered (.45 um) and stored in teflon stoppered sample vials at -20°C. Combined aqueous extracts were evaporated to a small volume in a round bottom flask at 40°C, transferred to a 50 ml volumetric flask and brought to volume with successive methanol rinses of the evaporator flask. After particulate matter was allowed to settle, 45 ml was removed to a round bottom flask and evaporated to dryness on the rotary evaporator at 40°C. The residue was dissolved in an aliquot of methanol, filtered (.45 um) and stored in teflon stoppered sample vials at -20°C.

Aliquots from samples stored in isooctane and methanol were analyzed for atrazine, DEA, DIA and DAA using a Hewlett-Packard Model 5790A gas chromatograph (GC). The GC was equipped with a Hewlett-Packard Ultra capillary column, cross-linked 5% phenylmethyl silicone (25 m by 0.31 mm i.d.), and a nitrogen-phosphorus detector operated in the split mode. Helium was used as carrier gas with a flow rate of 2 ml/min. Hydrogen and breathing air with flow rates of 4 ml/min and 80 ml/min respectively, were used with the detector. Injection and detector temperatures were both 300°C. Column temperature was 150°C. Residue concentrations were determined by comparing GC peak areas of samples against standards. Standard curves were prepared for each series of analyses; correlation coefficients (r^2) for these lines generally were greater than 0.95.

Aliquots from samples stored in methanol were analyzed for HYA using a Beckman modular HPLC system equipped with a Beckman Model 165 variable wavelength detector set at 240 nm and an Altex Ultrasphere-ODS C18 reversed phase column (25 cm by 4.6 mm i.d.). The flow rate of the mobile phase, methanol:water (7:3,v/v), was 1.7 ml/min. All analyses were performed at room temperature. Residue concentrations were determined using a Nelson Analytical System (Nelson Analytical, Inc.), comparing HPLC peak areas of samples against standards. Standard curves were prepared for each series of analyses; correlation coefficients (r^2) for these lines generally were greater than 0.95.

Extraction of ^{14}C-treated soils. Aliquots of treated and control moist soil were Soxhlet extracted with methanol:water (9:1,v/v) for 8 hr. Soil to solvent ratio was 1:2.5 (w/v). Solvent extracts were evaporated with a rotary evaporator at 40°C to a small volume, transferred to a volumetric 10 ml flask and the volume was brought to 10 ml with successive methanol rinses of the evaporator flask. Aliquots of solvent extracts were analyzed for radioactivity with a liquid scintillation counter and bound ^{14}C residues were determined by combustion to $^{14}CO_2$.

Bound and extracted ^{14}C residue analysis. An R. J. Harvey Model OX-400 biological material oxidizer was used to determine ^{14}C residues in extract and soil samples. Samples were oxidized and the $^{14}CO_2$ produced was absorbed in Carbon 14 Cocktail (R. J. Harvey Instrument Corp.). Trapping efficiency of the cocktail was 95.8% A ^{14}C standard, [^{14}C]methyl methacrylate (NEN Products), 1.65 x 10^{-4} uCi/mg, was used to determine instrument efficiency. Trapped $^{14}CO_2$ samples were analyzed by liquid scintillation counting using a Beckman Model LS-7000 scintillation counter and a quenched ^{14}C standard set (Nuclear-Chicago).

Results and Discussion

Atrazine dissipation and metabolite formation - microcosms. Atrazine concentration decreased exponentially over 180 d (Figure 2). The half-life for detectable residues of atrazine was 21 days. Decreasing atrazine concentration was accompanied by corresponding increases in HYA, DEA and DIA residues (Table II). Residues of these metabolites and their parent compound persisted in the soil at detectable levels through the 180-day studies. HYA was detected in higher concentrations than the phytotoxic monoalkylated metabolites at each sampling. These results are in agreement with Jones et al. (3) and Khan and Saidak (38). High HYA concentrations would be expected since hydroxylation is considered to be a major route of atrazine detoxification. Similar results were exhibited in irradiated microcosms at the termination of the studies. HYA concentration in irradiated soil, relative to initial atrazine concentration, was 68% greater than HYA concentrations in nonirradiated soil after 180 d. Presence of HYA residues in sterilized soils indicates hydroxylation is a significant route of abiotic detoxification. Also, the production of DEA and DIA in sterilized soils indicates the occurrence of N-dealkylation under abiotic conditions. This does not, however, exclude the possible presence of soil enzymes which may have mediated this process in the absence of viable organisms.

DEA and DIA were formed in soil microcosms; DEA concentrations were always greater than DIA. Production of these metabolites has been reported to occur in soils to which atrazine had been applied (38,39). Schiavon (32) and Muir and Baker (40) detected DEA and DIA in field leachates at concentrations greater than the parent. Although their soils were either sandy or clay loams, unlike the silt loam in this study, they found DEA in higher concentrations than DIA. DEA and DIA, when detected in this study, were measured in higher concentrations relative to initial parent application in irradiated microcosms than in nonirradiated microcosms on day 180. Only 180-day irradiated soils were examined for metabolite formation. DAA was not detected.

Atrazine dissipation and metabolite formation - Field study. Residue concentrations were measured in field soil, identical to soils used in the microcosms, to provide a comparison of parent dissipation rates and metabolite production. Atrazine concentration decreased exponentially during 180 d (Figure 2). The application rate in the field, 3.2 kg/ha, was higher than in microcosms, 2.2 kg/ha. The dissipation rate in field soil appears to be greater than in microcosms, because only the surface soil was examined in the microcosms. Any residues migrating into the soil horizon beyond 3 cm would not have been included in total residue concentrations. Also, microcosms were not subject to surface runoff which further increased parent dissipation rates in the field. The half-life for detectable residues of atrazine in the surface 3 cm of the field was 14 days. Metabolite

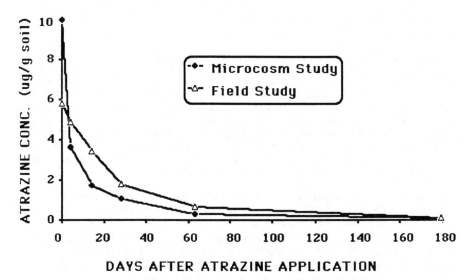

DAYS AFTER ATRAZINE APPLICATION

Figure 2. Comparison of atrazine disappearance in microcosms treated with 2.2 kg/ha of [U-ring-^{14}C] atrazine and the top three cm of field soil treated with 3.2 kg/ha commercial grade atrazine.

Table II.   Soil residues (μg/g soil, dry wt.) of atrazine, hydroxyatrazine, deisopropylatrazine and deethylatrazine in soil microcosms and irradiation sterilized soil microcosms dosed with [U-ring-^{14}C]atrazine at a rate of 2.2 kg/ha .   Values represent means of triplicate samples ± 95% C.I.

	Residues, μg/g, On Days After [^{14}C]-Atrazine Application							Irradiated Microcosms
			Microcosms					
Compound	Day 0	4	14	28	63	180	0	180
Atrazine	5.75[a]	4.85 ±0.36	3.42 ±0.38	1.77 ±0.27	0.66 ±0.09	0.06 ±0.01	4.98	0.13 ±0.03
Hydroxy- atrazine	-	0.08 ±0.05	0.14 ±0.03	0.28 ±0.05	0.53 ±0.04	0.37 ±0.04	-	0.54 ±0.05
Deisopropyl- atrazine	-	0.01 ±0.00	0.01 ±0.00	0.03 ±0.01	0.02 ±0.01	0.01 ±0.01	-	ND[b]
Deethyl- atrazine	-	0.13 ±0.01	0.16 ±0.03	0.14 ±0.02	0.11 ±0.01	0.03 ±0.01	-	0.04 ±0.01

[a]Means of duplicate samples.
[b]Not detectable; <0.01 ppm.

formation in field soils was similar to microcosm samples (Figure 3). The order of metabolite concentration in the field was HYA > DEA > DIA. This confirms similar results presented by Sirons *et al.* (*39*).

Metabolite dissipation in microcosms. Solvent-extractable metabolite concentrations decreased in microcosms during the assay period and followed first-order kinetics (Figure 4). Approximately 2.0% of applied DEA and DIA (0.02 μg/g soil) was detected after 180 d incubation. DAA was not detected after day 63. HYA concentration was equivalent to 33% of the applied compound (0.41 μg/g soil) after 180 d. Based on these data, metabolite half-lives fell into two groups relative to time. The first contained the chlorinated dealkylatrazines, DEA, DIA and DAA, which exhibited short half-lives, 26, 17 and 19 d respectively, similar to the parent herbicide which was approximately 21 d (*1*). The second group contained the hydroxylated metabolite, HYA, with a longer half-life of 121 d.

Metabolite mineralization in microcosms. Metabolism of atrazine and selected metabolites, resulting in the production of various triazine by-products and CO_2, has been shown to occur in several soils (*1,23,25-28*). Some of these studies using ^{14}C ring-labeled atrazine or one of its metabolites did not always produce results that clearly attributed mineralization to the radiolabeled molecule applied, because the percentage of ^{14}C evolved as [14]CO_2 was equal to or less than the concentration of ^{14}C impurities in the radiolabeled material.

In this study, mineralization rates of radiolabeled DEA, DIA, DAA and HYA were significantly greater in nonirradiated soil microcosms than in irradiated soil microcosms (Figures 5-8). Nonirradiated microcosms evolved approximately 10- to 600-fold more [14]CO_2 than irradiated microcosms. Evolution of [14]CO_2 from irradiated microcosms did not exceed 1%, except from microcosms dosed with [^{14}C]DIA (Figure 6), which evolved 1.77% of the radiolabel after 180 d of incubation. Since the radiolabel was 99.0% pure, it would appear that abiotic mineralization was a contributing factor. However, one of the irradiated replicates dosed with [^{14}C]DIA evolved less than 0.2% of the ^{14}C as [14]CO_2 after 180 d. The two other replicates evolved 2.0 and 3.1% during the same period. This increase in [14]CO_2 evolution may have resulted from bacterial contamination. Although irradiated microcosm soil showed no indication of viable organisms at the beginning of the study, the increased evolution of [14]CO_2 in the latter two microcosms after day 13, was more characteristic of the biotic mineralization seen in nonirradiated microcosms.

Since the radiolabeled metabolites were only 98 to 99% pure, it was not possible to attribute [14]CO_2 evolution in irradiated microcosms, other than those treated with [^{14}C]DIA, to actual degradation of the radiolabeled compound. The difference in [14]CO_2 evolution between nonirradiated and irradiated microcosms was an indication of the microbial mineralization of these compounds.

Mineralization rates of the two phytotoxic metabolites, DEA and DIA, in nonirradiated microcosms were similar to those measured for ^{14}C ring-labeled atrazine using the described microcosm system and soil from the same sites (*1*). DAA exhibited the highest mineralization rate (Figure 7), evolving 59% of the ^{14}C as [14]CO_2 after 180 d. These results are similar to those of Wolf and Martin (*25*) who reported 40% of the radioactivity as [14]CO_2 from ^{14}C ring-labeled DAA after 192 d. Schiavon (*33*) was not able to account for approximately 60% of the ^{14}C ring-labeled DAA applied to soils after 1 year, and attributed this loss to volatilization and degradation. Results from this study suggest that the majority of the loss was due to mineralization. HYA exhibited the lowest rate of dissipation

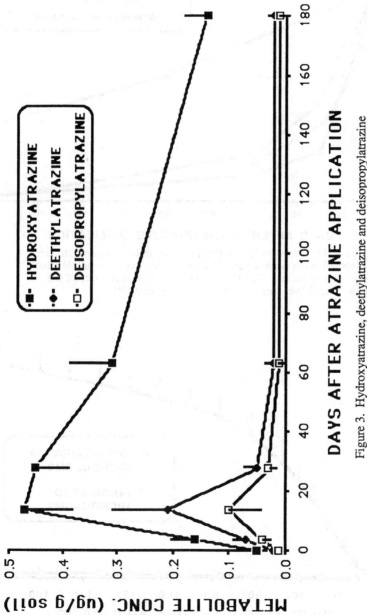

Figure 3. Hydroxyatrazine, deethylatrazine and deisopropylatrazine concentrations in the surface 3 cm of field samples during 180 d. Commercial grade atrazine, 3.2 kg/ha, was applied on day 0. Values represent means of triplicate samples ± 95% C.I.

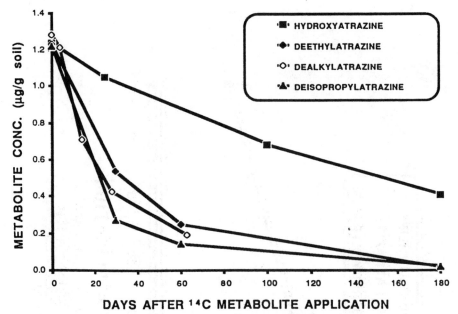

Figure 4. Hydroxyatrazine, deethylatrazine and deisopropylatrazine concentrations in nonirradiated microcosm soil for four radiolabeled assays during 180 d. All metabolites were applied at a rate of 0.5 kg/ha. Dealkylatrazine was not detected on day 180.

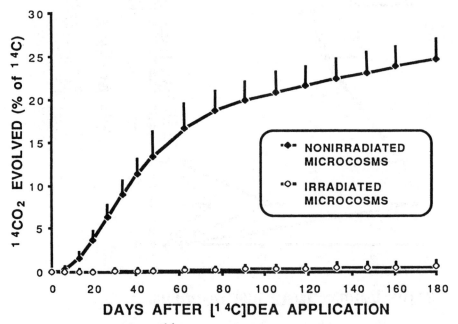

Figure 5. Evolved $^{14}CO_2$ in nonirradiated and irradiated soil microcosms dosed with 0.5 kg/ha of [U-ring-^{14}C]deethylatrazine during 180 d. Values represent means ± 95% C.I.

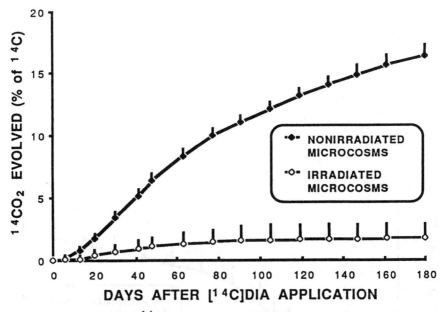

Figure 6. Evolved $^{14}CO_2$ in nonirradiated and irradiated soil microcosms dosed with 0.5 kg/ha of [U-ring-^{14}C]deisopropylatrazine during 180 d. Values represent means ± 95% C.I.

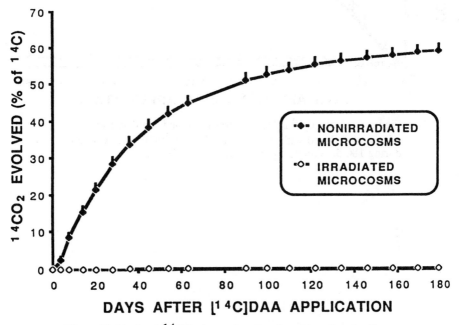

Figure 7. Evolved $^{14}CO_2$ in nonirradiated and irradiated soil microcosms dosed with 0.5 kg/ha of [U-ring-^{14}C]dealkylatrazine during 180 d. Values represent means ± 95% C.I.

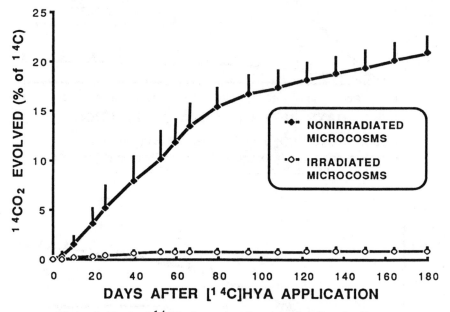

Figure 8. Evolved $^{14}CO_2$ in nonirradiated and irradiated soil microcosms dosed with 0.5 kg/ha of [U-ring-^{14}C]hydroxyatrazine during 180 d. Values represent means ± 95% C.I.

based on half-lives, however, $^{14}CO_2$ evolution from nonirradiated microcosms dosed with [^{14}C]HYA was relatively high, 21% of the applied ^{14}C after 180 d (Figure 8). Equivalent rates were measured by Goswami and Green (*26*) and Skipper and Volk (*23*). This suggests that the microbial mineralization of HYA in these soils proceeds at a rate comparable to the chlorinated monoalkylatrazines, despite the apparent longer half-life of HYA.

Soil bound residues and solvent extracts. Bound residues of atrazine and its metabolites have been shown to be persistent (*34*), and their formation influenced by dealkylation or hydroxylation of the parent herbicide (*33*). Schiavon (*33*), working with ^{14}C radiolabeled atrazine, DEA, DIA, DAA and HYA, found 49 to 67% of the ^{14}C in bound residues in the 0 to 6 cm level of soil columns incubated under field conditions for 1 year. He showed that the dealkylatrazines formed significantly higher amounts of bound residues in comparison to HYA which was adsorbed by soil organic matter in the surface soil. He found that the bidealkylated DAA formed the highest amounts of bound residues, followed by the monoalkylated metabolites, DEA and DIA. He also showed that HYA formed the lowest amounts of bound residues. Similar results were found in this study.

For atrazine, bound ^{14}C residue formation increased with time, and after 180 d of incubation, accounted for 43% of the ^{14}C added to microcosms (Table III). Capriel *et al.* (*34*) reported that approximately 50% of the initially applied [^{14}C]atrazine remained as bound residues 9 years after application. Bound residue formation has been shown to be influenced by alkyl group loss, with DAA forming the greater amount of bound residues (*33,34*). This may explain why DAA was not detected during these studies. Also, DAA was mineralized at a high rate in comparison to either parent, HYA, DEA or DIA, further explaining the lack of detection.

As radioactivity increased in the CO_2 and bound residue components in nonirradiated microcosms during incubation, it decreased in solvent extracts (Table III). The radioactivity in solvent extracts during these studies accounted for as little as 22% of the activity added to nonirradiated microcosms after 180 d. In comparison, solvent extracts from irradiated microcosms contained approximately twice as much radioactivity. Metabolite concentrations in irradiated soil, in general, were greater than in nonirradiated soil (Tables IV).

Bound ^{14}C residue concentrations of radiolabeled metabolites increased in all microcosm assays during incubation (Tables IV - VII). Bound ^{14}C residues accounted for 28 to 60% of the ^{14}C added to nonirradiated microcosm soil after 180 d. In comparison, irradiated microcosms exhibited as much as 72% of the radioactivity in the bound residue component. Dealkylated metabolites in irradiated microcosms produced high percentages of bound ^{14}C residues, with DAA forming the highest. However, DAA exhibited the highest mineralization rates in nonirradiated microcosms which contributed to the significant reduction of bound residue formation in these microcosms. Intermediate amounts of bound ^{14}C residues were formed by DEA and DIA, and after 180 d accounted for 48 to 60% of the radioactivity applied to nonirradiated and irradiated microcosms. Although these percentages were similar, and only negligible amounts of $^{14}CO_2$ were evolved in irradiated microcosms, solvent extracts from both microcosm groups contained less than 2% of the applied metabolites after 180 d, in comparison to the significantly higher percentages of ^{14}C remaining in the thermal extracts. This would indicate that a large portion of the radioactivity in thermal extracts was composed of degradation products of DEA and DIA, possibly hydroxylated compounds, which did not tend to form bound residues.

Table III. ^{14}C mass balance (percentage of total ^{14}C in various components) in soil microcosms and irradiation sterilized soil microcosms dosed with [U-ring-^{14}C]atrazine (2.2 kg/ha). Values represent means of triplicate samples ± 95% C.I.

	Percentage ^{14}C On Days After [^{14}C]Atrazine Application					
^{14}C Mass Balance	Microcosms					Irradiated Microcosms
Component	Day 4	14	28	63	180	180
$^{14}CO_2$ Evolved	0.02 ±0.01	0.61 ±0.24	2.47 ±0.44	11.97 ±1.39	28.31 ±1.70	0.05 ±0.01
Bound Residue	3.62 ±0.71	8.55 ±1.04	18.67 ±2.57	26.16 ±3.32	42.64 ±2.15	47.90 ±4.73
Solvent Extract	98.00 ±14.64	89.96 ±7.05	72.86 ±8.79	54.84 ±5.72	21.68 ±0.59	49.05 ±2.94
Total	101.64	99.12	94.00	92.97	92.63	97.00

Table IV. ^{14}C mass balance (percentage of total ^{14}C in various components) in soil microcosms and irradiation sterilized soil microcosms dosed with [U-ring-^{14}C]deethylatrazine (0.5 kg/ha) for various sampling days. Values represent means of duplicate samples or triplicate samples ± 95% C.I.

^{14}C Mass Balance Component	Percentage ^{14}C On Days After [^{14}C]Deethylatrazine Application			
	Microcosms			Irradiated Microcosms
	Day 30	60	180	180
$^{14}CO_2$ Evolved	7.47 ±1.40	16.20 ±2.28	24.72 ±1.96	0.63 ±0.29
Bound Residue	33.48[a]	45.44	60.30 ±5.32	48.47 ±3.41
Solvent Extract	56.89	35.66	12.92 ±2.13	47.14 ±4.32
Total	97.84	97.30	97.94	96.24

[a]Means of duplicate samples.

Table V. ^{14}C mass balance (percentage of total ^{14}C in various components) in soil microcosms and irradiation sterilized soil microcosms dosed with [U-ring-^{14}C]deisopropylatrazine (0.5 kg/ha) for various sampling days. Values represent means of duplicate samples or triplicate samples ± 95% C.I.

^{14}C Mass Balance Component	Percentage ^{14}C On Days After [^{14}C]Deisopropylatrazine Application			
	Microcosms			Irradiated Microcosms
	Day 30	60	180	180
$^{14}CO_2$ Evolved	3.37 ±0.63	7.94 ±0.45	16.38 ±0.99	1.76 ±1.28
Bound Residue	37.39[a]	50.14	57.68 ±4.86	55.38 ±5.30
Solvent Extract	59.35	37.08	13.56 ±1.05	32.98 ±5.31
Total	100.11	95.16	87.62	90.12

[a]Means of duplicate samples.

Table VI. ^{14}C mass balance (percentage of total ^{14}C in various components) in soil microcosms and irradiation sterilized soil microcosms dosed with [U-ring-^{14}C]dealkylatrazine (0.5 kg/ha) for various sampling days. Values represent means of triplicate samples ± 95% C.I.

Percentage ^{14}C On Days After [^{14}C]Dealkylatrazine Application						
^{14}C Mass Balance Component			Microcosms			Irradiated Microcosms
	Day 4	14	28	63	180	180
^{14}CO$_2$	2.46	15.36	28.56	45.06	59.22	0.10
Evolved	±0.46	±0.21	±0.92	±0.75	±0.68	±0.01
Bound Residue	4.14 ±0.28	12.41 ±0.70	15.78 ±0.79	22.89 ±2.44	29.34 ±5.02	71.66 ±1.94
Solvent Extract	91.26 ±7.17	73.83 ±9.16	46.71 ±4.95	23.08 ±1.85	4.33 ±1.82	31.42 ±4.11
Total	97.86	101.60	91.05	91.03	92.89	103.18

Table VII. ^{14}C mass balance (percentage of total ^{14}C in various components) in soil microcosms and irradiation sterilized soil microcosms dosed with [U-ring-^{14}C]hydroxyatrazine (0.5 kg/ha) for various sampling days. Values represent means of duplicate samples or triplicate samples ± 95% C.I.

Percentage ^{14}C On Days After [^{14}C]Hydroxyatrazine Application				
^{14}C Mass Balance Component		Microcosms		Irradiated Microcosms
	Day 25	100	180	180
^{14}CO$_2$	5.21	17.27	20.99	0.87
Evolved	±2.60	±1.64	±1.63	±0.13
Bound Residue	10.26[a]	25.83	27.56 ±2.62	23.89 ±1.88
Solvent Extract	81.04	54.85	40.15 ±5.42	68.03 ±4.68
Total	96.51	97.95	89.50	92.79

[a]Means of duplicate samples.

HYA formed the lowest amounts of bound ^{14}C residues in both nonirradiated and irradiated microcosms. Approximately 28 and 24% of the radioactivity was in the bound residue component after 180 d in nonirradiated and irradiated microcosms respectively. HYA is adsorbed by soil organic matter (*41,42*), but unlike the dealkylatrazines, does not form bound residues as readily. The majority of HYA residues 1 year after application were found in the plow zone of a 60 cm soil column where soil organic matter concentrations were highest (*33*). Intensive extraction procedures are not required to dissociate a large proportion of adsorbed residues of HYA from soil organic matter. Bound residues in irradiated microcosms would indicate the potential for their formation in the absence of biological activity. Based on the potential to form bound residues, the order was: DAA > DEA ≥ DIA > HYA.

The implication of these data is that the parent and chlorinated monoalkylatrazines (the phytotoxic metabolites), which had lower extractable residue concentrations, exhibited shorter half-lives than the hydroxylated metabolite due to the increased formation of bound residues. In normal ambient extraction procedures, these residues would go undetected and results would tend to underestimate their true soil burden. Although bound ^{14}C residues do not identify the applied metabolite or the extent of its degradation, they do indicate the presence of the applied compound or a degradation product.

Solvent-extractable radioactivity decreased during 180 d incubation in all studies. The solvent extract fraction for chlorinated dealkylatrazines in nonirradiated microcosms at the termination of the assays contained the lowest amounts of activity. In comparison, this component in the HYA study contained the highest amount of radioactivity in both nonirradiated and irradiated microcosms, again indicating a lower potential for bound residue formation.

The activity of herbicides has been shown to be influenced by several soil properties. Soil organic matter has been shown to exert a significant effect on the phytotoxicity of herbicides (*43,44*). Rahman and Matthews (*43*) demonstrated a significant correlation between the inverse relationship of organic matter and phytotoxicity. Although there is a positive association of bound residue formation with soil organic matter, which reduces phytotoxicity, bound residues have been shown to be bioavailable, and should be considered when evaluating herbicide persistence.

Acknowledgements

This research was partially supported by the Ciba-Geigy Corporation, Sigma Xi, The Scientific Research Society and the State of Tennessee, Department of Health and Environment, Division of Water Management. The irradiated samples were prepared through the courtesy of Vicki Spaet at the Research Reactor Facility, University of Missouri, Columbia, Missouri.

Literature Cited

1.   Frank, R.; Sirons, G. J. *Bull. Environ. Contam. Toxicol.* **1985**, *35*, 541-548.
2.   Khan, S. U., Marriage, P. B., Hamill, A. S. *J. Agric. Food Chem.* **1981**, *29*, 216-219.
3.   Jones, T. W., Kemp, M. W., Stevenson, J. C., Means, J. C. *J. Environ. Qual.* **1982**, *11*, 632-638.
4.   Wolf, D. C., Martin, J. P. *J. Environ. Qual.* **1975**, 4:134-139.
5.   Skipper, H. D., Volk, V. V. *Weed Sci.* **1972**, *20*, 344-347.
6.   Skipper, H. D., Gilmour, C. M., Furtick, W. R. *Soil Sci. Soc. Am. Proc.* **1967**, *31*, 653-656.

7.   Wagner, G. H. Chahal, K. S. *Soil Sci. Soc. Am. Proc.* **1966**, *30*, 752-754.
8.   Hall, J. K. Hartwig, N. L. *J. Environ. Qual.* **1978**, *7*, 63-68.
9.   Ritter, W. F., Johnson, H. P., Lovely, W. G.*Weed Sci.* **1973**, *21*, 38-384.
10. Goswami, K. P., Green, R. *Soil Sci. Soc. Am. Proc.* **1973**, *37*, 702-707.
11. Burnside, O. C., Fenster, F. R., Wicks, G. A.*Weed Sci.* **1971**, *19*, 290-293.
12. Scheunert, I., Zhang Oiao, Korte, F. *J. Environ. Sci. Health.* **1986**, *6*, 457-485.
13. Jachetta, J. J., Radosevich, R. *Weed Sci.* **1981**, *29*, 37-44.
14. Winkle, M. E., Leavitt, J. R. C., Burnside, O. C. *Weed Sci.* **1981**, *29*, 405-409.
15. Harvey, R. G. *Weed Sci.* **1973**, *21*, 204-206.
16. Parochetti, J. V.*Weed Sci.* **1973**, *21*, 157-160.
17. Kaufman, D. D., Kearney, P. C. *Residue Rev.* **1970**, *32*, 235-265.
18. Shimabukuro, R. H., Walsh, W. C., Lamoureux, G. L., Stafford, L. E. *J. Agric. Food Chem.* **1973**, *21*, 1031-1036.
19. Armstrong, D. E., Chesters, G., Harris, R. F. *Soil Sci. Soc. Amer. Proc.* **1967**, *31*, 61-66.
20. Giardina, M. C., Giardi, M. T, Filacchioni, G. *Agric. Biol. Chem.* **1980**, *44*, 2067-2072.
21. Kaufman D. D., Blake, J. *Soil Biol Biochem.* **1970**, *2*, 73-80.
22. Behki, R. M., Khan, S. U. *J. Agric. Food Chem.* **1986**, *34*, 746-749.
23. Skipper, H. D., Volk, V. V. *Weed Sci.* **1972**, *20*, 344-347.
24. Wagner, G. H. Chahal, K. S. *Soil Sci. Soc. Am. Proc.* **1966**, *30*, 752-754.
25. Wolf, D. C., Martin, J. P. *J. Environ. Qual.* **1975**, *4*, 134-139.
26. Goswami, K. P., Green, R. E. *Environ. Sci. Technol.* **1971**, *5*, 426-429.
27. Führ, R., Mittelstaedt, W. *J. Agric. Food Chem.* **1980**, 28, 122-125.
28. Fuhreman, R. W., Lichtenstein, E. P. *J. Agric. Food Chem.* **1978**, 26, 605-610.
29. Khan, S. U., Hamilton, H. A. *J. Agric. Food Chem.* **1980**, 28, 126-132.
30. Scheunert, I., Zhang Oiao, Korte, F. *J. Environ. Sci. Health* **1986**, 6, 457-485.
31. Federal Register. **1975**. *40(123)*, 26802-26928.
32. Schiavon, M. *Ecotoxicol. and Environ. Safety* **1988**, *15*, 46-54.
33. Schiavon, M. *Ecotoxicol. and Environ. Safety* **1988**, *15*, 55-61.
34. Capriel, P., Haisch, A., Khan, S. U. *J. Agric. Food Chem.* **1985**, *33*, 567-569.
35. Führ, R., Mittelstaedt, W. *J. Agric. Food Chem.* **1980**, 28, 122-125.
36. Fuhreman, R. W., Lichtenstein, E. P. *J. Agric. Food Chem.* **1978**, *26*, 605-610.
37. Khan, S. U. *J. Agric. Food Chem.* **1980**, *28*, 1096-1098.
38. Khan, S. U.,Saidak, W. J. *Weed Res.* **1981**, *21*, 9-12.
39. Sirons, G. J., Frank, R., Sawyer, T. *J. Agric. Food Chem.* **1973**, *21*, 1016-1020.
40. Muir, D. C., Baker, B. E. *J. Agric. Food Chem.* **1976**, *24*, 122-125.
41. Armstrong, D. E., Chesters, G. *Environ. Sci. Technol.* **1968**, *2*, 683-689.
42. Hayes, M. H. B. *Residue Rev.* **1970**, *32*, 131-174.
43. Rahman, A., Matthews, L. J. *Weed Sci.* **1979**, *27*, 158-161.
44. Harrison, B. W., Weber, J. B., Baird, J. V. *Weed Sci.* **1976**, *24*, 120-126.

RECEIVED December 17, 1990

Chapter 7

Factors Affecting the Degradation of 3,5,6-Trichloro-2-Pyridinol in Soil

Kenneth D. Racke and Susan T. Robbins

Environmental Chemistry Laboratory, DowElanco, 9001 Building, Midland, MI 48641–1706

The degradation and sorption of 3,5,6-trichloro-2-pyridinol (TCP), a primary metabolite of the insecticide chlorpyrifos and the herbicide triclopyr, was examined in 25 different soils in order to better predict its environmental fate and significance. TCP exhibited sorption (K_d) coefficients of between 0.3 and 20.3 ml/g (mean 3.1) and calculated mean K_{oc} coefficients for the neutral and anionic forms of 3344 and 54 ml/g, respectively. Mineralization was used as an indicator of the degradation in soil of TCP and similar organic compounds chosen for comparative study (3-chloro-5-trifluoromethyl-2-pyridinol, 2-hydroxypyridine, 2,4-dichlorophenol). Although rates of TCP degradation varied between soils, multiple regression analyses revealed poor correlation of degradation with commonly measured soil properties. However, inclusion of soil degradation rates of related organic compounds or glucose, and sorption information in the regression model resulted in a significant improvement in the ability to predict TCP degradation. The mineralization of TCP is microbially-mediated, yet it is unclear whether catabolic or cometabolic processes are predominant. Only 2 soils contained microbial populations that could utilize TCP as a sole carbon source in mixed culture.

Most environmental fate studies have focused strongly if not exclusively on the parent pesticide molecule and have devoted little attention to the fate of pesticide metabolites. A number of pesticides contain halogenated pyridine ring systems, and various halogenated pyridinols may be produced from these compounds during degradation. One of these pyridinols of interest is 3,5,6-trichloro-2-pyridinol (TCP). TCP is a primary metabolite of

0097–6156/91/0459–0093$06.00/0

the insecticide chlorpyrifos and the herbicide triclopyr in soil and
water and on plant surfaces. Formation of TCP occurs via both
hydrolytic and photolytic mechanisms, and measurable quantities of
TCP can be produced in soil following application of either of these
pesticides. Maximum TCP concentrations in surficial soils following
application of chlorpyrifos or triclopyr have been measured as
0.07-0.7 and 0.4-11 $\mu g/g$, respectively (1-3).

Although many halogenated phenols have been studied quite
extensively, few halogenated pyridinols have been as thoroughly
investigated. In contrast, TCP has been fairly well characterized.
It is an acidic compound with a pka of 4.55, a water solubility of
117 $\mu g/ml$ (pH 2-3), and a negligible vapor pressure of $<1 \times 10^{-7}$ at
pH 7 (4). Photolysis half-lives of TCP range from 4 to 15 minutes in
aqueous solution (5). TCP is moderately mobile in soil with reported
K_{oc} coefficients of 19 and 281 for a soil column and a batch
equilibrium study, respectively (4,6). Soil degradation half-lives
for TCP have been found to vary greatly between soil types, with
reported laboratory-determined values ranging from 10 to 325 days
(7). Microbially-mediated mineralization appears to be the primary
degradative pathway. Racke et al. (8) reported that <10 to >80% of
the applied dose of TCP was mineralized within a 2-week period in
various soils. Degradation has been shown to be concentration
dependent (9). Minimal mineralization of TCP applied at 50 $\mu g/g$ was
noted in soils which had displayed rapid mineralization of a dose of
5 $\mu g/g$ (8).

The purpose of the present investigation was to determine the
factors that are important in governing the dissipation and sorption
of TCP in the soil environment. Therefore, the sorptive and
degradative behavior of TCP and degradative behavior of several
related compounds were studied in a number of different soils so that
better predictions of the environmental fate of TCP can be made.

Materials and Methods

Chemicals. Five radiolabeled test compounds were used for this
study. These compounds were: [14]C-(2,6-ring)3,5,6-trichloro-
2-pyridinol (TCP), [14]C-(3,5-ring)3-chloro-5-trifluoro-
methyl-2-pyridinol (TFP), [14]C-(2,6-ring)2-hydroxypyridine (HPYR),
[14]C-(U-ring)2,4-dichlorophenol (DCPH), and [14]C-(U-ring)glucose (not
pictured).

TCP   TFP   HPYR   DCPH

Nonradioactive analytical standards of each compound were used to
dilute the radiolabeled compounds for soil treatment. Non-

radiolabeled TCP was also used as a chromatographic reference standard.   All other laboratory chemicals and solvents were of reagent grade.

Soils. The soils used and their physical properties are listed in Table I.   These soils were chosen for their variety of properties and locations of origin.   Soils M177, M242 and M271 were three yearly samples of the same soil, collected in 1986, 1987, and 1988, respectively.   They were included to determine if any loss of microbial activity had occurred upon prolonged storage.   This information was important for choosing other soils to be used.   Soils from one site in Nebraska (M284, M285) and two sites in Illinois (M245, M247; M254, M255) were selected so that at each site one soil came from a plot that had received previous chlorpyrifos applications and another soil came from an adjacent plot that had never been treated.   These soils were included to determine if prior treatment with chlorpyrifos, and presumed exposure of the soil microbial community to TCP, would influence the degradation rate of TCP in these soils.

Degradation of TCP and Related Compounds in Soil. The objective for this portion of the study was to screen many soils over a relatively short period of time to determine the differences in degradation rates of TCP among the soils.   Samples of soils were also treated with TFP, HPYR, and DCPH to compare the capability of each soil for degradation of similar pyridinyl and phenolic compounds.   Some soil samples were treated with glucose in order to obtain an estimate of the general microbial activity present (10).   Because TCP is mineralized quite readily to CO_2 with minor accumulation of intermediates, cumulative $^{14}CO_2$ evolution was used as an indicator of the extent of TCP degradation (8).   A strong inverse relationship has been demonstrated to exist between percent mineralization of TCP and its degradation half-life in soil (7).   The relationship is not linear, but on log transformation a reasonable correlation is achieved ($R^2 = 0.84$).   Percent mineralization was also used as a degradation indicator for the other test compounds.

    To facilitate the examination of the degradation of TCP, TFP, HPYR, DCPH, and glucose in many soils, a simple soil incubation assay was used.   The method chosen was similar to that used by Racke and Coats (11).   For each compound tested, duplicate 25-g (dry weight) samples of each soil were weighed and placed into individual 8-oz French square bottles.   Samples of soils were treated with the following compounds (approximately 0.1 μCi): ^{14}C-TCP (1 μg/g), ^{14}C-TFP (0.08 μg/g), ^{14}C-HPYR (0.01 μg/g), ^{14}C-DCPH (1 μg/g), ^{14}C-glucose (10 μg/g).   Soils M177, M242, M245, M247, M254, and M255 were treated only with TCP.   All compounds, excluding glucose, were applied in 300 μl of acetone, with glucose applied as an aqueous solution.   After the acetone was evaporated, so as not to interfere with the microbial activity of the soil, distilled water was added to each sample to maintain a soil moisture tension of approximately 0.3 bar.   Scintillation vials containing 10 ml of 0.2N NaOH were placed in each bottle to serve as CO_2 traps.   The bottles were then capped with single-hole rubber stoppers through which a 4-inch length of 2 mm (ID) glass capillary tubing had been inserted to serve as a

Table I.   Soil Physical and Chemical Properties

SOIL	ORIGIN	pH	%O.C.	%H$_2$O at 1/3 bar	C.E.C.	%S	%Si	%C	Texture
M177[a]	ND	6.8	3.1	30.0	15.9	40	38	22	L
M242[a]	ND	7.5	3.2	31.5	16.8	34	46	20	L
M271[a]	ND	7.8	3.2	30.4	20.7	40	38	22	L
M261	MI	7.5	1.9	15.4	9.0	73	18	9	SL
M269	FL	5.4	0.7	4.4	2.4	90	6	4	S
M273	PA	4.2	2.4	30.4	6.0	32	44	24	L
M262	OH	6.1	1.7	30.3	15.3	26	30	44	C
M264	CA	7.3	0.5	15.0	5.7	58	30	12	SL
M266	MS	7.5	0.5	16.3	7.0	44	44	12	L
M268	MS	5.7	1.7	31.7	20.2	18	38	44	C
M272	GA	6.1	0.5	9.1	2.8	79	12	9	SL
M274	FL	7.5	1.9	5.6	9.3	90	4	6	S
M234	BRAZIL	5.7	1.4	20.0	3.7	56	22	22	SCL
M244	CANADA	8.0	3.1	18.2	10.6	48	35	17	L
M256	SD	5.6	2.1	27.2	17.5	32	43	25	L
M275	AZ	8.3	0.9	18.5	14.5	60	22	18	SL
M284[b]	NE	5.8	2.2	22.9	14.0	18	66	16	SiL
M285[b]	NE	5.5	2.0	22.2	15.3	20	62	18	SiL
M236	BRAZIL	5.9	0.8	9.5	3.0	74	12	14	SL
M258	ENGLAND	7.6	2.4	19.8	16.9	64	20	16	SL
M265	MT	6.5	1.1	13.1	7.5	68	22	10	SL
M267	GA	5.6	0.6	9.1	2.7	70	14	16	SL
M270	KS	7.9	0.9	28.9	18.6	26	44	30	CL
M277	HI	5.7	5.9	67.2	13.4	76	15	9	SL
M283	TX	8.0	1.2	23.2	21.6	32	38	30	CL
M245[b]	IL	5.8	1.6	26.4	14.0	14	58	28	SiCL
M247[b]	IL	6.0	2.0	27.0	14.9	16	56	28	SiCL
M254[b]	IL	5.2	2.6	30.9	22.2	20	58	22	SiL
M255[b]	IL	6.2	3.0	31.9	20.6	18	56	26	SiL

[a] Soils M177, M242, and M271 were all collected from the same field in 1986, 1987, and 1988, respectively.

[b] Soils M284 and M285 (NE), M245 and M247 (IL), and M254 and M255 (IL) were companion soils that came from adjacent plots with one plot (M284, M245, M254) untreated and one plot treated with chlorpyrifos for last 5 years (M285, M247, M255) at each site.

[c] Abbreviations: %O.C. = % organic carbon; C.E.C. = cation exchange capacity (meq/100 g); %S = sand, % Si = silt, %C = clay; L = loam, SL = sandy loam, SiL = silt loam, SCL = sandy clay loam, CL = clay loam, SiCL = silty clay loam.

support for the vial, a sampling pathway, and an avenue of aeration. Preliminary experiments indicated that this ventilation did not result in significant loss of evolved CO_2. The samples were then placed in an incubation chamber at $25^{\circ}C$ in the dark for up to three weeks during which time the traps were periodically sampled for evolved [14]CO_2.

Preliminary experiments showed that opening the bottles and replacing the vials of NaOH resulted in significant losses of CO_2 during sampling. Therefore, a 10 ml syringe with a 6-inch needle was inserted through the glass tubing into the scintillation vial to draw out the used trapping solution and replace it with new solution. Duplicate 1-ml samples of each trapping solution along with 15 ml of Ultima Gold cocktail were analyzed by liquid scintillation counting (LSC) for [14]CO_2 content using a Packard Tri-Carb Liquid Scintillation Analyzer.

Sorptive Behavior of TCP. The sorption coefficient of TCP was determined in 25 soils. The method used for these analyses was similar to that described by Felsot and Dahm (*12*). Four treating solutions of ^{14}C-TCP in methanol were made up such that 15 μl contained either 1, 5, 10, or 100 μg of total TCP. For each soil, four 2-g (dry-weight) samples were placed in 25 ml glass centrifuge tubes with teflon-lined caps. To each tube 10 ml of 0.01N $CaCl_2$ solution was added. For each soil, individual tubes were then treated with 15 μl of ^{14}C-TCP solution for initial aqueous TCP concentrations of 0.1, 0.5, 1.0, and 10.0 μg/ml. The tubes were shaken on a horizontal Eberbach shaker for 2 hours. A preliminary experiment demonstrated that this time was sufficient for sorptive equilibrium to be reached. The tubes were then centrifuged for 15 minutes at 2000 rpm to separate solution and soil phases, and duplicate 1 ml samples of aqueous phases were analyzed for nonsorbed ^{14}C by LSC. Sorbed ^{14}C-TCP was estimated as the difference between applied and nonsorbed material. Sorption coefficients were calculated using the Freundlich equation:

$$S = K_d C^{1/N}$$

wherein S and C are the sorbed-phase (μg/g) and solution-phase (μg/ml) concentrations of the sorbate, respectively, N is an empirical power term, and K_d (ml/g) is the sorption coefficient. The Freundlich K_d for each soil was also corrected for soil organic carbon content (K_{oc}=K_d/organic carbon %).

TCP Degradation in Soil Microbial Cultures. These incubations were carried out to determine if there were organisms present in the test soils that could use TCP as a sole carbon source in aqueous culture. An initial kinetic study was conducted using 2 soils, M269 and M285, that represented opposite ends of the spectrum in their TCP degradation rates.

A shake flask culture technique was used to monitor TCP degradation over time. A sterile basal salts medium was prepared as described by Krieg (*13*). A ^{14}C-TCP treating solution was prepared in acetone and 100 μl (0.01 μCi) was added to each flask and the solvent allowed to evaporate. To each flask was added 5 ml of sterile basal salts medium, bringing the initial TCP concentration to 5 μg/ml.

Samples of each soil (0.01 g) were added to each of 6 flasks.
Samples of previously sterilized (autoclaved) soils were also added
to several flasks to serve as sterile controls. The flasks were then
plugged with loose cotton stoppers and placed in a Fisher water bath
shaker at 25°C and an agitation rate of 68 strokes/minute for up to
four weeks.

One sample from each soil inoculum was sacrificed at 0, 4, 7, 14,
and 21 days and analyzed for TCP remaining. The contents of the
flasks were transferred to 25 ml centrifuge tubes and centrifuged for
15 minutes at 2000 rpm. Two 0.5 ml aliquots were taken for
quantitative LSC analysis of solution phase ^{14}C, and a 1 ml aliquot
was taken for qualitative analysis by High-Pressure Liquid
Chromatography (HPLC). HPLC analyses were performed with a Waters
600E Powerline System that included a μ-Bondapak C18 column.
Detection of analytical standards was by UV absorption at 300 nm λ,
and reconstructed radiochromatograms of samples were generated by
collecting eluent fractions and analyzing them by LSC. The mobile
phase consisted of a gradient of two solutions: A) 890 ml distilled
water, 9.0 ml glacial acetic acid, 0.45 ml N,N-dimethyl-octylamine
and B) 890 ml methanol, 9.0 ml glacial acetic acid, 0.45 ml
N,N-dimethyloctylamine. Samples were injected under 1.5 ml/min flow
conditions under a linear gradient that went from 100% solvent A to
100% solvent B in 30 minutes.

After the initial incubation with 2 soils and the sterilized
control, additional fixed-length incubations of the remaining soils
with TCP were conducted. Based on the response time for the first
experiment, samples were incubated for 4 weeks at which time they
were analyzed for remaining TCP.

Data Analyses. Statistical analyses of the relationships between TCP
degradation data and soil factors (properties, degradation of other
compounds) were conducted using SAS. Correlation procedures (PROC
CORR) and regression procedures (PROC RSQUARE, PROC REG) were used to
investigate interrelationships.

Results

Degradation of TCP in Soil. The degradation rates of TCP varied
between different soils (Table II). Cumulative percent
mineralization during the 21-day incubation ranged from 2.4 to 45.1%
of applied. The level of mineralization observed was slightly less
than previously reported by Bidlack (7) and Racke et al. (8) who
respectively noted up to 56% and 85% TCP mineralization over a 2-week
period. This inconsistency was most likely due to experimental
differences in application methods, application rates, and soil
incubation procedures. In the current study, most soils displayed a
rather constant flux of ^{14}CO2 during the incubation, whereas a few
soils with more rapid rates of degradation showed a decrease in rate
over time (Figure 1). Only soils M258, M268, M284, and M285
displayed acceleration in TCP mineralization rate over the course of
the incubation. A preliminary examination of the properties of soils
with rapid and slow TCP mineralization rates revealed few general
trends to explain the variation observed.

The mineralization of TCP in the three annually collected soils
was examined to determine the effect of soil storage on the microbial

Table II.   Mineralization of Organic Compounds in Soil

SOIL	TCP 21 DAYS	TFP 21 DAYS	HPYR 14 DAYS	DCPH 7 DAYS	GLUCOSE 3 DAYS
	CUMULATIVE $^{14}CO_2$ AS % OF APPLIED ^{14}C				
M177	13.1	NM[a]	NM	NM	NM
M242	10.2	NM	NM	NM	NM
M271	15.4	1.6	54.6	12.6	16.0
M261	14.4	1.5	54.2	12.4	22.2
M269	2.4	0.1	13.5	6.2	32.8
M273	30.9	0.7	19.0	19.0	21.8
M262	24.5	0.6	54.8	17.2	22.5
M264	13.0	2.1	56.5	11.6	25.5
M266	10.3	3.9	56.4	14.1	31.1
M268	36.8	0.4	56.4	13.5	18.7
M272	21.7	0.5	54.8	12.1	25.8
M274	7.2	1.1	52.8	8.8	29.4
M234	7.6	0.1	38.5	8.4	28.3
M244	7.6	1.7	51.1	11.2	21.3
M256	15.5	0.6	58.4	12.7	18.6
M275	15.7	6.6	58.5	12.7	21.6
M284	35.4	0.6	55.7	14.5	21.1
M285	45.1	0.7	55.6	15.2	20.6
M236	8.9	0.1	38.8	6.1	31.2
M258	29.8	1.5	55.7	11.9	17.6
M265	25.0	1.3	57.1	13.1	17.5
M267	17.3	0.3	56.3	10.7	24.4
M270	11.7	4.6	56.1	11.3	18.4
M277	40.7	3.6	51.8	11.0	24.7
M283	13.6	1.2	52.8	12.5	16.8
M245	16.1	NM	NM	NM	NM
M247	14.7	NM	NM	NM	NM
M254	28.0	NM	NM	NM	NM
M255	27.1	NM	NM	NM	NM

[a] NM: Not measured

mineralization of TCP. Soils M177, M242, and M271 had been stored
for 32, 19, and 7 months, respectively, prior to use. The most
recently collected sample, M271, displayed the fastest degradation
rate for this compound, but overall the rates were not inversely
correlated with storage length. In fact, the rate of TCP
mineralization was quite constant in these 3 soil samples considering
they were collected in different years from somewhere within the same
field. It is concluded from these results that storage and handling
of these soil samples did not alter their activity toward TCP. This
is corroborated by the poor regression relationship for TCP
mineralization rate versus soil storage length (months) noted for all
25 soils used (R^2 = 0.15). Therefore, use of some of the older soil
samples for this study was valid.

Results for incubation of TCP with the pairs of untreated
(control) and chlorpyrifos-history soils are shown in Figure 2.
Previous treatment with chlorpyrifos did not enhance TCP
mineralization in 2 of the pairs of soil (M245, M247; M254, M255).
However, in the chlorpyrifos-history soil of one pair, the treated
soil (M285) mineralized TCP at a more rapid rate than the untreated
soil (M284). This could be explained by slight differences in soil
types within a single field. These results support previous work
concluding that chlorpyrifos and TCP are not subject to enhanced
degradation in soil (*14,15*).

Degradation of Other Test Compounds. The objective in investigating
the degradation of TFP, HPYR, and DCPH in the same soils was to
determine whether any correlation existed between TCP mineralization
rates and mineralization of similar compounds. Degradation of
glucose was also investigated in these soils to account for any
correlation between general microbial metabolic activity and TCP
mineralization.

TFP. Cumulative mineralization of TFP over a 21-day period
ranged between 0.06% and slightly over 6% of applied (Table II).
Apparently, elimination of one chlorine atom and substitution of
another with a trifluoromethyl group renders TFP much less
mineralizable than TCP. Although most soils displayed a constant,
albeit slow rate of TFP mineralization, soils M256, M270, and M266
showed some acceleration in mineralization rate with time.

HPYR. The rate of HPYR degradation was more rapid than that for
either TCP or TFP. Between 13.5 and 58.5% of applied HPYR was
mineralized within 14 days (Table II). Rather than a wide
distribution of mineralization rates, as was the case with TCP,
nearly all the soils degraded HPYR at a similar rate. Only soils
M273, M269, M236, M234, and M244 displayed somewhat slower HPYR
mineralization rates. The rapid mineralization of HPYR observed in
this study support previous conclusions on the biodegradability of
HPYR in soil suspensions (*16*).

DCPH. Mineralization of DCPH also proceeded rapidly, but with
less variation between soils than for TCP. Quantities of DCPH
mineralized within a 7-day period represented from 6.1 to 19.0% of
applied material (Table II). At low concentrations DCPH is easily
degraded by soil microorganisms, which have shown the ability to
adapt to rapidly catabolize DCPH as a sole carbon source (*17*).

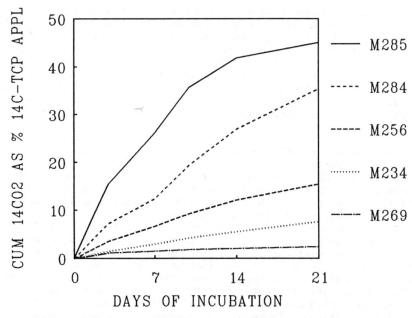

Figure 1. Cumulative Mineralization of TCP in Selected Soils.

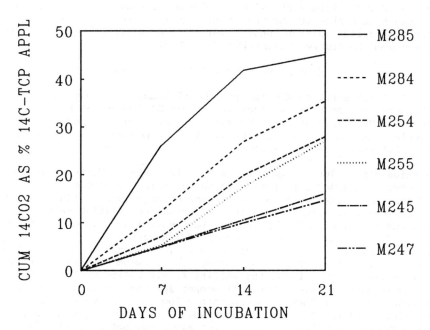

Figure 2. Cumulative Mineralization of TCP in Paired Chlorpyrifos History and Untreated Soils from 3 Sites.   Site 1: M285 (Hist), M284 (Untr).   Site 2: M255 (Hist), M254 (Untr).   Site 3: M247 (Hist), M245 (Untr).

Glucose. Glucose mineralization has been used to estimate the general microbial activity of soil and as a relative measure of soil biomass (*10*). For the soils in this study, between 16.0 and 32.8% of the applied glucose was mineralized within 3 days (Table II). All soils were quite active in degrading glucose, although some relative differences were observed. The fact that the degradation of an easily utilized carbon source such as glucose may be accompanied by incomplete mineralization underscores the observed balance between mineralization and incorporation that is characteristic of soil microbial activity (*18*).

Sorptive Behavior of TCP. Freundlich sorption coefficients for TCP are listed in Table III. The TCP sorption coefficients (K_d) varied between 0.3 and 20.3 ml/g (average K_d = 3.1), and K_{oc} values ranged between 27 and 389 ml/g (average K_{oc} = 168). According to the soil mobility classes proposed by Hamaker (*19*), TCP would be considered as low to moderately mobile depending on soil type.

To elucidate the factors that influence the sorption of TCP, multiple regression of soil organic carbon and soil pH on K_d was conducted. The relationship between sorption and organic carbon content was poor (R^2 = 0.11), and indicates that without consideration of soil pH, the K_{oc} expression is not of much value in predicting TCP mobility (note: soil M277 was an outlier with very high organic carbon and was excluded from the regression and the predictive equation). There was an inverse relationship between TCP adsorption and pH (R^2 = 0.58) that is probably due to the acidic nature of TCP. With a pK_a of 4.55, TCP would tend to be more ionic in nature at higher pH and thus less tightly sorbed. Inclusion of both soil organic carbon content and pH in a multiple regression with K_d improved predictive capability (R^2 = 0.75):

$$K_d = 10.80 + 0.94 \ O.C.\% - 1.49 \ pH$$

A plot of TCP sorption K_{oc} versus soil pH demonstrates the pH dependency of the sorption process (Figure 3). Sorption K_{oc} values for TCP (3344 ml/g) and the TCP anion (54 ml/g) were separately calculated using a curve-fitting model developed by Fontaine et al. (*20*) that takes pK_a into account.

Relationship of TCP Degradation to Soil Variables. Correlation and regression analyses were performed on the soil properties (Table I) and the mineralization data for TCP and other compounds (Table II) to determine any interrelationships. A partial correlation matrix obtained from the PROC CORR program (not shown) demonstrated that no single factor explained the behavior of TCP very well. The only soil property that was at all correlated with TCP degradation was K_d (R = 0.61). There was a measure of correlation between the ability of a soil to mineralize TCP and the ability to mineralize DCPH (R = 0.60). It should also be noted that soil properties such as organic carbon percent, pH, or texture also had rather low correlation with the degradation of any of the test compounds including glucose. Thus, soil physical and chemical properties alone may not be very valuable in predicting rates of some microbially-mediated degradation processes (i.e. TCP mineralization).

Table III.   Freundlich Soil Sorption Coefficients for TCP

SOIL	Kd (ml/g)	percent organic carbon	Koc (ml/g)
M177	3.05	3.08	99
M242	1.96	3.18	62
M271	1.65	3.20	52
M261	1.01	1.87	54
M269	2.72	0.70	389
M273	5.47	2.35	233
M262	2.67	1.71	156
M264	0.37	0.52	71
M266	0.30	0.45	67
M268	4.37	1.74	251
M272	1.15	0.47	245
M274	0.84	1.92	44
M234	3.64	1.42	256
M244	0.84	3.06	27
M256	6.74	2.12	318
M275	0.35	0.88	40
M284	6.52	2.20	296
M285	6.28	2.00	314
M236	2.08	0.75	277
M258	1.33	2.43	55
M265	1.67	1.10	152
M267	1.71	0.59	290
M270	0.29	0.89	33
M277	20.29	5.90	344
M283	0.80	1.20	67
AVG =	3.12	1.83	168

Note: Calculated Koc values for the neutral and anionic forms of TCP were 3344 and 54 ml/g, respectively (Figure 3).

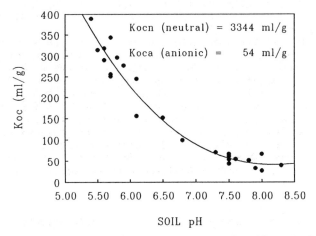

Figure 3. Relationship Between Freundlich Sorption Coefficient for TCP (Koc) and Soil pH.

To determine whether soil factors may be interacting to influence the observed TCP mineralization rates a multiple regression procedure was employed. The PROC REG procedure was set up in such a way that for each number of independent variables in the model (i.e. soil properties) the best 3 combinations were selected for output. As shown in Table IV, one of the best fits that could be obtained (R^2 = 0.58) was with a model that contained 4 variables: Kd, silt content, pH, and CEC. Although inclusion of 5 variables slightly increased the regression fit, little progress was made by including greater than 4 independent variables. However, the level of predictability provided by even the best regression model which included only soil physical and chemical properties was poor.

Additional multiple regression analyses were performed to determine whether inclusion of both soil property information and biodegradation of other test compounds provided a better model for explaining TCP degradation. As shown in Table IV, models that included soil mineralization of other test compounds provided much better fits of the data than models that included soil physical and chemical properties alone. The best model (R^2 = 0.78) included 6 independent variables: Kd, DCPH mineralization, HPYR mineralization, pH, CEC, and clay content. Thus, inclusion of soil mineralization of other test compounds, most notably DCPH and pyridinol, led to an increased ability to predict soil response to TCP. However, the practicality of obtaining this biodegradation information solely to predict TCP degradation rates would be rather low.

Degradation of TCP in Soil Microbial Cultures. To determine whether microorganisms capable of using TCP as a sole carbon source were present in soils in which TCP was rapidly mineralized, cultures inoculated with either soil M269 (2.43% TCP mineralization), soil M285 (45.11% TCP mineralization), or sterilized soil M285 were incubated. As evident from a plot of TCP concentration over time in the cultures (Figure 4), there was no degradation of TCP as a result of inoculation with either soil M269 or sterilized samples of soil M285. However, after a lag of about 7 days, TCP was rapidly degraded in cultures inoculated with soil M285. The lag phase followed by rapid degradation is characteristic of microbial adaptation for catabolism. This is the first report of adaptation for microbial catabolism of TCP.

Additional culture incubations were conducted in which TCP-treated cultures inoculated with samples of each of the remaining soils were incubated for standard length incubations of 3 weeks, at which time the quantities of TCP remaining were assayed. Of all the cultures, only the 2 inoculated with soils M285 and M247 substantially degraded the added TCP, with 6% and 37% TCP remaining, respectively. Both soils M285 and M247 came from fields that had been treated with chlorpyrifos for up to 5 years previously. Several incubations in which mannitol at 1280 μg/l was added as a supplemental carbon source to the cultures failed to enhance TCP degradation in the cultures inoculated with any of the soil samples.

Discussion

The rates of TCP degradation in soil are variable between different

Table IV.   Multiple Regression Models and Goodness of Fit for
             Predicting TCP Mineralization in Soil

# PARAMETERS	R^2	VARIABLES IN MODEL
Soil Physical/Chemical Properties Only		
1	0.2246	%O.C.
1	0.3297	1/3 BAR
1	0.3680	Kd
2	0.4875	pH, CEC
2	0.5120	Kd, %SILT
2	0.5200	Kd, %SAND
3	0.5403	pH, CEC, %CLAY
3	0.5490	Kd, CEC, pH
4	0.5733	1/3 BAR, pH, CEC, %CLAY
4	0.5765	Kd, %SILT, pH, CEC
5	0.5949	1/3 BAR, pH, CEC, %CLAY, %SAND
5	0.5949	1/3 BAR, pH, CEC, %CLAY, %SILT
Soil Physical/Chemical Properties and Degradation Data		
1	0.3633	DCPH
2	0.5533	Kd, GLUCOSE
2	0.6635	Kd, DCPH
3	0.6873	Kd, DCPH, HPYR
3	0.6914	Kd, DCPH, GLUCOSE
4	0.7301	Kd, DCPH, HPYR, pH
4	0.7443	Kd, DCPH, GLUCOSE, 1/3 BAR
5	0.7560	Kd, DCPH, GLUCOSE, pH, HPYR
5	0.7561	Kd, DCPH, GLUCOSE, 1/3 BAR, HPYR
6	0.7762	DCPH, 1/3 BAR, pH, HPYR, %CLAY, CEC
6	0.7764	Kd, DCPH, HPYR, pH, CEC, %CLAY

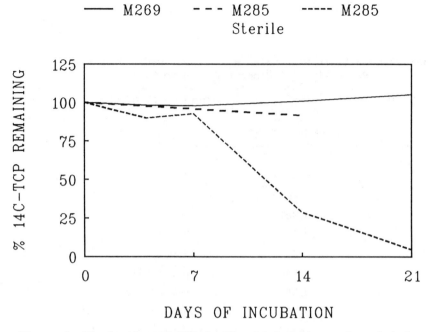

Figure 4. Dissipation of TCP in Microbial Cultures Inoculated
with Chlorpyrifos-History (M285) or Untreated (M269) Soil.

soil samples, and are poorly predicted by regression models based on
soil physical and chemical properties.   Thus, it is currently
difficult to anticipate accurately the persistence of this compound
under field conditions.
        Results of the present study indicate that there likely is an
important interaction between soil microbial populations and TCP that
determines its persistence. A fair regression model for predicting
TCP persistence could only be generated by inclusion of both soil
properties and biodegradation data for other organic compounds.   This
interaction is apparently not strictly associated with gross numbers
or activity of soil microorganisms, as evidenced by the lack of
correlation between soil mineralization of TCP and glucose.   However,
the correlation observed between the ability of a given soil to
mineralize TCP and DCPH suggests that there are specific
subpopulations of xenobiotic-degrading microorganisms that must be
present and active for rapid TCP degradation to occur.   It is unclear
whether the biodegradation of TCP is generally associated with
catabolic or cometabolic processes.   Both catabolism and cometabolism
have been shown to be important in the microbial mineralization of
similar compounds such as p-nitrophenol and 2,4-D, and only one or
both processes may be important in a given soil (21,22).   Only 2
soils with long histories of chlorpyrifos use and TCP exposure
contained populations of microorganisms that adapted to degrade TCP
as a sole carbon source in mixed culture.

Literature Cited

1. Woods, W. J.; McKellar, R. L.   The Dow Chemical Company, unpublished report, **1979**.
2. Fontaine, D. D.; Wetters, J. H.; Weseloh, J. W.; Stockdale, G. D.; Young, J. R.; Swanson, M. E.   The Dow Chemical Company, unpublished report, **1987**.
3. Moore, B. C.; Danesh, B. *The Environmental Fate of Triclopyr in Soils Following Forest Site Preparation Applications.* Washington State University: Pullman, WA, **1989**.
4. Meikle, R. W.; Hamaker, J. W.   The Dow Chemical Company, unpublished report, **1981**.
5. Dilling, W. L.; Lickly, L. C.; Lickly, T. D.; Murphy, P. G. *Environ. Sci. Technol.* **1984**, *18*, 540-543.
6. McCall, P. J.   The Dow Chemical Company, unpublished report, **1985**.
7. Bidlack, H. D.   The Dow Chemical Company, unpublished report, **1976**.
8. Racke, K. D.; Coats, J. R.; Titus, K. R.   *J. Environ. Sci. Health* **1989**, *B23*, 527-539.
9. Bidlack, H. D.   The Dow Chemical Company, unpublished report, **1980**.
10. Anderson, J. P. E.; Domsch, K. H. *Soil Biol. Biochem. 10*, **1991**, 215-221.
11. Racke, K. D.; Coats, J. R. *J. Agric. Food Chem.* **1987**, *35*, 94-99.
12. Felsot, A.; Dahm, P. A.   *J. Agric. Food Chem.* **1979**, *27*, 557-563.
13. Krieg, N.R.   1981.   In *Manual of Methods for General Bacteriology*; Gerhardt, P., Ed.; American Society for Microbiology: Washington, DC, **1981**; 112-142. 42.
14. Harris, C. R.; Chapman, R. A.; Tolman, J. H.; Moy, P.; Henning, K.; Harris, C. *J. Environ. Sci. Health* **1989**, *B23*, 1-32.
15. Racke, K. D.; Laskowski, D. A.; Schultz, M. R.   *J. Agric. Food Chem.* **1990**, *38*, 1430-1436. 36.
16. Sims, G. K.; Sommers, L. E. *Environ. Toxicol. Chem.* **1986**, *5*, 503-509.
17. Tyler, J. E.; Finn, R. K. *Appl. Microbiol.* **1974**, *28*, 181-184.
18. Baldock, J.A.; Oades, J.M.; Vassallo, A.M.; Wilson, M.A. *Environ. Sci. Technol.* **1990**, *24*, 527-530.
19. Hamaker, J. W.   In *Environmental Dynamics of Pesticides*; Haque, R.; Freed, V. H., Ed.; Plenum Press: New York, NY, **1975**; 115-133.
20. Fontaine, D. D.; Lehmann, R. G.; Miller, J. R. *Weed Sci.* (in press) **1990**.
21. Fournier, J. C. *Chemosphere* **1980**, *9*, 169-174.
22. Ou, L. T.; Sharma, A. *J. Agric. Food Chem.* **1989**, *37*, 1514-1518.

RECEIVED November 8, 1990

Chapter 8

Bound (Nonextractable) Pesticide Degradation Products in Soils

Bioavailability to Plants

Shahamat U. Khan

Land Resource Research Centre, Research Branch, Agriculture Canada,
Ottawa, Ontario K1A 0C6, Canada

A significant proportion of degradation products from
certain pesticides applied in agriculture remains in
soils as bound (nonextractable) residues. Soil organic
matter is largely responsible for the formation of bound
residues. In addition to chemical binding, the
degradation products of pesticides are also firmly
retained by soil organic matter fractions to form bound
residues by a process that more likely involves
adsorption on external surfaces and entrapment in the
internal voids of molecular sieve-type structural
arrangements. The bound residues of pesticides
degradation products in soil are bioavailable to plants.

Although the formation of bound (nonextractable) pesticide residues
in soil and plants have been known to occur for over two decades,
their significance has been critically addressed only recently when
it became obvious that these residues are not excluded from
environmental interactions. The true nature of bound residues of
pesticides and their degradation products in soil and plants is
still poorly understood. However, it is well established that
various pesticides and/or metabolites can form appreciable amounts
of bound residues. This was revealed primarily by use of
radiolabeled pesticides, which after application, led to the
detection and quantitation of residues undetectable by any
conventional analytical techniques. Thus, for a long time the
possible soil or plant burden of total pesticide and/or its
degradation products residues has been underestimated.

The question of the significance of residues of bound
pesticides and/or their degradation products can be evaluated in
terms of their availability to plants if they should be released
from soil. Recent evidence indicates that bound residues may be
released from soil and absorbed by plants.

The formation of bound residues of pesticides in soil and
plants, and their bioavailability has been reviewed by Klein and

0097–6156/91/0459–0108$06.00/0
Published 1991 American Chemical Society

Scheunert (14), Khan (7), Roberts (16) and Calderbank (3). The present paper summarizes the information on bound residues of the degradation products of certain pesticides in soil and their uptake by plants. The mechanisms(s) of the formation of bound residues in soil is also discussed. Most of the data reported in this paper are based on research efforts in our laboratory at the Land Resource Research Centre, Ottawa.

Formation of Bound Residues in Soil

Figure 1 illustrates several of the direct and indirect routes by which pesticides and their degradation products form bound residues in soil (3). The majority of these bound residues in soil originate from deliberate application of pesticides to the soil or to the foliage of crop plants and weeds. Very often substantial quantities of the applied pesticide reach the soil by either missing the target or by run-off from leaves and stems. Death of plants or root exudation may also contribute to the incorporation of pesticide in soil. Although leaching, volatilization and biota accumulation may result in the loss of pesticide from soil, a proportion of many of these chemicals on or in soil is subjected to biological, chemical and photochemical degradation of the parent chemical. A portion of the pesticide and/or its degradation products then becomes much more firmly held by the soil component than the average and is now referred to as a "bound" residues. The mechanism of binding processes by which the pesticide and/or metabolites become bound in soil has been discussed by Calderbank (3).

In studying bound residues in soil it often becomes very difficult to differentiate between bound residues of parent pesticides and their degradation products. Drastic experimental methods utilized to extract the bound radiolabeled residues or combustion to $^{14}CO_2$ to determine the total bound ^{14}C residues destroy or alter the chemical nature of the residues. Nevertheless, identification of bound residues in soil as degradation products of the parent pesticide has been reported in many instances. Katan et al (6) and Katan and Lichtenstein (5) demonstrated rapid binding of the amino analogue of parathion. These soil bound residues of aminoparathion were unextractable and therefore not detected in routine residue analyses. Wheeler et al (20) also observed a significant relationship between the amount of binding and the nature of substitution on the amino nitrogen. These authors postulated that some of the metabolites containing secondary or primary amino functional groups may have become part of the bound residue in soil. Spillner et al (19) implicated 2-methylhydroquinone, an oxidative product of 3-methyl-4-nitrophenol, as the precursor to the formation of bound residues of fenitrothion in aerobic soil. However, under anaerobic conditions binding was thought to proceed through the intermediate. Golab et al (4) suggested that α,α,α-trifluorotoluene-3,4,5-triamine, a degradation product of trifluralin, may be a key compound in the formation of soil bound residues. In other studies concerned with the formation of bound residue in soil treated with fenitrothion it was suspected that part of the bound residue was 3-methyl-4-nitrophenol (15). Klein and Scheunart (14) presented data on bound residues of pesticides which have been shown to form anilines and phenols in soil.

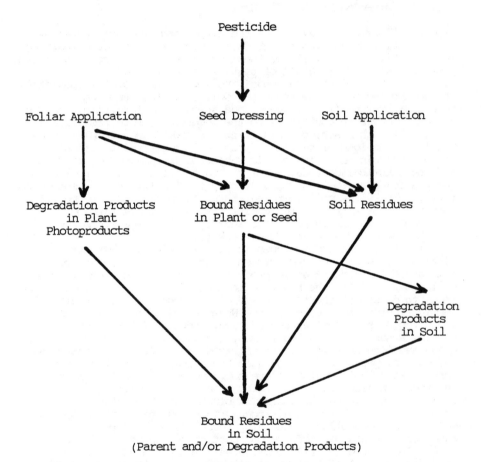

Figure 1.   Possible routes for the formation of bound residues in soil following pesticide application.

In an earlier study (10) it was shown that bound residues of the degradation products of prometryn were formed when an organic soil was incubated with the herbicide for one year (Table I). At the end of incubation period it was observed that 57.4% of the ^{14}C initially added was not extractable with solvents, constituting a bound residue. While more than 50% of the total bound ^{14}C residues constituted the parent herbicide, measurable amounts of the mono-N-dealkylated and hydroxy analogues of prometryn were also present in the form of bound ^{14}C residues (Table I). It was shown later that incubation of this soil containing bound residues with a fresh soil inoculum resulted in a release of [14]C (27%) which had been initially unextractable (11). Examination of the extractable material indicated the presence of prometryn and several metabolites including hydroxypropazine (9%) and partially N-dealkylated compounds (2%). The radioactivity which remained bound consisted of prometryn (31%), hydroxypropazine (7%) and mono-N-dealkylated prometryn (<4%) expressed relative to the initially bound prometryn residue. It is possible that some of the metabolites formed during the incubation period may have become part of the bound portion of the residues in the soil (Table II).

In another study (8) it was demonstrated that soil-bound ^{14}C residues were absorbed by oat plants grown in an organic soil previously incubated with ^{14}C prometryn for one year. Analysis of plant tissues indicated the presence of extractable ^{14}C hydroxypropazine in the form of conjugates. It was also observed that some of the radioactivity absorbed by the plants from the soil containing bound ^{14}C residues became again bound in the plant tissues. These bound plant residues were present in the form of mono-N-dealkylated compounds, namely 2-(methylthio)-4-amino-6-(isopropylamino)-s-triazine and traces of 2-(methylthio)-4,6-diamino-s-triazine (Table III).

In Germany, uniformly ^{14}C ring-labeled atrazine was applied to a mineral soil under field conditions (2). Nine years after application of the herbicide soil samples were collected for analysis. The soil contained about 50% [14]C residues in the bound form. The bound ^{14}C residues were distributed among the various soil humic fractions. It was observed that, in addition to the parent herbicide a considerable proportion of these residues consisted of the hydroxy analogues of atrazine and their dealkylated products (Table IV). These data further suggest that in addition to the parent compound bound residues in soil may also include several degradation products.

In a similar study the formation of bound ^{14}C residues in an organic soil treated with ^{14}C-fonofos was investigated under field conditions (12). It was observed that fonofos was metabolized in the field plot soil and the products formed were also bound (Table V).

In a recent study we investigated the release of bound residues from soil treated with ^{14}C-atrazine by incubating the solvent extracted soil with two species of Pseudomonas capable of metabolizing atrazine (13). The soil initially contained 25 ppm atrazine and a total radioactive count of 1.4×10^6 dpm/100g and was incubated for one year in the laboratory (Figure 2). Distilled water was added as necessary to maintain the initial moisture content of the sample during incubation. It was observed that, after an incubation period of one year, the soil contained 54% [14]C

Table I. Bound ^{14}C residues of prometryn and its degradation products in an organic soil incubated with the herbicide for one year

Compound Identified	% ^{14}C of the total bound ^{14}C residues
Prometryn	54
Hydroxypropazine	8
Mono-N-dealkylated prometryn	<2
Mono-N-dealkylated hydroxypropazine	traces
Unidentified methanol soluble products	16

Table II. Bound and extractable ^{14}C residues after incubating soil containing only ^{14}C bound residues with a fresh inoculum

Bound ^{14}C		Extractable ^{14}C	
compound identified	%*	compound identified	%*
Prometryn	31	Prometryn	14
Hydroxypropazine	7	Hydroxypropazine	9
Mono-N-dealkylated prometryn	4	Mono-N-dealkylated prometryn	2

* % of initially bound after one year incubation

Table III. ^{14}C Residues in oat plants grown in soil containing bound ^{14}C residues

Bound ^{14}C	Extractable ^{14}C
Mono-N-dealkylated prometryn	Hydroxypropazine
Di-N-dealkylated prometryn	(conjugate)

Table IV. Bound residues in soil and humic material nine years after atrazine application in the field

Compound	Soil	Humic acid	Fulvic acid	Humin
		ppm[a]		
Atrazine	0.11	0.30	ND	ND
Deethylatrazine	T	T	ND	ND
Deisopropylatrazine	T	0.12	ND	ND
Hydroxyatrazine	0.10	0.30	T	0.11
Deethylhydroxyatrazine	0.13	0.20	T	T
Deisopropylhydroxyatrazine	0.07	0.20	ND	T

$_{ND}$ = nondetectable; <0.01 ppm.

T = trace amount; <0.05 ppm.

Table V. Extractable and bound (nonextractable) ^{14}C residues in a soil and onion following the application of ^{14}C-fonofos to small organic soil field plots (% of the initially applied ^{14}C)

Sampling time (days)	Soil			Onion							
	Extractable	Bound*	Total recovered	Extractable				Bound**			
				Shoot	Peel	Bulb	Total	Shoot	Peel	Bulb	Total
0	96.5	1.7	98.2								
31	69.6	6.3	75.9								
89	51.0	14.1	65.1	0.4	0.8	0.5	1.7	0.3	0.7	1.0	2.0
130	41.6	21.8	63.3	0.4	0.9	0.3	1.6	0.3	0.9	1.5	2.7

* Fonofos, Methylphenylsulfoxide, fonofos oxon

** Fonofos oxon

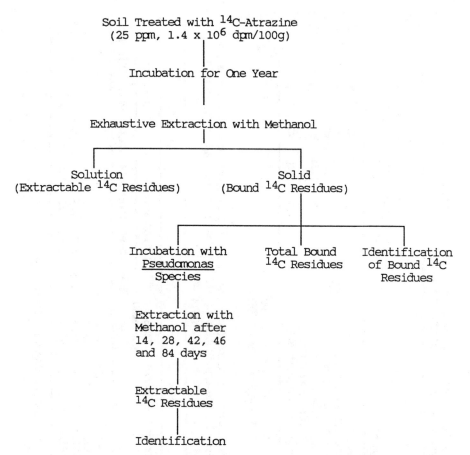

Figure 2. Schematic diagram for the analysis of bound ^{14}C residues from the ^{14}C-atrazine treated soil incubated with Pseudomonas species.

of the originally applied radioactivity in bound form. Analysis of bound ^{14}C residues revealed the presence of atrazine and its degradation products as shown in Table VI.

The extracted soil containing bound ^{14}C residues was incubated with two Pseudomonas species (13). These species were isolated by modified enrichment culture technique from a soil which had a long history of atrazine application. In our earlier studies these bacteria were shown to be capable of metabolizing atrazine (1). It was observed that, at the end of incubation period, the amounts of released (extractable) ^{14}C residues ranged from 30 to 35% of its initially bound ^{14}C. The nature of the released ^{14}C residues in methanol extracts from the incubated soils was determined by GC and GC-MS. The identity and the amount of each compound released is shown in Figure 3. The data suggest that the initial increase in the extractable ^{14}C residues of atrazine up to 28 days was followed by a decrease in concentration of the herbicide over the incubation period of 84 days. In contrast, the amounts of extractable ^{14}C residues of hydroxyatrazine, dealkylated atrazine and their hydroxy analogues increased during incubation. It was demonstrated in our earlier study (1) that very little dechlorination of the parent compound atrazine occurred with the Pseudomonas species used in this study. It was suggested that the presence of both alkyl groups on the atrazine molecule may be inhibitory for bacterial dechlorination. Thus, it is apparent that the extractable hydroxyatrazine determined during the incubation period was mainly released from the bound form present in the soil rather than resulting from the dechlorination of the released parent compounds (13).

Analysis of soil extracts after different incubation periods revealed a relatively high proportion of deisopropylatrazine and its hydroxyanalogue. The amounts of the extractable compound present (Figure 3) cannot be solely attributed to the release of such residues from the bound form. It appears that some of the released bound atrazine was also further metabolized preferentially by N-dealkylation of isopropyl moiety by the two Pseudomonas species (1). It also appears that the mono-N-dealkylated products were readily subjected to dechlorination resulting in the formation of the respective hydroxy analogues (Figure 3).

The foregoing studies clearly demonstrate that often a considerable proportion of pesticide degradation products may be present in soil in the form of bound residues. These residues may become available for uptake by plants and may be carried over into succeeding crops as they may persist in soil for long periods. It is possible that the release of these residues in soil solution may also allow their leaching into ground water.

Mechanism of Formation of Bound Residues

Several researchers have reported that in the formation of bound residues, the pesticides and/or their metabolites are chemically bound to the soil organic matter. However, our work provides evidence that support the contention that physical binding may also play on important role in the formation of soil-bound residues.

As described earlier an organic soil treated with ^{14}C-prometryn contained 57.4% of the total applied radioactivity in bound form

Table VI. Bound ^{14}C residues in soil incubated with ^{14}C atrazine
            (25 ppm) for one year

Bound ^{14}C residues	Concentration (ppm)
Atrazine	3.7
Hydroxyatrazine	1.5
Deethylatrazine	2.1
Deisopropylatrazine	1.1

The soil contained 54% bound ^{14}C expressed as percent of
originally applied.

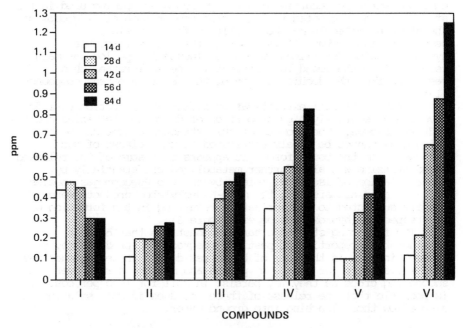

Figure 3. Compounds released from soil containing bound ^{14}C
residues by incubation with <u>Pseudomonas</u> species.   I -
Atrazine; II - Hydroxyatrazine; III - Deethylatrazine; IV -
Deisopropylatrazine, V - Deethylhydroxyatrazine and VI -
deisopropylhydroxyatrazine.

following an incubation period of one year (9). In a subsequent
study it was shown that the bound residues became distributed among
the various soil organic matter fractions, predominantly humic
substances (9). These observations were consistant with several
other studies reported in the literature. Table VII shows the
distribution and identification of bound ^{14}C in humic materials.
The presence of the 2-hydroxy metabolite in fulvic acid(FA) is of
special interest. FA is known to be present in surface waters and
imparts a yellow to brown colour in natural water (17). Thus, FA
bound degradation products of pesticides would be expected to become
bioavailable to both plants and exposed aqueous or soil fauna.

   In our study we employed thermoanalytical methods in order to
obtain information on the nature of the formation of bound residues
of pesticides and their degradation products. These techniques have
been used for investigating the mechanism of thermal decomposition
of organic matter (18). The experiments were carried out by using
the High Temperature Distillation technique isothermally (10). The
amount of ^{14}C released from humic material as a function of
temperature under a helium stream was determined (Figure 4). It was
observed that about 1% of the radioactivity was released at 150°C,
about half of the total bound ^{14}C was released at 325° to 350°C and
by 700°C the recovery of ^{14}C was nearly quantitative. Differential
thermogravimetry of humic materials (Figure 5) showed elimination of
all COOH and OH groups between 200° to 400°C and decomposition of
humic "nuclei" at 450° to 550°C (18). Table VII shows that in
addition to hydroxy and N-dealkylated analogues, a considerable
portion of the bound residue in the humic materials was present in
the form of prometryn. One can postulate that bound residue
formation may be linked with phenolic-OH and -COOH groups of soil
organic matter involving chemically stabilizing reactions between
these functional groups and the pesticide and/or metabolites. Under
these conditions the presence of unchanged pesticides or unreactive
degradation products in the bound residue is not expected. However,
earlier studies have demonstrated the formation of such bound
residues. Thus, conceivably, in addition to the chemical binding of
the reactive metabolites of pesticides, physical binding of the
parent molecules and/or unreactive metabolites may also play an
important role in the formation of soil-bound residues.

   It has been suggested by Schnitzer and his coworkers (17) that
humic materials consist of phenolic and benzenecarboxylic acids
joined by hydrogen bonds to form a molecular sieve-type polymeric
structure of considerable stability (Figure 6). One of the
characteristics of this proposed structure is that it would contain
voids or holes of different molecular dimensions which could trap
organic molecules such as pesticides and/or their degradation
products. This may be considered analogous to the "clathrate
compounds". The application of this structural concept to the bound
residues of pesticide and their metabolites in organic matter or
humic materials is still a matter of conjecture. However, judging
from the thermoanalytical data, the thermal decomposition of organic
matter or humic substances appears to weaken the structure by
eliminating the functional groups and eventually decomposing the
nuclei. This, in turn, may permit the release of bound residues
that are trapped in the molecular-sieve type structure.

Table VII. Bound residues of the herbicide and its metabolites in humic materials from an organic soil treated with ^{14}C-prometryn

Humic fraction	Prometryn	Hydroxypropazine	Deisopropylprometryn
		ppm	
Humic Acid	0.3	<0.1	0.2
Fulvic Acid	0.3	0.9	<0.1
Humin	1.9	ND[a]	1.1

[a]Nondetectable; <0.01 ppm

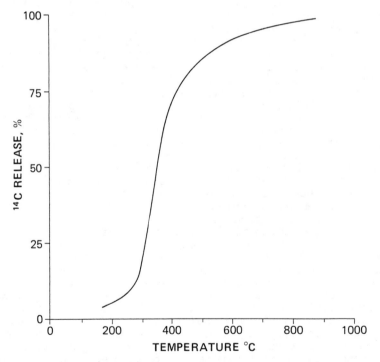

Figure 4. Thermal profile of bound ^{14}C released from humic materials.

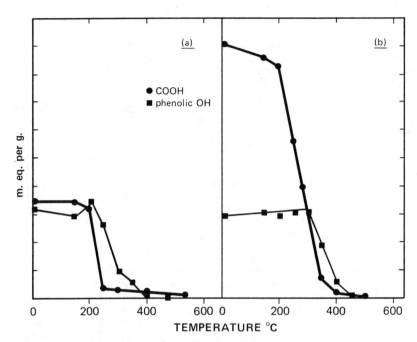

Figure 5. Changes in carboxylic and phenolic hydroxyl groups during pyrolysis (a) – Humic acid, (b) – Fulvic acid.

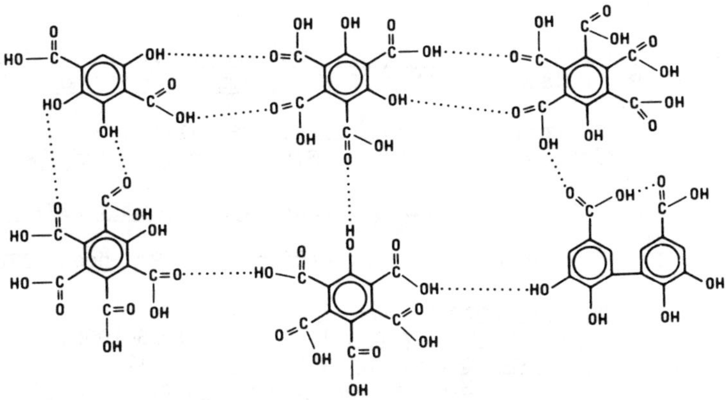

Figure 6. Partial structure of humic materials. (Reproduced with permission from reference 17. Copyright 1982 Marcel Dekker.)

Conclusion

The use of ^{14}C-labeled pesticides has made us aware of the existence of bound residues of pesticides and their degradation products in soil. These residues would escape detection by the conventional analytical methods and would result in an underestimation of the soil burden of total pesticides and/or their degradation products. Considerable evidence is presented in the literature to demonstrate the bioavailability of soil-bound pesticides and/or their degradation products to plant and soil fauna. However, availability of residues to plants is considerably lower from bound residues than from freshly pesticide-treated soil. It has been suggested that the uptake ratio of pesticides and their degradation products from freshly-treated soils compared to those from soils containing bound residues was of the order of about 5:1 (7).

The organic fraction of a soil appears to have the potential for forming bound residues with pesticides or products arising from their degradation. It is suggested that in addition to chemical binding, the pesticide and/or their metabolites are also firmly retained by humic materials by a process that more likely involves adsorption on external surfaces and entrapment in the internal voids of a molecular sieve-type structural arrangement.

In future it will be important to obtain more information about the mechanism(s) of binding of pesticides and their degradation products to form bound residues. This will possibly shed some light on the mechanism of their potential release and the conditions by which this release might occur. Since not much information is available pertaining to the nature and potential biological activity of the degradation products that are bound in soil more research in this field is desirable.

Literature Cited

1. Behki, R.M.; Khan, S.U. J. Agric. Food Chem., 1986, 34, 746-749.
2. Capriel, P., Haisch, A.; Khan, S.U. J. Agric. Food Chem., 1985, 33, 567-569
3. Calderbank, A. Residue Reviews, 1989, 108, 71-103
4. Golab, T.; Althaus, W.A.; Wooten, H.L. J. Agric. Food Chem. 1979, 27, 163-179.
5. Katan, J.; Lichtenstein, E.P., J. Agric. Food Chem. 1977 25, 1404-1408.
6. Katan, J.; Fuhremann, T.W. Lichtenstein, E.P. Science, 1976 193, 891-894
7. Khan, S.U. Residue Reviews 1982, 84, 1-24
8. Khan, S.U. J. Agric. Food Chem., 1980., 28, 1096-1098.
9. Khan, S.U. J. Agric. Food Chem., 1982, 30, 175-179.
10. Khan, S.U.; Hamilton, H.A. J. Agric. Food Chem. 1980, 28, 126-132.
11. Khan, S.U.; Ivarson, K.C. J. Agric. Food Chem. 1981, 29, 1301-1303.
12. Khan, S.U., Bélanger, A. Chemosphere, 1987, 16, 167-170.
13. Khan, S.U., Behki, R.M. J. Agric. Food Chem., 1990, 38, 2090-2093

14. Klein, W.; Scheunert, I. In Agrochemicals: Fate in Food and Enviroment. Proc. Interm. Symp. IAEA Vienna, 1982, pp.177-205.
15. Mac Rae, I.C. Soil Biol. Biochem., 1986 18, 221-225.
16. Roberts, T.R. Pure and Appl. Chem., 1984, 945-956.
17. Schnitzer, M.; Khan, S.U. Humic Substances in the Environment, Marcel Dekkar, New York, 1982.
18. Schnitzer, M.; Hoffman, J. Geochemica et Cosmochimica Acta, 1965, 29, 859-866.
19. Spillner, C.J.; DeBaum, J.R.; Menn, J.J. J. Agric. Food Chem. 1979, 27, 1054-1060.
20. Wheeler, W.B.; Stratton, G.D.; Twilley, R.R.; Ou, L.T.; Carlson, D.A.; Davidson, J.M. J. Agric. Food Chem. 1979, 27, 702-705.

RECEIVED December 13, 1990

Chapter 9

Enzymatic Binding of Pesticide Degradation Products to Soil Organic Matter and Their Possible Release

Jean-Marc Bollag

Laboratory of Soil Biochemistry, The Pennsylvania State University, University Park, PA 16802

Many pesticides are hydrolyzed or otherwise transformed to phenol or aromatic amine intermediates. These intermediates are frequently incorporated into humic material through covalent binding. A reaction causing this binding is oxidative coupling catalyzed by oxidoreductive enzymes or abiotic agents. The reaction forms free radicals which are easily observed if pesticide intermediates are substituted phenols or anilines. The free radicals bind to reactive groups of humic substances or polymerize to become part of humus. The chemical linkages formed are quite stable and resistant. Nevertheless, the question arises to what extent the bound pesticide residues resist subsequent release during microbial or chemical attack. Our investigation with chlorinated phenols showed that the release of these xenobiotics is very slow and the released compounds are further degraded by microorganisms. Therefore, it appears that pesticides bound to humus usually do not present a hazard to the environment.

Many industrial and agricultural chemicals are structurally similar to humus constituents. Consequently, these pollutants can become incorporated into soil organic matter during the humification process (1-3). The binding of xenobiotics to soil constituents decreases the availability of the compound for interaction with the biota, and consequently decreases their toxicity. The incorporation of these chemicals into humus may provide a convenient tool for neutralizing pollutants, provided that the subsequent release of xenobiotics is limited.

Many methods used for the decontamination of soils are effective. However, they may also be costly when employed for large-scale or long-term clean-ups. Therefore, new technologies are

0097–6156/91/0459–0122$06.00/0
© 1991 American Chemical Society

being developed for soil detoxification. Since oxidoreductive enzymes catalyze oxidative coupling reactions and initiate the covalent binding of xenobiotics to humic substances, this observation has led to the idea that enzymatic coupling may represent a means for the detoxification of pollutants in the environment (4).

The research presented in this paper deals with the enzymatic coupling of chlorinated phenols and aromatic amines to humic substances. We have characterized some oxidative coupling reactions and the resulting products. We also present data on the fate of the bound chemicals and provide further insight into the use of enzyme-induced polymerization of xenobiotics as a decontamination method.

Oxidative Coupling of Phenols

Binding of phenolic compounds to organic matter can occur by several mechanisms, but covalent bonding usually produces the most persistent complexes. The cross-linking reaction is mediated by a number of biological and abiotic catalysts including microbial oxidoreductases (2) (e.g. laccase, peroxidase, and tyrosinase), clay minerals, and metal oxides (5,6).

Several studies have been conducted to investigate the covalent binding of halogenated phenols to humic constituents. Initial experiments were performed with 2,4-dichlorophenol, a degradation product of the herbicide 2,4-dichlorophenoxyacetic acid (2,4-D). 2,4-Dichlorophenol was incubated in the presence of a phenoloxidase with humic-derived compounds such as orcinol, syringic acid, vanillic acid, and vanillin, and the formation of cross-coupling products was determined (7). The initial step is the enzymatic formation of an aryloxy radical that can then react with the humus constituents. The hybrid products formed from these reactions ranged from dimers to pentamers. Subsequent experiments demonstrated that a wide variety of halogen-substituted phenols could be cross-coupled to naturally occurring humic materials.

In additional experiments, phenols containing one to five chlorines (4-chlorophenol, 2,4-dichlorophenol, 2,4,5-trichlorophenol, 2,3,5,6-tetrachlorophenol and pentachlorophenol) were incubated with syringic acid in the presence of a laccase from the fungus <u>Rhizoctonia</u> <u>praticola</u> (8). The resulting products were isolated by thin-layer chromatography (TLC) or high-performance liquid chromatography (HPLC) and then characterized by mass spectroscopy (Table I). Two types of cross-coupling products were formed: (1) quinonoid oligomers in which chlorophenols were bound by ether linkages to orthoquinoline products of syringic acid; and (2) phenolic oligomers in which chlorophenols were bound by ether linkages to decarboxylated products of syringic acid.

The enzymatic coupling of phenolic compounds to soil substances seems to be a naturally occurring phenomenon. Martin et al. (9) have demonstrated that ^{14}C-catechol in soil is rapidly polymerized by soil enzymes or by autooxidation. Cheng et al. (10) have reported that 2,4-dichlorophenol is degraded in soil to a lesser extent than other chlorinated phenols and is gradually incorporated into the soil organic matrix.

TABLE I. Formation of cross-coupling products of syringic acid and chlorophenols with a laccase from _Rhizoctonia praticola_ as catalyst

Reaction of syringic acid with:	Quinoid oligomers		Phenolic oligomers			
	Dimer	Trimer	Dimer	Trimer	Tetramer	Pentamer
4-Chlorophenol	264^a $C_{13}H_9O_4Cl^b$	---	280 $C_{14}H_{13}O_4Cl$	432 $C_{22}H_{21}O_7Cl$	584 $C_{30}H_{29}O_{10}Cl$	736 $C_{38}H_{37}O_{13}Cl$
2,4-Dichlorophenol	298 $C_{13}H_8O_4Cl_2$	434 $C_{21}H_{16}O_6Cl_2$	314 $C_{14}H_{12}O_4Cl_2$	466 $C_{22}H_{20}O_7Cl_2$	618 $C_{30}H_{28}O_{10}Cl_2$	770 $C_{38}H_{36}O_{13}Cl_2$
2,4,5-Trichlorophenol	332 $C_{13}H_7O_4Cl_3$	---	348 $C_{14}H_{11}O_4Cl_3$	500 $C_{22}H_{29}O_7Cl_3$	652 $C_{30}H_{27}O_{10}Cl_3$	804 $C_{38}H_{35}O_{13}Cl_3$
2,3,5,6-Tetrachlorophenol	---	---	382 $C_{14}H_{10}O_4Cl_4$	534 $C_{22}H_{18}O_7Cl_4$	686 $C_{30}H_{26}O_{10}Cl_4$	838 $C_{38}H_{34}O_{13}Cl_4$
Pentachlorophenol	---	---	416 $C_{14}H_9O_4Cl_5$	568 $C_{22}H_{17}O_7Cl_5$	720 $C_{30}H_{25}O_{10}Cl_5$	872 $C_{38}H_{33}O_{13}Cl_5$

SOURCE: Adapted from ref. 8
[a] m/z value of molecular ion
[b] molecular composition based on number of chlorine atoms and m/z value of a given compound

In order to further study the coupling reaction, we have examined the polymerization of phenols to humus under conditions which approximate the natural habitat. ^{14}C-labeled 2,4-dichlorophenol was incubated with stream fulvic acid in the presence of various oxidoreductases (11). The binding of chlorophenol to humic material was determined by measuring the amount of radioactivity incorporated into fulvic acid. Chromatographic analysis indicated that although a large portion of the radioactivity remained in the solution, no unbound ^{14}C-2,4-dichlorophenol was present in the supernatant. Two types of polymers were formed during the coupling reaction: hybrid molecules containing fulvic acid and 2,4-dichlorophenol and polymers of 2,4-dichlorophenol alone. Importantly, the presence of fulvic acid in the reaction mixture essentially doubled the removal of 2,4-dichlorophenol.

Other parameters which affect the enzymatic polymerization process have also been analyzed. When substituted phenols were incubated with peroxidase, tyrosinase, and laccases from R. praticola and Trametes versicolor, the removal of phenol through polymerization depended on the chemical structure and concentration of substrate, pH of the reaction mixture, activity of the enzyme, length of incubation, and temperature (12).

Recent work has focused on the stabilization of phenoloxidases through their immobilization on solid supports. Laccase immobilized on kaolinite and silt loam soil was shown to effectively promote the polymerization of 2,4-dichlorophenol (13). The immobilization of enzymes enhances their thermostability, inhibits their degradation by proteolytic enzymes, and increases their half-life. Because of their enhanced stability, immobilized enzymes may provide a more effective method for promoting phenolic detoxification reactions in soils.

Furthermore, in several studies we have found that the addition of a highly reactive humic monomer, such as guaiacol, vanillic acid, ferulic acid or syringic acid, to a phenoloxidase-containing medium can effectively initiate the binding of a molecule which by itself is only poorly, if at all, transformed. For example, when syringic acid was added to 2,4-dichlorophenol, the amount of xenobiotic removed by the laccase of R. praticola increased by more than two-fold (14). The quantity of 2,4-dichlorophenol removed was dependent on the concentration of syringic acid added. This observation may be potentially very important, as by the appropriate choice of co-substrate(s), various xenobiotics which are relatively inert to oxidative coupling enzymes could be efficiently removed from contaminated soil or water.

Oxidative Coupling of Anilines

Anilines (aromatic amines) are released to the environment during the industrial production of dyes and pigments and during microbial and/or chemical decomposition of pesticides. These compounds are transformed in the soil by a variety of processes. One of the major transformation reactions involves the binding of substituted anilines to soil organic matter, namely the humic acid fraction. As with the phenols, the nature of the binding can range from sorptive forces such as charge-transfer, hydrogen bonding, and hydrophobic interactions to irreversible covalent bonds that are more resistant to acid and base hydrolysis, thermal treatment, and microbial degradation.

Studies were conducted to identify the chemical products formed as a result of covalent reactions occurring between anilines and humic materials. The laccase from the fungus R. praticola was shown to catalyze the cross-coupling of anilines (4-chloroaniline, 3,4-dichloroaniline, and 2,6-diethylaniline) (Figure 1) and humic monomers (ferulic, protocatechuic, vanillic, and syringic acids) (15). The reactions yielded a number of hybrid oligomers ranging from dimers to tetramers.

To further investigate the binding mechanism, studies were carried out on the polymerization of guaiacol and 4-chloroaniline (16). Polymerization reactions were catalyzed by a variety of inorganic and biological soil components, including manganese dioxide, horseradish peroxidase, tyrosinase, and laccases from the fungi T. versicolor and R. praticola. Each of these catalysts promoted the formation of the same oligomeric products. During the initial stages of polymerization, five co-oligomeric compounds and six guaiacol-derived compounds were formed. The co-oligomers were found to have aminoquinone, carbazole, and iminodiphenoquinone structures. Figure 2 shows the proposed pathway for the formation of these co-oligomers. Overall, two different mechanisms for the incorporation of anilines in humic substances can be proposed: (1) formation of addition products with quinones; and (2) condensation of hydroquinones with anilines leading to the formation of heterocyclic structures. The reactions are believed to occur via nucleophilic additions and through free-radical coupling reactions.

The use of enzymatic coupling and polymerization for soil decontamination purposes may be hindered by the relative inertness of some pollutants to enzymatic action. For example, while the laccase from Rhizoctonia readily oxidizes halogenated phenols, the enzyme does not oligomerize anilines under certain conditions (17). However, when 2,4-dichlorophenol was added to a medium containing a halogenated aniline (e.g. 2,4-dichloroaniline) and laccase, the aniline was effectively transformed. The coupling of the aniline to the products of 2,4-dichlorophenol is thought to be of a chemical nature. In fact, further studies have shown that the quinone products generated from 2,4-dichlorophenol react with various anilines in the absence of enzyme. Therefore, the addition of easily oxidized substrates may enhance the reactivity of some anilines and may be a useful strategy for the removal of inert pollutants.

Fate of Bound Xenobiotics

The implications surrounding the binding of xenobiotic compounds to humic materials are not easy to assess. It is clear, however, that once bound to humic materials, xenobiotic residues often persist in the soil for long periods. In one study, 46% of the applied 3,4-dichloroaniline remained in the soil in the form of bound residues 2 years after treatment (18). Theoretically, the binding of pesticides or their derivatives to humic materials should decrease

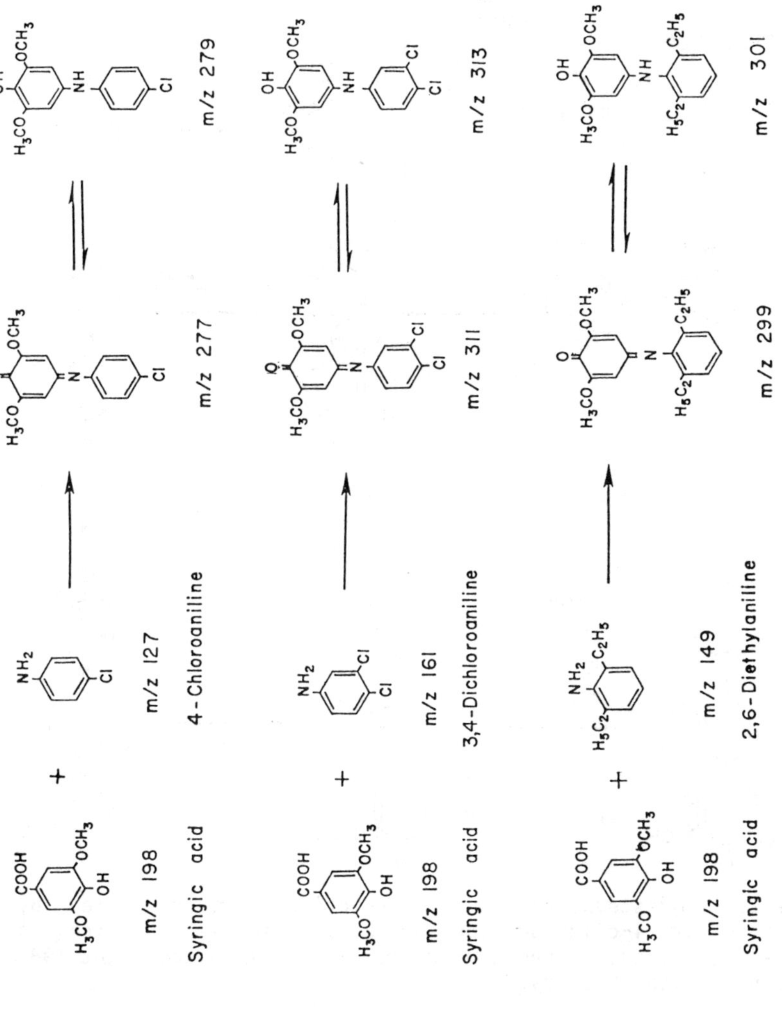

Figure 1. Structures of cross-coupling products between a lignin derivative (syringic acid) and variously substituted anilines. (Reproduced from ref. 15. Copyright 1983 American Chemical Society.)

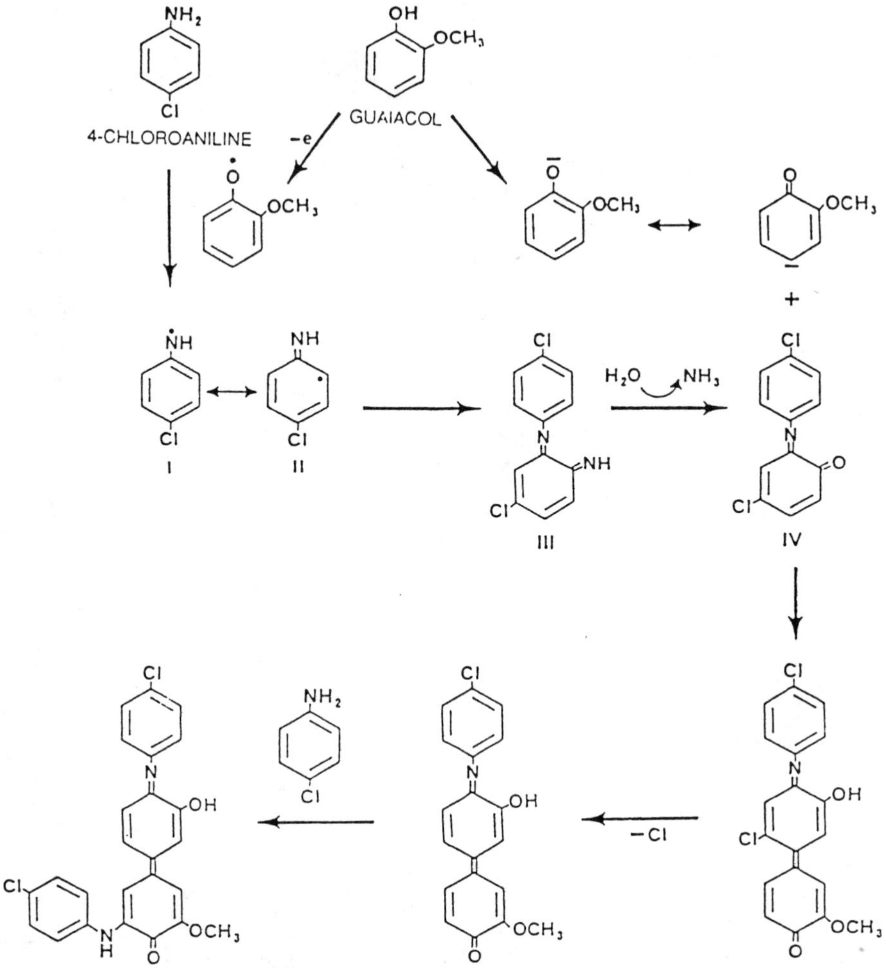

Figure 2. Suggested pathway for reaction products formed by incubation of 4-chloroaniline and guaiacol in the presence of various catalysts. (Reproduced from ref. 16. Copyright 1989 American Chemical Society.)

the toxic effects of these substances. Binding reduces the amount
of compound available to interact with the biota and as the amount
of available xenobiotic is reduced, the toxicity of the compound
also declines.

In order to demonstrate the decrease in toxicity of bound
xenobiotics, a variety of phenolic pesticides were incubated with
laccase and a naturally occurring phenol in the presence of the
fungus R. praticola (19).   The amount of pesticide added was based
on the minimum concentration that would inhibit the growth of the
fungus.   The addition of laccase to p-cresol and to 2,6-xylenol and
the addition of laccase and syringic acid to o-cresol allowed growth
of the previously inhibited R. praticola.   The toxicity of the
phenolic pesticides decreased as a result of the transforming and/or
polymerization reactions catalyzed by the laccase.

Any acceptable detoxification procedure based on the
polymerization and binding of xenobiotics must take into account the
possibility of release of the pollutant from the humic material. If
environmentally significant concentrations of the bound material are
released at a later time, the accumulation of released residues
would constitute a potential health hazard.   The mechanism of
residue release from soil and organic matter is not yet well
understood.   Experimental data suggest that the release of bound
xenobiotics is mediated primarily by microbial activity.   Certain
soil fungi, such as Penicillium frequetans, are capable of degrading
humic substances and in so doing free bound xenobiotics for plant
uptake or further mineralization (20).

Experiments were conducted in order to determine the extent to
which bound residues were released from humic complexes.   [14]C-
labeled chlorophenols (4-chlorophenol, 2,4-dichlorophenol, 2,4,5-
trichlorophenol, and pentachlorophenol) were covalently bound to a
synthetic humic acid polymer and incubated with microbial soil
cultures (21).   Insignificant quantities of ^{14}C were released during
a 10-week incubation period, and this release was accompanied by a
simultaneous mineralization of 1.2 to 6% of the initially bound
material to [14]CO_2.   Most of the radioactivity remained bound to the
humic material (Table II).

In a second study, ^{14}C-labeled-2,4-dichlorophenol was bound to
synthetic and natural humic materials or polymerized by enzymes
(22).   After 12 weeks of incubation with forest soil microorganisms,
the amount of radioactive substance released into the media was very
small (a maximum of 2.2% of the initially bound ^{14}C).   Differences
were observed between the humic acid bound and the polymerized forms
of 2,4-dichlorophenol.   The radioactivity that remained bound to the
2,4-dichlorophenol-polymer at the end of the incubation was much
higher (91.7%) than that observed for the 2,4-dichlorophenol bound
to humic material (64.3%).   The rate of [14]CO_2 evolution from the
2,4-dichlorophenol-polymer was minute (0.5%) as compared to that for
the 2,4-dichlorophenol-synthetic humic acid complex (4.8%).   The
mineralization of free and humic bound 2,4-dichlorophenol was also
studied.   The pattern of [14]CO_2 release indicated that the major
source of [14]CO_2 production may not be 2,4-dichlorophenol but rather
a derivative of this compound.   It was concluded that the binding of
xenobiotics to humic material may enhance their ability to be

TABLE II. Percentage distribution of radioactivity after a 10-week
incubation of synthetic humic acid polymers containing
^{14}C-labeled substituted chlorophenols

| | | Aqueous phase | | | |
Bound compound	Humic acid (precipitate)	Extracted with CH_2Cl_2	Remaining	$^{14}CO_2$	Total recovered
4-Chlorophenol	79.0	0.9	11.7	6.0	97.6
2,4-Dichlorophenol	77.2	0.5	7.9	3.9	89.5
2,4,5-Trichlorophenol	83.8	2.1	7.7	1.8	95.4
Pentachlorophenol	78.9	12.4	6.3	1.2	98.8

SOURCE: Adapted from ref. 21

mineralized, which may actually be an advantage in the
decontamination process.
    Our results are in agreement with the findings of other
laboratories (23-25).   To date, most available data indicate that
the release of bound xenobiotics is minimal.   Furthermore, once the
xenobiotics are released they are mineralized by the activities of
microorganisms and therefore do not accumulate to large levels in
the soil.

Conclusions

Results obtained in our laboratory have shown that xenobiotics, such
as phenols and substituted anilines, can be covalently cross-linked
to humic constituents.   The reactions can be catalyzed by both
microbial enzymes and by abiotic agents.   Two mechanisms by which
xenobiotics can become incorporated into soil organic matter have
been proposed: (1) direct chemical attachment of the substance to
reactive sites on colloidal organic surfaces; and (2) incorporation
of the substance into the structure of newly formed fulvic and humic
acids during the process of humification.   Given that xenobiotics
can be covalently bound to soil constituents, the question arises
whether this binding can be used and applied as a method of
detoxification.
    The incorporation of toxic xenobiotics into soil humus has
three important consequences: (1) bound xenobiotics form frequently
insoluble precipitates, and therefore leaching of these substances
is reduced; (2) the amount of bioavailable compound is diminished;
and (3) the polymerized xenobiotic appears to be less toxic than the
parent compound.   Therefore, the enzyme-induced immobilization of
xenobiotic pollutants would appear to be an effective method for
soil decontamination.   However, the primary problem associated with

this method is the potential for the release of the bound xenobiotic. The release of bound xenobiotics appears to occur very slowly and to a minimal extent. Furthermore, once released, the xenobiotics are mineralized by microorganisms and abiotic factors. All available information indicates that the release of humus-bound xenobiotics does not appear to pose a health hazard. Thus, enhanced rates of enzyme-induced binding of xenobiotics continues to provide a promising technique for the decontamination of environmental sites.

Literature Cited

1. Stevenson, F. J. In Bound and Conjugated Pesticide Residues; Kaufman, D. D.; Still, G. G.; Paulson, G. D.; Bandal, S. K., Eds.; ACS Symposium Series 29, 1976; pp. 180-207.
2. Bartha, R. ASM News 1980; 46, 356-360.
3. Bollag, J.-M. In Aquatic and Terrestrial Humic Substances; Christman, R. F.; Gjessing, E. T., Eds.; Ann Arbor Science Publishers, Inc.: Ann Arbor, MI, 1983; pp 127-141.
4. Sjoblad, R.D.; Bollag, J.-M. In Soil Biochemistry; Paul, E. A.; Ladd J. N., Eds.; Marcel Dekker, NY, 1981, Vol. 5, pp. 113-152.
5. Wang, T. S. C.; Huang, P. M.; Chou, C.H.; Chen, J.H. In: Interactions of Soil Minerals with Natural Organics and Microbes; Spec. Pub. No. 17; Soil Science Society of America, Madison, WI, 1986; pp. 251-281.
6. Shindo, H.; Huang, P. M.. Soil Sci. Soc. Am. J. 1984; 48, 927-934.
7. Bollag, J.-M.; Liu, S.-Y.; Minard, R. D. Soil Sci. Soc. Am. J. 1980; 44, 52-56.
8. Bollag, J.-M.; Liu, S.-Y. Pestic. Biochem. Physiol. 1985; 23, 261-272.
9. Martin, J. P.; Haider, K.; Linhares, L. F. Soil Sci. Soc. Am. J. 1979; 43, 100-104.
10. Cheng, H. H.; Haider, K.; Harper, S. S. Soil Biol. Biochem. 1983; 15, 311-317.
11. Sarkar, J. M.; Malcolm, R. L.; Bollag, J.M. Soil Sci. Soc. Am. J. 1988; 52, 688-694.
12. Dec, J.; Bollag, J.-M. Arch. Environ. Contam. Toxicol. 1990; 19, 534-541.
13. Ruggiero, P.; Sarkar, J. M.; Bollag, J.-M. Soil Sci. 1989; 147, 361-370.
14. Shuttleworth, K. L.; Bollag, J.-M. Enzyme Microbiol. Technol. 1986; 8, 171-177.
15. Bollag, J.-M.; Minard, R. D.; Liu, S. Y. Environ. Sci. Technol. 1983; 17, 72-80.
16. Simmons, K. E.; Minard, R. D.; Bollag, J.-M. Environ. Sci. Technol. 1989; 23, 115-121.
17. Liu, S.Y.; Minard, R. D.; Bollag, J.-M. J. Agric. Food Chem. 1981; 29, 253-257.
18. Viswanathan, R.; Scheunert, I.; Kohli, J.; Klein, W.; Korte, F. J. Environ. Sci. Health B 1978; 13, 243-259.
19. Bollag, J.-M.; Shuttleworth, K. L.; Anderson, D. H. Appl. Environ. Microbiol. 1988; 54, 3086-3091.

20. Mathur, S. P.; Paul, E. A. <u>Can. J. Microbiol</u>. 1967; <u>13</u>, 573-580.
21. Dec, J.; Bollag, J.-M. <u>Soil Sci. Soc. Am. J</u>., 1988; <u>52</u>, 1366-1371.
22. Dec, J.; Shuttleworth, K. L.; Bollag, J.-M. <u>J. Environ. Qual</u>. 1990; <u>19</u>, 546-551.
23. Hsu, T. S.; Bartha, R. <u>Soil Sci</u>. 1974; <u>118</u>, 213-220.
24. Khan, S. U.; Ivarson, K. C. <u>J. Agric. Food Chem</u>., 1981; <u>29</u>, 1301-1303.
25. MacRae, I. C. <u>Soil Biol. Biochem</u>. 1981; <u>18</u>, 221-225.

RECEIVED November 8, 1990

Chapter 10

Mineralization of Pesticide Degradation Products

Cathleen J. Hapeman-Somich

Pesticide Degradation Laboratory, Beltsville Agricultural Research
Center, Agricultural Research Service, U.S. Department of Agriculture,
Beltsville, MD 20705

Pesticides and pesticide degradation products, which are not
mineralized to carbon dioxide, ammonia, water and inorganic
salts, can leach and contaminate water supplies. Altering
pesticides by photolysis or ozonation has been shown to enhance
significantly the rate of mineralization. Photodegradation
products of s-triazines, chloroacetanilides and paraquat were
dechlorinated and/or oxidized. Ozonation of these same
herbicides afforded products in which the alkyl side chains were
oxidized or removed and the aromatic ring was oxidized or
cleaved, however, dechlorination did not occur. In some cases,
the matrix in which these compounds were found affected
microbial activity.

Pesticides applied to soil may dissipate through a variety of processes
including volatilization, adsorbtion to soil particles and/or chemical
transformation. In addition, pesticides may be degraded by indeginous soil
microorganisms to structurally simple compounds or eventually mineralized to
water, CO_2, NO_3^- and other ions. Microbial degradation is dependent on
microbial populations and a number of environmental conditions including
temperature, soil moisture, pesticide toxicity, and bioavailability of cometabolic substrates and other nutrients. The rate of pesticide degradation is
often times slow and under less than ideal conditions, the rate of pesticide
transport through the soil may exceed rate of degradation, thus posing a
groundwater contamination risk.

Chemical alteration of pesticides by photolysis or ozonation affords
products that are more amenable to biological degradation. Photolysis
requires that the emission spectrum of the light source overlap with the
absorbtion spectrum of the compound to be degraded. For most organic
pesticides, high energy UV (wavelengths less than 254 nm) lamps must be

used. Aqueous ozonation of organic compounds generally involves strong oxidants, such as hydroxy radicals and superoxide, which are formed from the degradation of ozone, as opposed to direct ozonolysis *(1-3)*. The addition of hydrogen peroxide can increase the reaction rate under certain conditions.

This paper discusses some recent developments in the combined use of chemical pretreatment and microbial degradation to mineralize four widely used herbicides: alachlor, metolachlor, paraquat and atrazine. Much of the work described was initially conducted as part of an overall scheme to prevent point source pollution arising from pesticide storage areas, contained spills and collected equipment rinsates. However, the information obtained may also be useful in predicting the fate of these pesticides and their degradation products in the environment.

Experimental

Photolysis and ozonolysis were carried out in a 220-mL reactor equipped with a Hanovia-Conrad medium pressure lamp in a quartz water cooled immersion well, a sintered glass disk in the bottom for introducing gases, a gas outlet and liquid sampling valve at the bottom *(4)*. Some ozonation product studies were conducted in a low efficiency reaction chamber *(5)*.

Product isolation and characterization studies were carried out using HPLC, direct insertion probe or gas chromatography mass spectroscopy, and ^{1}H-NMR techniques. MS data were acquired by William R. Lusby, Insect Hormone Laboratory, USDA/ARS, and NMR data by Rolland Waters, Environmental Chemistry Laboratory, USDA/ARS. Specific methodologies are referenced accordingly.

Biodegradation of pesticide degradation products on soil was carried out in biometer flasks *(6)*. Mineralization of ^{14}C-labelled pesticides and pesticide degradtion products was determined by trapping released $^{14}CO_2$ in KOH added to the side arm flask. Samples were removed, counted and fresh base added at specific intervals. Experimental details are referenced.

Results and Discussion

Alachlor. The major products resulting from irradiation of alachlor in water were hydroxyalachlor (**1**) and the lactam **2** *(4)*. Three other products, norchloralachlor (**3**), 2',6'-diethylacetanilide (**4**) and 2-hydroxy-2',6'-diethyl-*N*-methylacetanilide (**5**), were also identified. A suggested mechanism for the formation of these products is shown in Scheme I. Absorbtion of a photon by the carbonyl moiety presumably gives rise to homolytic cleavage of chlorine and the resonance stabilized radical **6**. Trapping of this radical with water affords **1** and in some cases **3**. The radical **6** can also undergo intramolecular hydrogen abstraction which upon collapse of the resultant radical leads to the formation of **2**. Other secondary reactions, such as loss of a methyl, methoxy or methylmethoxy moiety, afforded a variety of products. All of these compounds underwent further photodegradation to compounds which eluted near the void volume on reverse phase HPLC. Studies using either [*carbonyl*-^{14}C]alachlor or [U-*ring*-^{14}C]alachlor as starting material showed that over 70% of the radiolabel was retained in these secondary compounds. This suggests

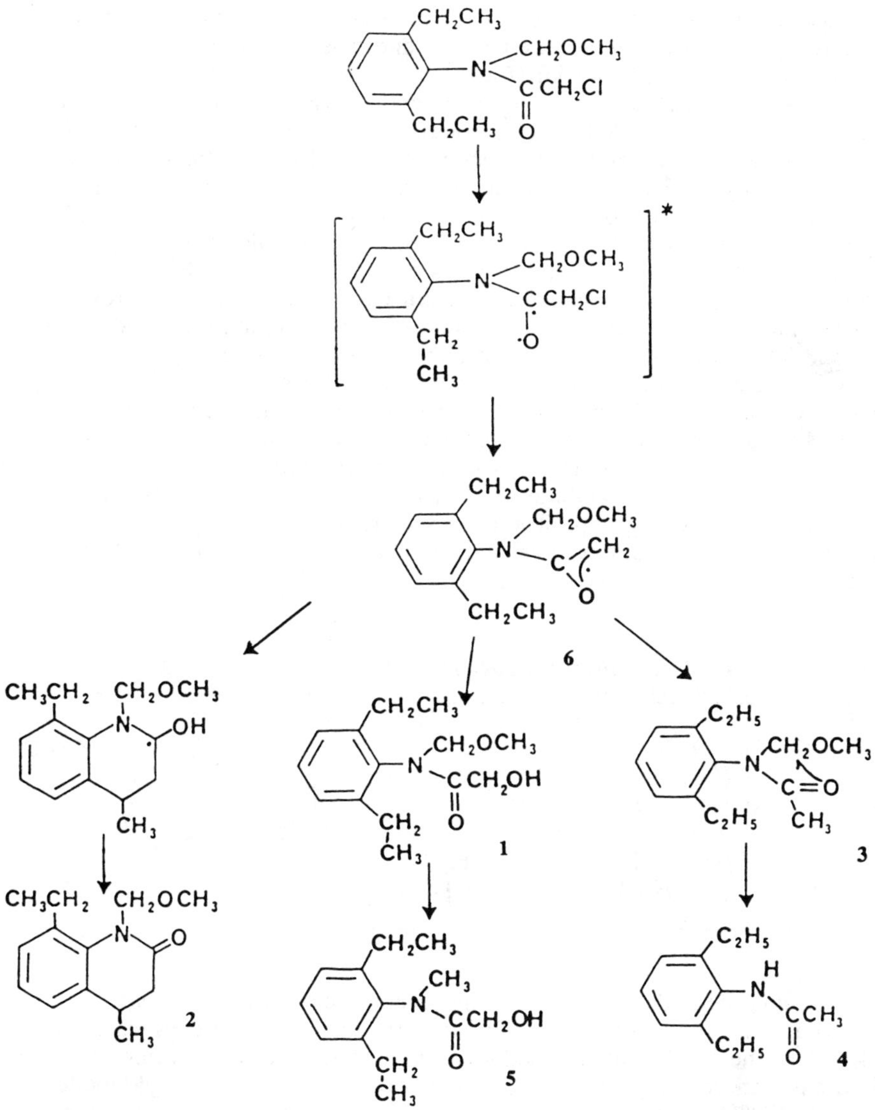

Scheme I (Reproduced from ref. 4. Copyright 1988 American Chemical Society.)

that during photolysis, the overall acetanilide structure remained intact in the majority of the products. This is in sharp contrast to the microbial degradation of alachlor where cleavage of the amide bond was observed in most of the metabolities (7), although **3** has been observed in anaerobic degradation studies (8).

Experiments were conducted to determine if partial breakdown of alachlor by UV light resulted in enhanced microbial degradation in soil. Solutions of formulated alachlor (Lasso) with carbonyl- or ring-labelled alachlor were irradiated. These solutions and similar solutions that had not been irradiated were added to soil in biometer flasks and the evolution of $^{14}CO_2$ monitored to determine the extent of mineralization (Figure 1). After two days nearly 50% of the label was released in the flasks with irradiated alachlor. A continued slow release of $^{14}CO_2$ was observed over the next 33 days to give a 60% total label recovery from irradiated alachlor. Only 3% of the label from nonirradiated alachlor was mineralized during the same period (35 d). Carbonyl- and ring-labelled alachlor gave virtually the same result.

Alachlor was also treated with ozone (9). GC/MS data indicated that the chlorine was not removed in any of the initial products. Ozonation was carried out using carbonyl- and ring-labelled alachlor as above. Ninety-seven percent of the carbonyl label remained over the course of the reaction, whereas 83% of the ring label was retained, indicating that the amide bond remained intact and that in some products the aromatic ring was cleaved. The most abundant intermediate was **7** where one of the ethyl side chains was converted to an acetyl group. The mass spectrum of this compound showed a base peak at 174 m/z and a molecular ion peak of 283 (285) m/z. Other ions observed are listed in Table I.

Table I. Ion Listing from Mass Spectrum of 7	
Ion Mass	*Fragment Lost*
248	Cl
240 (242)	$C(O)CH_3$
206	$C(O)CH_2Cl$
188	CH_2Cl, CH_2OCH_3, H
174	$C(O)CH_2Cl$, $HOCH_3$
160	$C(O)CH_2Cl$, CH_2OCH_3, H
132	$C(O)CH_2Cl$, CH_2OCH_3, CH_2CH_3

Microbial degradation studies using organic rich soil showed that ozonated labelled-alachlor was rapidly mineralized in several days whereas only 3% of the $^{14}CO_2$ was released from the untreated alachlor after 35 days (Figure 2). Interestingly, more $^{14}CO_2$ was recovered using carbonyl-labelled substrate (80%) as compared to ring-labelled compounds (60%). Perhaps, the ring carbons are more resistant to degradation than the carbonyl carbon, but more probable is that the ring carbons are more likely to be incorporated into biomass than the carbonyl carbon.

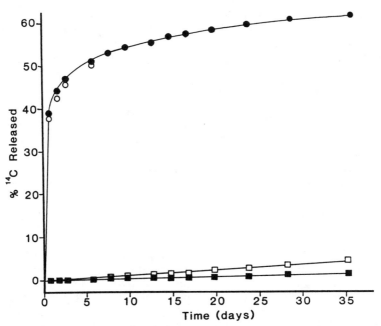

Figure 1. Metabolism of irradiated and nonirradiated alachlor in soil. ■, Nonirradiated [U-*ring*-^{14}C]alachlor; □, nonirradiated [*carbonyl*-^{14}C]alachlor; ●, irradiated [U-*ring*-^{14}C]alachlor; ○, irradiated [*carbonyl*-^{14}C]alachlor. (Reproduced from ref. 4. Copyright 1988 American Chemical Society.)

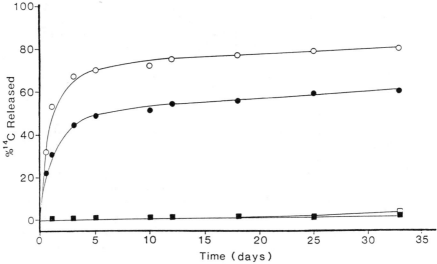

Figure 2. Metabolism of ozonated and non-ozonated alachlor in soil. ■, Non-ozonated [U-*ring*-^{14}C]alachlor; □, non-ozonated [*carbonyl*-^{14}C]alachlor; ●, ozonated [U-*ring*-^{14}C]alachlor; ○, ozonated [*carbonyl*-^{14}C]alachlor. (Reproduced from ref. 4. Copyright 1988 American Chemical Society.)

Metolachlor. The ozonation of metolachlor has been investigated and the products shown in Figure 3 *(10)*. Oxidation generally occured at the electron rich secondary carbons such as the ethyl side chain (8), the methoxy carbon (9) and the carbon attached to the methoxy group (10). The chlorinated carbon which is electron deficient was not oxidized. In addition, NMR data and MS data of several unidentified products suggested that the aromatic ring was partially oxidized. All the initial products contained chlorine except for 11. Formation of this compound can easily be rationalized as an intramolecular displacement reaction not involving ozone or another oxidizing species, followed by oxidation of the cyclized intermediate (Scheme II). Only 9 has also been identified as a metabolic intermediate *(11)*. All these products underwent further oxidation before introduction to soil and, as with alachlor, were mineralized more rapidly when compared to parent material *(12)*. Eighty percent of the label was recovered as $^{14}CO_2$ after two days using ozonated [U-*ring*-^{14}C]metolachlor as substrate compared to no mineralization after nine days using untreated material (Figure 4).

Paraquat. Slade *(13)* and Funderburk et al. *(14)* have shown that photolysis of paraquat (1,1'-dimethyl-4,4'-bipyridinium dichloride) afforded 4-carboxy-1-methylpyridinium ion (12) and methylamine through a somewhat transient ketone or aldehyde (Scheme III). Kearney et al. *(15)* found that when paraquat was irradiated in the presence of oxygen, 4,4'-bipyridyl (13) and 4-picolinic acid (14) were also formed. Sequential loss of methyl from paraquat was presumed to give rise to monoquat (1-methyl-4,4'-bipyridinium ion) (15) and 13. 4-Picolinic acid was postulated to form either by oxidative ring cleavage of 13 or by demethylation of 12. Further irradiation of these paraquat degradation products afforded oxalate, succinate, malate and *N*-formyl glycine.

The major microbial degradation product of paraquat was found to be monoquat, which was further metabolized to 12 *(16)*. Wright and Cain *(17-19)* demonstrated that *Achromobacter* D, isolated from soil, utilized 12 as a sole carbon and nitrogen source and produced CO_2, formate, methylamine and succinate. The mineralization of 2-picolinamide was found to proceed through hydroxylation of the pyridine ring of 2-picolinic acid. Subsequent ring opening and hydrolysis yielded maleic acid *(20)*. These studies did not include 4-picolinic acid; however, mineralization of this isomer would be expected to proceed via a similar pathway *(21)*.

Oxidation of paraquat in the presence of hydrogen peroxide and sodium hydroxide occured at the 2- and/or 2'-carbons to afford 12, 16 and 17 (Scheme IV) *(22)*. Ozonation of paraquat at high pH in the presence of hydrogen peroxide has been examined cursorily examined as part of a disposal effort for pesticide rinsate *(12)*. No products were isolated in this investigation; however, soil metabolism studies were conducted using [*methyl*-^{14}C]paraquat. Sixty percent of the radiolabel was recovered from ozonated paraquat after 13 days compared to less than 5% for untreated parent material (Figure 5).

Atrazine. Many dialkylated-*s*-triazines, including atrazine, have been shown to undergo rapid photodehalogenation in aqueous solution to afford the

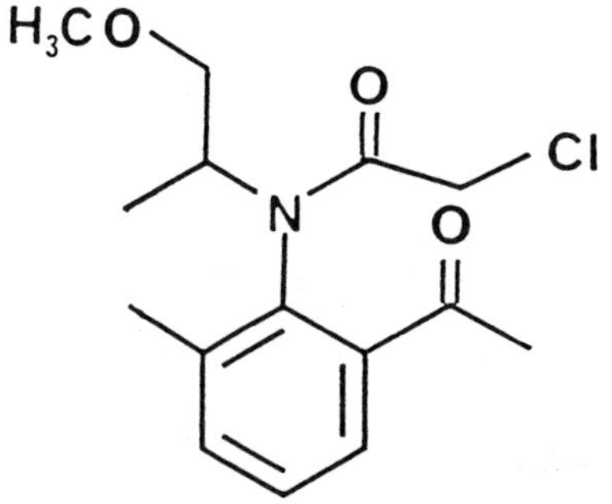

Figure 3. Metolachlor ozonation products.

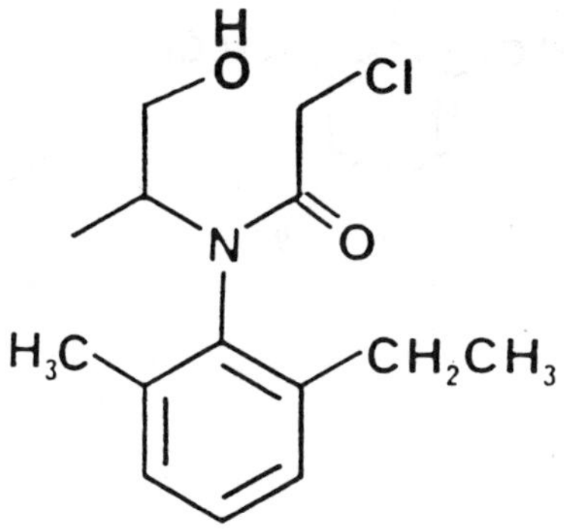

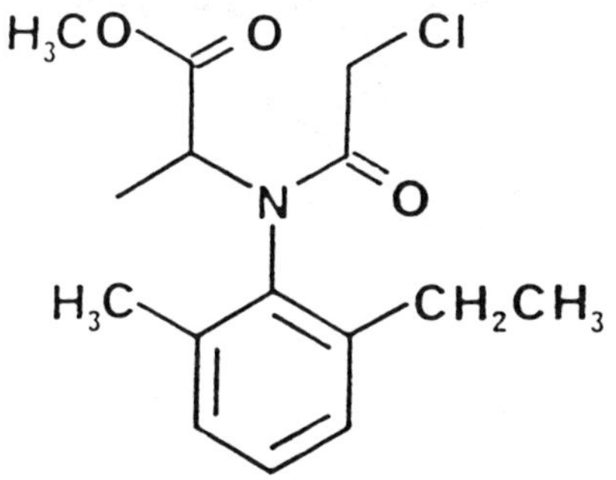

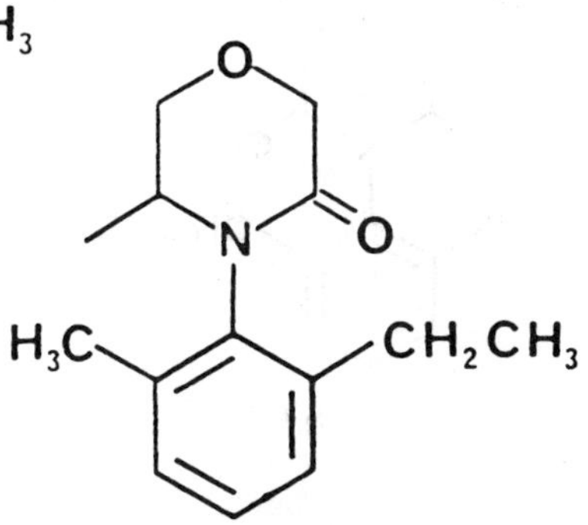

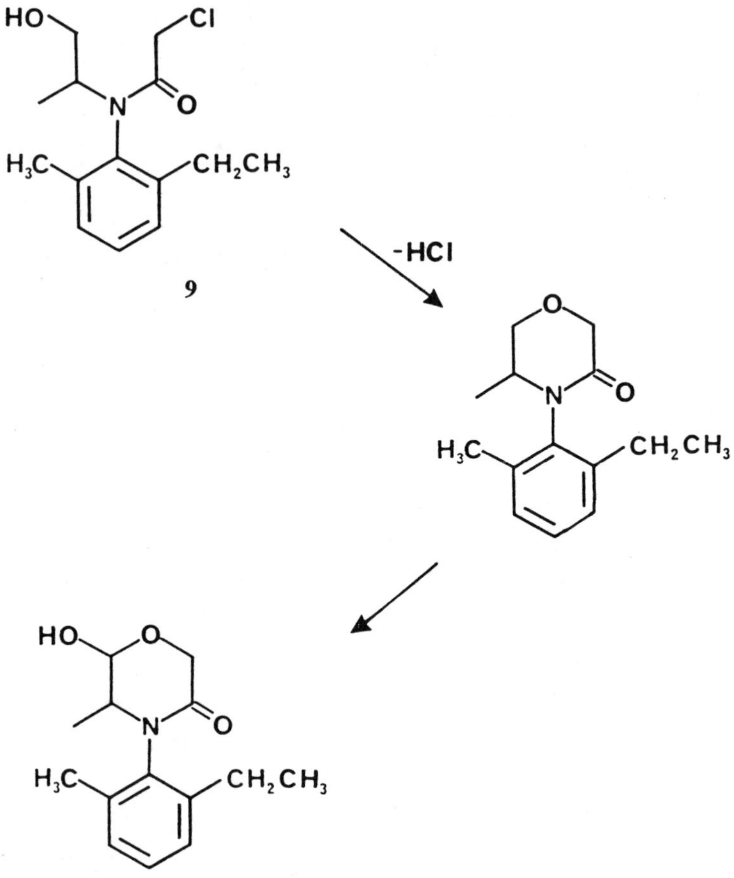

Scheme II

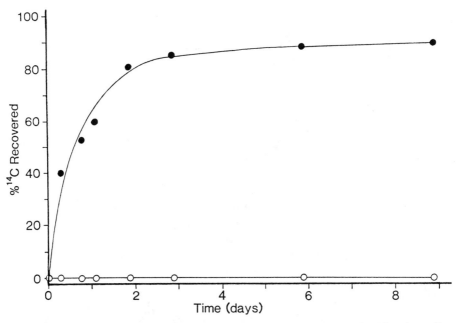

Figure 4. Metabolism of ozonated and non-ozonated metolachlor in soil. ●, Non-ozonated [*carbonyl*-14C]metolachlor; ○, ozonated [*carbonyl*-14C]metolachlor. (Reproduced from ref. 12. Copyright 1990 American Chemical Society.)

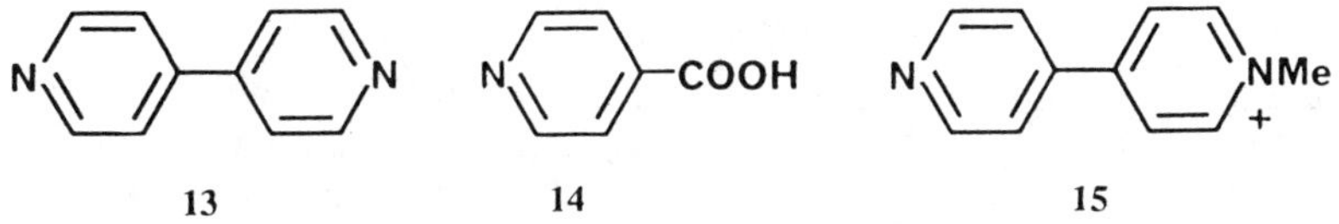

Scheme III

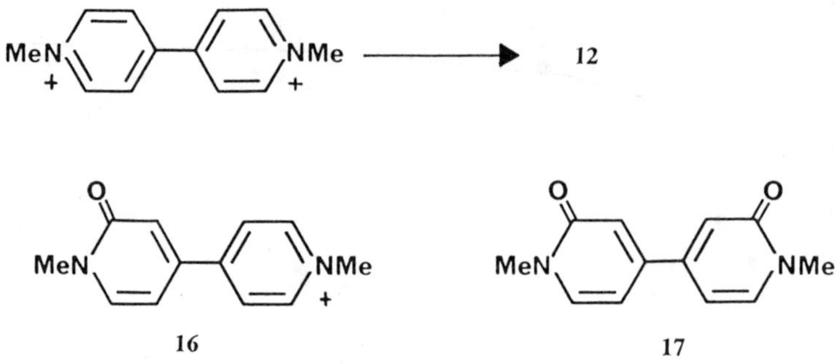

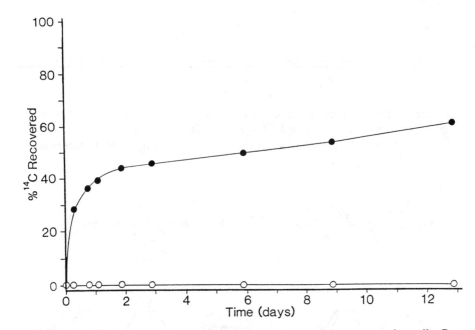

Scheme IV

Figure 5. Metabolism of ozonated and non-ozonated paraquat in soil. ○, Non-ozonated [*methyl*-^{14}C]paraquat; ●, ozonated [*methyl*-^{14}C]paraquat. (Reproduced from ref. 12. Copyright 1990 American Chemical Society.)

corresponding hydroxy-dialkylated-s-triazine *(23)*. This reaction proceeds in quantitative fashion as shown for atrazine in Figure 6 *(24)*. Dehalogenation was also found to be a major metabolic process *(25)*. Subsequent photodealkylation of hydroxylated compounds to give the mono- and nonalkylated species has been demonstrated *(26)*. A series of [U-*ring*-^{14}C]- and [*ethyl*-^{14}C]di- and mono-N-alkylated-hydroxy-s-triazines and [U-*ring*-^{14}C]-hydroxy-s-triazines were subjected to microbial degradation in soil *(27)*. The results showed that the alkyl substituents strongly retard degradation. Cyanuric acid, ammelide and ammeline were nearly completely degraded within 30 days, whereas, after 90 days, mono-N-alkylated ammeline and ammelide were only mineralized ca. 5% and less than 1% hydroxysimazine was mineralized.

Ozonation of atrazine did not yield dechlorinated compounds. Rather the alkyl groups were either removed or oxidized to the acetamide giving rise to four primary products 18-21 and three secondary products 22-24 (Scheme V) *(5)*. The monoalkylated-chloro-s-triazines, 18 and 19, have also been observed as atrazine metabolities *(25)*. The dealkylated products 22-24 are presumed, based on the above study with hydroxy-s-triazines, to be more biolabile than 18-21, which possess an N-alkyl moiety. Initial experiments *(28)* using indigenous soil microorganisms indicated that ozonation enhanced mineralization (Figure 7), but at a slower rate than observed for the other herbicides *(12)*. Further work has shown that the biodegradation rate of these final ozonation products can be increased by innoculation of the soil with a *Pseudomonas sp.* *(28)*; however, these organisms were metabolicly inhibited under field conditions *(12)*. Several organisms have recently been isolated which degrade 24 more efficiently, and characterization of these strains is in progress *(21)*.

Conclusion

The chemical degradation products of four pesticides have been characterized and in a few cases were found to be similar to products observed in microbial processes. Ozonation reactions typically did not involve chlorine removal. Rather ozone and other oxidants formed during the reaction cleaved double bonds, opened aromatic rings, removed or oxidized alkyl groups giving rise to alcohols, carbonyls or carboxylic acids. Similar oxidation products were obtained in photolytic reactions although dechlorination readily occured. In all cases, chemical pretreatment was found to enhance the rate of microbial mineralization, although the s-triazines were more slowly mineralized than paraquat or the chloroacetanilides, alachlor and metolachlor. It may be necessary, therefore to provide organisms capable of carrying out the desired transformations as opposed to relying on indigenous soil microbes to mineralize these chemical degradation products. Development of methods to insure rapid mineralization of pesticides in aqueous matrices will reduce the threat of contamination to surface and groundwater supplies.

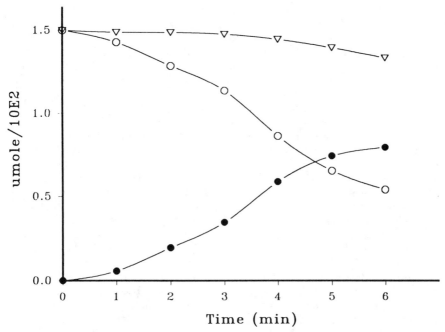

Figure 6. Photolysis of atrazine and the formation of hydroxyatrazine. ○, Moles of atrazine; ●, moles of hydroxyatrazine; ∇, sum of moles of atrazine and hydroxyatrazine.

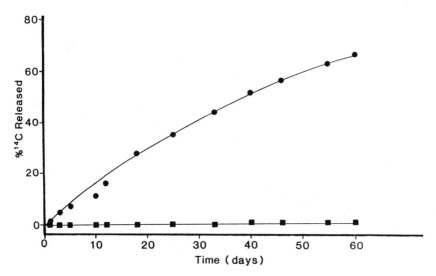

Figure 7. Metabolism of ozonated and non-ozonated atrazine in soil. ■, Non-ozonated [U-*ring*-^{14}C]atrazine; ● ozonated [U-*ring*-^{14}C]atrazine. (Reproduced from ref. 28. Copyright 1988 American Chemical Society.)

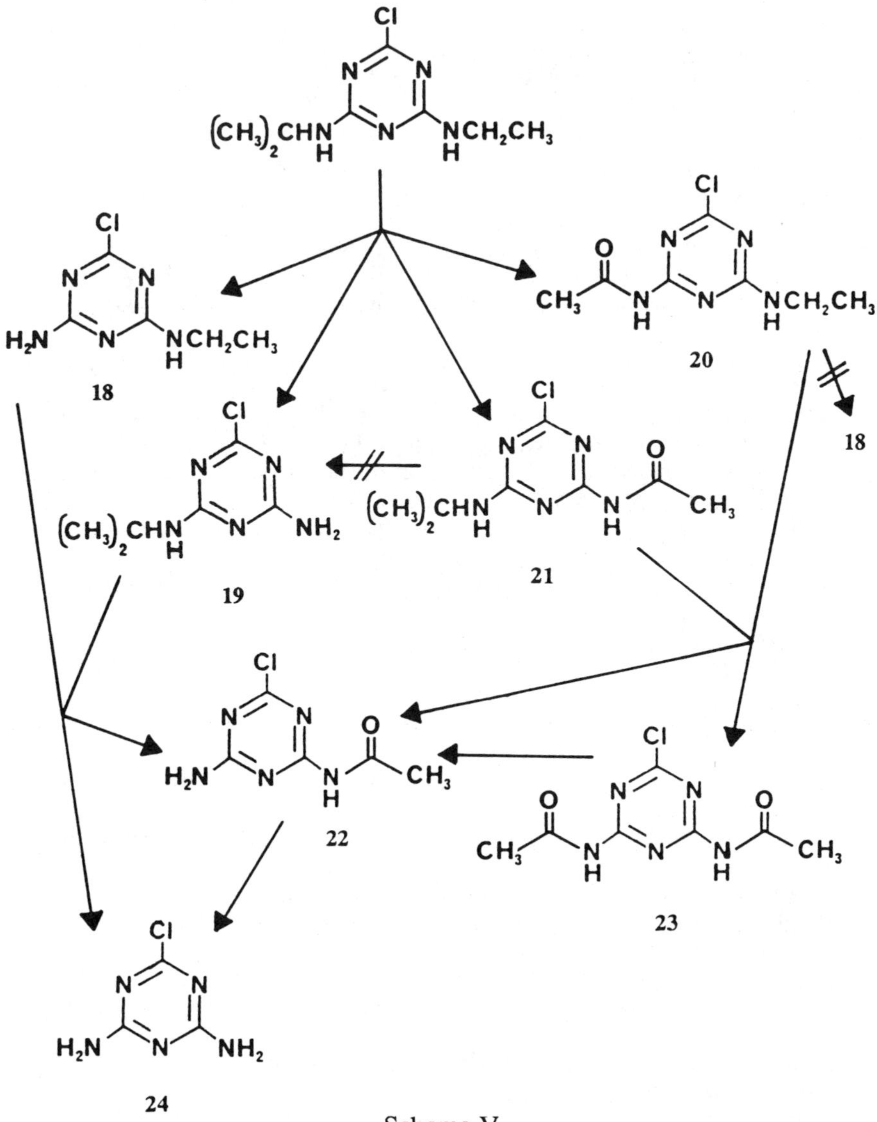

Scheme V

Literature Cited

1. Staelin, J.; Hoigne, J. *Environ. Sci. Technol.* **1985**, *19*, 1206-1213.
2. Glaze, W. H. *Environ. Sci. Technol.* **1987**, *19*, 224-230.
3. Peyton, G. R.; Glaze, W. H. In *Photochemistry of Environmental Aquatic Systems*; Zika, R. G.; Cooper, W. J., Eds.; ACS Symposium Series No. 327; American Chemical Society: Washington, DC, 1987; pp 76-88.
4. Somich, C. J.; Kearney, P. C.; Muldoon, M. T.; Elsasser, S. *J. Agric. Food Chem.* **1988**, *36*, 1322-1326.
5. Zong, G. -M.; Lusby, W. R.; Muldoon, M. T.; Hapeman-Somich, C. J. "Aqueous Ozonolysis of *s*-Triazines. I. Description of Atrazine Degradation Pathway and Product Identification". (manuscript to be submitted to Environ. Sci. Technol. 1990).
6. Bartha, R.; Pramer, D. *Soil Sci.* **1965**, *100*, 68-69.
7. Tiedje, J. M.; Hagedorn, M. L. *J. Agric. Food Chem.* **1975**, *23*, 77-81.
8. Bollag, J.-M.; McGahen, L. L.; Minard, R. D.; Lin, S.-Y. *Chemosphere* **1986**, *15*, 153-162.
9. Somich, C. J.; Kearney, P. C.; Muldoon, M. T., Elsasser, S. "A Product Study of Ozonated Alachlor". *Abstracts of Papers*, 195th National Meeting of the American Chemical Society, Toronto, Canada, June 5-10, 1988; American Chemical Society: Washington, DC, 1988, AGRO 161.
10. Ludwig, M.; Somich, C. J. "Ozonization of the Herbicide Metolachlor". *Abstracts of Papers*, 198th National Meeting of the American Chemical Society, Miami, FL, September 10-15, 1989; American Chemical Society: Washington, DC, 1989, AGRO 18.
11. Sexena, A.; Zhang, R.; Bollag, J.-M. *Appl. Environ. Microbiol.* **1987**, *53*, 390-396.
12. Somich, C. J.; Muldoon, M. T.; Kearney, P. C. *Environ. Sci. Technol.* **1990**, *24*, 745-749.
13. Slade, P. *Nature*, **1965**, *207*, 515-516.
14. Funderburk, H. H., Jr. ; Negi, N. S.; Lawrence, J. M. *Weeds* **1966**, *14*, 240-243.
15. Kearney, P. C.; Ruth, J. R.; Zeng, Q.; Mazzocchi, P. *J. Agric. Food Chem.* **1985**, *33*, 953-957.
16. Funderburk, H. H., Jr. ; Bozarth, G. A. *J. Agric. Food Chem.* **1965**, *15*, 563-567.
17. Wright, K. A.; Cain, R. B. *Soil Biol. Biochem.* **1969**, *1*, 5-14.
18. Wright, K. A.; Cain, R. B. *Biochem. J.* **1972**, *128*, 543-559.
19. Wright, K. A.; Cain, R. B. *Biochem. J.* **1972**, *128*, 561-568.
20. Orbin, C. G.; Knight, M.; Evans, W. C. *Biochem. J.* **1972**, *128*, 819-831.
21. Shelton, D. R. Pesticide Degradation Laboratory, USDA/ARS, personal communication, 1990.
22. Calderbank, A.; Charlton, D.F.; Farrington, J.A.; James, R. *J. Chem. Soc. Perkin Trans. I.* **1972**, 138-140.
23. Pape, B. E.; Zabik, M. T. *J. Agric. Food Chem.* **1970**, *18*, 202-207.
24. Hapeman-Somich, C. J.; Rouse, B. J. Pesticide Degradation Laboratory, USDA/ARS, unpublished data.
25. Muir, D. C. G.; Baker, B. E. *Weed Res.* **1978**, *18*, 111-120.

26. Crosby, D.G. In *Herbicides. Chemistry, Degradation, and Mode of Action*; Kearney, P. C.; Kaufman, D. D., Eds.; Marcel Dekker: New York, New York, 1976, Vol. 2; pp 825-890, and references therein.
27. Loos, M. A.; Kearney, P. C. Pesticide Degradation Laboratory, USDA/ARS, unpublished data.
28. Kearney, P. C.; Muldoon, M. T.; Somich, C. J.; Ruth, J. R.; Voaden, D. J. *J. Agric. Food Chem.* **1988**, *36*, 1301-1306.

RECEIVED December 12, 1990

Chapter 11

Computer-Assisted Molecular Prediction of Metabolism and Environmental Fate of Agrochemicals

Mahmoud Abbas Saleh

Department of Chemistry, Texas Southern University, Houston, Texas 77004 and Environmental Research Center, University of Nevada, Las Vegas, NV 89154

Knowledge about metabolism and environmental degradation of agrochemicals is the key to evaluating their toxicological effects and environmental fate. It is also a part of the requirement for registration of the chemical by the U.S. EPA for field applications. Laboratory identification of metabolites and studies of environmental degradation are very costly and time-consuming. Several artificial intelligence software packages capable of predicting metabolism and their hazardous effects and environmental fate have recently become available. In this paper, the results of using computer-assisted modeling for the prediction of metabolism in target and non-target organisms and their hazardous effects are discussed and compared to those actually obtained from data derived by laboratory techniques for selected groups of agrochemicals.

Computer-aided artificial intelligence has recently been demonstrated to have great potential applications in the fields of chemistry and environmental research *(1)*. Artificial intelligence normally deals with situations that require symbolic rather than numerical data and whose solutions rely more on logical reasoning than calculation. Within the agrochemical systems there is great opportunity for the use of artificial intelligence techniques to predict structures and biological or environmental degradation products along with their properties and associated data. A possible application of artificial intelligence to pesticide chemistry and biochemistry is the utilization of the technology of "expert" systems. Such systems are characterized by their ability to extrapolate from information that may be inexact or incomplete, operating on a system of rules containing the knowledge required to solve a particular problem. Recently, several software packages targeted at specific areas of prediction have been developed for possible application in the field of pesticide metabolism and environmental degradation. This paper deals with the application of Computer-Aided Molecular Prediction (CAMP) software for prediction of pesticide metabolism, estimation of metabolites partitioning between water and octanol calculated as log P, and their possible environmental hazardous effects. These programs were developed by CompuDrug USA Inc. of Austin, Texas. The software systems encode the knowledge of an expert in the form of rules, which can be used to reach a conclusion based on the user's input into a database.

Prediction of Pesticide Metabolism

Exogenous chemicals entering a living system can undergo a number of chemical modifications by a wide array of enzymes which use these chemicals as substrates. Rarely does a compound simply produce a single metabolite. In general, complex metabolic patterns of competitive and sequential reactions occur. Metabolic pathways can be classified into two groups, Phase I and Phase II. Reactions in Phase I involve the transformation of specific functional groups in the molecule, thus introducing new reactive functional groups. As a general rule, the resulting metabolite displays an increased water solubility as compared to the parent molecule. On the other hand, Phase II enzymes bring about conjugations to various endogenous substrates, such as sugars and amino acids, thus forming exceedingly water-soluble products that are readily excreted. The metabolic fate of a molecule is highly dependent upon its structural elements. This fact provides the basis for treating molecules as sets of substructures when formulating the rules of possible metabolic transformations. For,example, although direct conjugation is not infrequent, Phase I reactions usually precede Phase II reactions in the biotransformation of most compounds. While products of Phase II reaction are not usually metabolized further, a metabolite generated through a Phase I reaction may undergo subsequent biotransformations. In general, closely related substances have rather similar metabolic fates. Therefore, it is reasonable to make use of the accumulated data in the literature or from direct experimentation to predict the metabolic fate of unstudied compounds. Data concerning species specificity and prioritization of the predicted metabolites can also be used for similar prediction.

The most important requirement of a Computer-Assisted Molecular Prediction (CAMP) system is that it should significantly contribute to the realization of the long-range scientific objective of the field, the formation of a unified theory of structure-metabolism relationship. Based on a reliable theoretical foundation, CAMP is expected to meet the most important practical aim of many metabolism researchers by providing substitution or partial substitution of time consuming and expensive experimentation with economical calculations.

Two programs were used in this study: 1) AGROMETABOLEXPERT, which was developed for predicting metabolic pathways of agrochemical compounds in plants; and 2) METABOLEXPERT, which predicts metabolites in several organisms such as microorganisms, insects, and mammals, including man. They also allows the user to modify and complete the metabolism transformation knowledge base by drawing from his own experiences or from updated literatures. The knowledge base and the inference rules of these programs are based on literature data of the metabolic fate of many important pesticide types in various plant species. METABOLEXPERT might be used to improve the predictive ability of the system in a specific case. During the generation of metabolite predictions, METABOLEXPERT tries to match generalized metabolic transformation rules listed in the knowledge base with substructures of the parent compound or its metabolites. In addition to this mainly retrosynthetic approach, the program can also "reason by analogy" on the basis of a representative collection of compounds with known metabolic trees.

The knowledge base of both programs is composed of the following four elements: 1) the substructure changed during the metabolic transformation, 2) the new substructure formed, 3) a list of substructures from which at least one should be present in the molecule for the transformation to occur, and 4) a list of those substructures whose presence prevents the metabolic transformation. An example of one of the basic transformations of the program knowledge data base, the acylation of glycine is shown in Figure 1.

TRANSFORMATION NAME: F23_CONJUGATION WITH GLYCINE
REAGENT: Glycine-N-Acylase
REACTION CLASS: Phase II
REACTION TYPE: 8_Amino acid conjugations
ACTIVE SUBSTRUCTURE: 2-O4
REPLACEMENT SUBSTRUCTURE: 2-N5-6-9 = O10,9-O11
POSITIVE SUBSTRUCTURES: 1-2 = O3
NEGATIVE SUBSTRUCTURES: O4-X12 :1-N13-14 = O15
14-X16 :O17-1-18-O19

Figure 1. Example of basic transformation of the knowledge maintenance data base.

In this example the hydroxyl group (2-O4) of the active substructure (acyl group) are replaced by -NH-CH$_2$COOH group. The presence of a structural part described in positive substructures (C=O) is essential for the performance of the transformation, i.e., the hydroxyl group must be a part of a carboxylic acid group. The presence of any of the negative substructures will prevent the reaction from occurring. In Figure 1, the presence of any atom other than hydrogen connected to the oxygen of the carboxyl group will prevent the reaction from proceeding.

The description of a transformation rule is completed by its classification as a Phase I or Phase II reaction and its type. The product of a

Phase II reaction will not be metabolized further, while a metabolite generated through a Phase I reaction may undergo subsequent biotransformation. In the example shown in Figure 1, the glycine conjugate will not be metabolized further, since its high hydrophilicity will most likely result in rapid elimination from the body rather than further metabolism. The Agrochemical program contains 60 pesticides stored in the data base with their metabolites. The metabolexpert contains 200 compounds in the analog knowledge base, all of which are pharmaceutical drugs.

The predictability of metabolites of three pesticides (unknown to the programs data base) belonging to three groups were examined. The first compound is the herbicide 2,4,5-trichlorophenoxy acetic acid (2,4,5-T). The second compound is one of the toxic components (toxicant B) of the toxaphene mixture *(2)*, an example of chlorinated insecticide. The third compound is malathion, an organophosphate. The AGROMETABOLEXPERT program was used to predict metabolites in plants, while the METABOLEXPERT program was used to predict metabolites in animals. Predicted metabolites of the three selected compounds are shown in Figures 2, 3 and 4 respectively. The results for the major metabolites of the 2,4,5-T (figure 2) are identical to those obtained in laboratory measurement, except in the case of conjugation reactions (Phase II). The program predicted conjugation with glutamic acid rather than the laboratory observed conjugation with aspartic acid *(3)*. Metabolites of toxaphene component B (figure 3) were predicted in general mammals using the METABOLEXPERT program, and the predicted products were similar to those reported in laboratory measurements on mammals *(2,4,5)*. Predicted metabolites of malathion (figure 4) agree with those reported by laboratory experiments *(6)*.

The metabolic pathways may be generated automatically or under manual control. Transformation possibilities are searched in the order in which they appear in the knowledge base. After finishing the prediction of the first-order metabolites, second- and higher-order products are predicted. It is possible to limit the maximum number of predicted metabolites. By searching manually, it is possible to exclude the prediction of a metabolite and its higher-order metabolites. The program does not look up the metabolites in a data base. Rather, the database stores inference rules based on literature data of the metabolic fate of some pesticides in various plant species. The program examines the structure of the compound of interest at every possible transformation site and uses the stored rules of metabolism to predict what will happen. The software also allows users to modify, teach, and complete the metabolism transformation knowledge base. This teaching ability is very useful for improving and updating the prediction ability of the system. The ability to modify and expand the knowledge data base can also provide a means for building new data bases for nonbiological transformations such as photochemical and chemical degradation of compounds. The user simply provides the active functional group, the product of photochemical or chemical reactions and the functional groups in the rest of the molecule that allow or prevent the transformation. Metabolites can also be listed in one or more

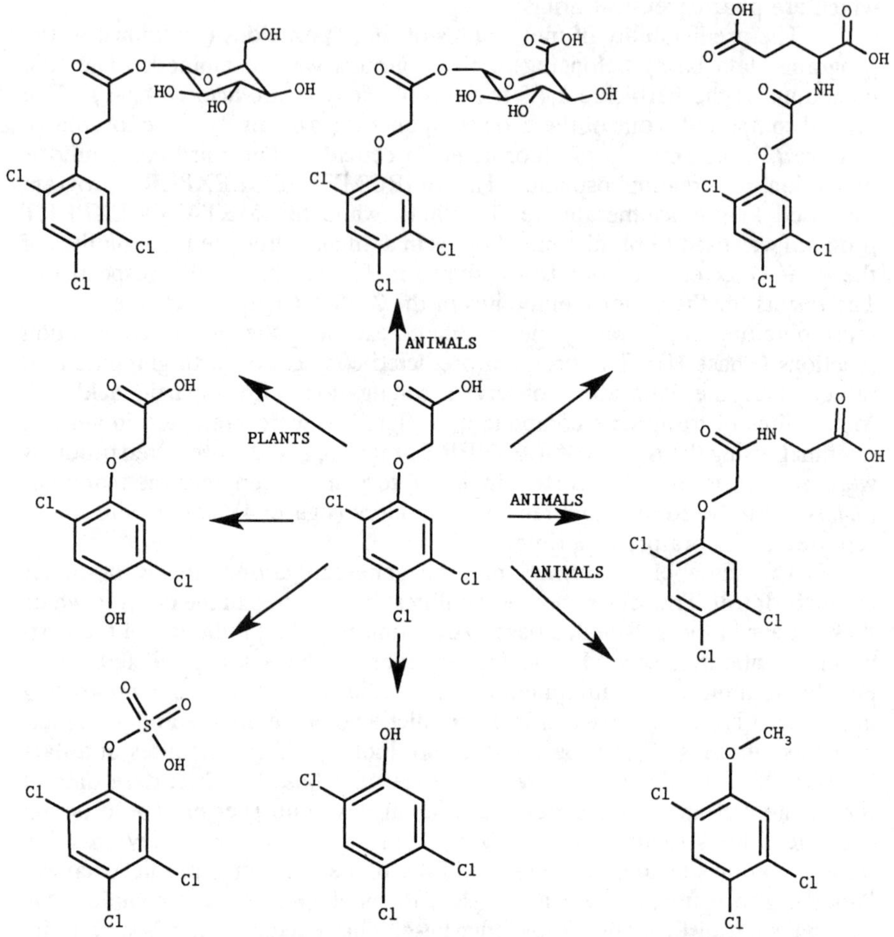

Figure 2. Predicted metabolites of 2,4,5-trichlorophenoxy acetic acid.

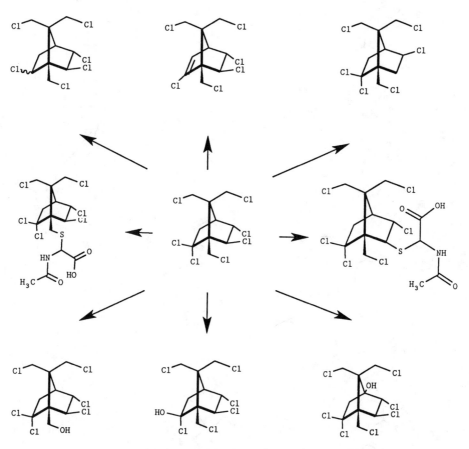

Figure 3. Predicted metabolites of toxaphene toxicant B.

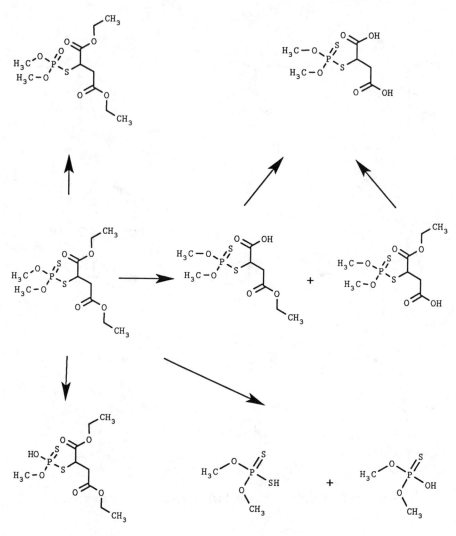

Figure 4. Predicted metabolites of malathion.

selected species, along with their respective excretion percentages. These percentages represent the total amount detected in the various excreted fluids and waste products following administration of the parent substance.

In our experience the system has the following theoretical and practical problems:

1. Current formulation of the rules treats compounds as independent sets of substructures. This approach makes it difficult to draw conclusions when a property assigned to the whole compound controls the metabolic transformation. A good example is the p-hydroxylation of aromatic rings in non-condensed polyaromatic compounds, where normally only one or a few benzene rings will be hydroxylated.

2. Stereoselectivity is not handled by the system, while a number of metabolic reactions cannot be treated without considering stereoselectivity.

3. Controversial difficulties arise on attempting to formalize rules when selectivity within and between species, or dependency on experimental conditions may seriously influence metabolite formation.

Prediction of Bioconcentration and Environmental Hazards of Metabolites

Metabolites that were predicted by METABOLEXPERT or AGROMET-ABOLEXPERT or determined experimentally can be evaluated for their physical and toxicological properties by using another package which calculates lipophilicity as the partition of the compound between octanol and water as a $\log (pK_{o/c})$ as well as bioavailability and bioconcentration factors. Lipophilicity is a very important molecular description because it is often well correlated with the bioactivity of chemical entities. There are numerous difficulties in the experimental determination of real log P, often due to hydrolysis, chemical degradation or extremely low solubility of the tested compound. For these reasons, a wide range of partition coefficients may be measured depending upon the solvent systems, the selectivity of the analytical method, and the method of detection. Since published data are measured on a wide range of temperatures, the likelihood of comparing published versus experimental data is not good.

A number of theoretical methods have been developed to predict log P, the Rekker's method *(7)* is the most widely used method. The program PRO-LOGP, a logic based expert system, calculates hydrophobicity of organic compounds using the Rekker's algorithm. Another expert system, HAZARDEXPERT is capable of predicting the overall health hazard effect of the metabolites relative to the parent compound and also calculates log P and pK_a. It can estimate both chronic and acute toxicity of organic chemicals in different animals, plants, and microorganisms. The log P and pKa calculated by this program are not quantitative and can only be used for comparing a set of chemicals of the same family. The PRO-LOGP program is

more accurate. However, its results were still found to diverge widely from observed values recorded in the literature. Toxicological evaluation of the metabolite malaoxon for a medium oral dose to mammals as compared to the parent compound malathion, are shown in Figs. 5 and 6. The program predicted a stronger hazardous effect for maloxon relative to malathion. It also showed that the metabolite is more water soluble than the parent compound. The columns on the right side of the figure represent different toxicodynamic effects (i.e., oncogenicity, mutagenicity, teratogenicity, and membrane irritation as well as other chronic, acute, and neurotoxic characteristics). The two columns on the left side of the figure provide information on the modifying effects of the internal intensive parameters of the molecule on the toxic effects. The areas above and below the horizontal axis express the ratio of arguments supporting and opposing the hypothesis that the molecule is toxic. The three characteristics of the compound being investigated (MW, log P and pKa) are shown in the bottom right corner of each illustration. The striped areas on the right side of the figure can be larger or smaller than the respective columns corresponding to the toxico-dynamic effects, according to the deviations of the external factors from the basic interpretations. The striped part of the column represents the role of bioaccumulation which increases with exposure. The integrated prediction shown at the bottom summarizes the partial judgements corresponding to the properties of the molecule and all its inherent toxic characteristics, and the effects originating from the interaction with the organism. The top right column refers to the greatest toxico-dynamic component in the molecule. The left column represents the average of the intensive internal factors which support the toxicity of the given substance. The bottom left column represents the average of the intensive internal, which opposes the toxicity of the given substance. The bottom right column represents the uncertainty of the judgment expressed in the top right column. This uncertainty is calculated from a ratio of the unsubstantiated answers of the experts surveyed in the referenced EPA report *(8)*. The dotted line is calculated from the toxico-dynamic effect (i.e., the top right column), modified by the results of the intensive factors (represented by the left-hand side of the diagram). The solid line that marks the final result of the estimation is derived from the result above, taking into consideration the effect of the external parameters. Its value is a number between 0 and 100. A percentage shown in the lower portion of the screen in brackets (following a statement concerning the measure of the overall health hazard effect of the compound) lies in the following ranges:
(0-4) NO Hazardous Effect, (5-14) Low, (15-29) weak, (30-44) Medium, and (45-100) Strong.

Conclusion

CAMP offered great advantages in organizing and updating databases for biotransformation of agrochemicals and to predict their metabolism, toxicological properties and environmental hazardous effects. The compound

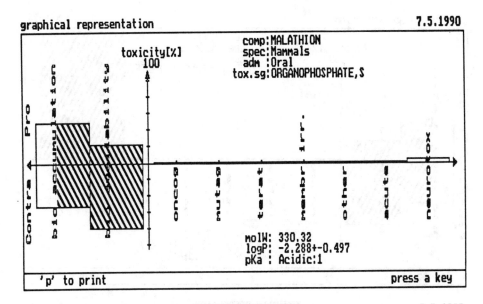

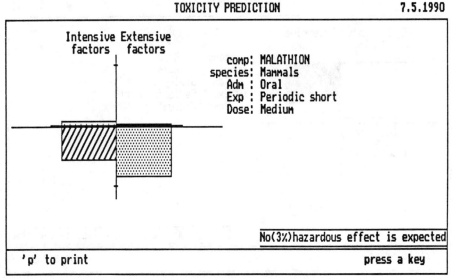

Figure 5. Predicted toxicity of malathion.

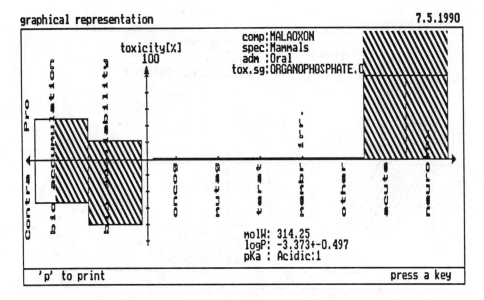

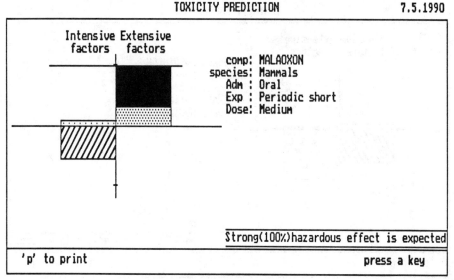

Figure 6. Predicted toxicity of malaoxon.

can be known, or it can be a new structure. The programs do not look up the products or effects in the database; they examine the structure at every possible transformation site and use stored rules to predict its effects. While CAMP can provide the means for suggesting possibilities, it will not totally replace laboratory experiments. The learning ability of the programs allows further application for predicting nonbiological transformation such as photochemical and chemical degradation. This capability will greatly enhance the storage and knowledge information of pesticides degradation.

Acknowledgements

This work was partially supported by U.S EPA grant #CR814342-01. Although the research described in this article has been supported by the U.S EPA, it has not been subjected to Agency review and, therefore, does not necessarily reflect the view of the Agency, and no official endorsement should be inferred. The author thanks Marie Schnell of the UNLV/ERC for helpful suggestions.

Literature Cited

1.   Hopkinson, G.A.; Judson, P.N. *Chemistry in Britain.* **1989**, pp 112.
2.   Saleh, M.A.; Turner, W.V.; Casida, J.E. *Science.* **1977**, 198, 1256.
3.   Mullison, W.R. In *Fate of Pesticides in the Environment;* Biggar, J.W., Seiber, J.N., eds., University of California Publication 3320. **1986**, pp 121-131.
4.   Saleh, M.A.; Skinner, R.F.; Casida, J.E. *J. Agric. Food Chem.* **1979**, 27, 731.
5.   Saleh, M.A. *Rev. Environ. Cont. Toxic.,* **1990**, in press.
6.   Matsumura, F.; Hogendijk, C.J. *J. Agric. Food Chem.* **1964** 12, 447.
7.   Rekker, R.F. *The Hydrophobic Fragmental Constant, Its Derivation and Application. A Means of Characterizing Membrane Systems;* Publisher: Elsevier, Amsterdam 1977.
8.   Brink, R.H.; Walker, J.D., *EPA TSCAITC Interim Report*, Rockville, MD,   June 4, 1987.

RECEIVED November 20, 1990

SIGNIFICANCE

Chapter 12

Interactions Between Pesticides and Their Major Degradation Products

L. Somasundaram and Joel R. Coats

Pesticide Toxicology Laboratory, Department of Entomology, Iowa State University, Ames, IA 50011

The interactions between parent pesticide compounds and their degradation product(s) can influence the fate of both pesticides and degradation products. The role of pesticide degradation products in influencing the persistence and degradation of parent compounds, and the enhanced degradation of degradation products are discussed. The significance of these interactions in crop protection is also addressed.

Pesticides applied to agricultural lands are degraded through biological, chemical, and physical mechanisms (1). Some pesticides have very low persistence levels, resulting in rapid transformation to degradation products. For example, 50% of applied terbufos was oxidized to its sulfoxide within three days of application (2). More than 90% of applied metham-sodium was transformed to methyl isothiocyanate within 3 hrs (3). In the environmental matrices, parent compounds and their degradation products have been detected simultaneously. The interactions between pesticides and their degradation products could be synergistic, antagonistic, or additive. This paper focuses on the interactions between some of the commonly used pesticides and their major degradation products (Table I).

Degradation Products as Inducers of Pesticide Degradation

The failure of some soil-applied pesticides to control their target pests (4) has resulted in significant research on microbial adaptation to pesticides (5). The enhanced biodegradation of some pesticides has been attributed to the substrate value of their degradation products (6-8). Degradation products of pesticides belonging to different classes such as phenoxyacetic acids, carbamothioates, N-methyl carbamates, and organophosphates have been reported to condition soils for rapid degradation of their respective parent compounds (Table II).

0097–6156/91/0459–0162$06.00/0
© 1991 American Chemical Society

Table I. Structures of Some Commonly Used Pesticides and their Degradation Products

Pesticide	Structure	Degradation product	Structure
2,4-D		2,4-Dichlorophenol	
EPTC		EPTC sulfoxide	
Butylate		Butylate sulfoxide	
		Butylate sulfone	

Continued on next page

Pesticide	Structure	Degradation product	Structure
Chlorpyrifos		3,5,6 Trichloro-2-pyridinol	
Fonofos		Methyl phenyl sulfone	
Diazinon		2-Isopropyl-6-methyl-4-hyrdoxypyrimidine	

Carbofuran phenol

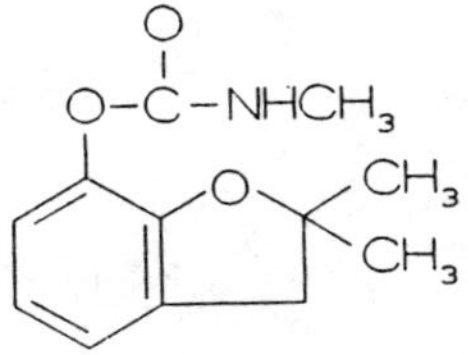

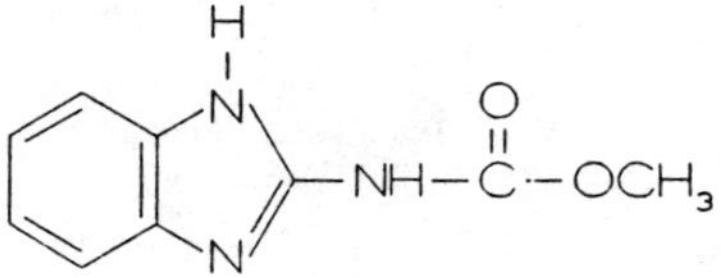

2-Aminobenzimidazole

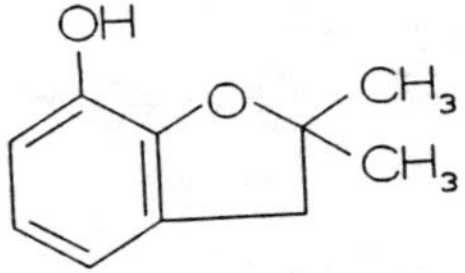

Benzimidazole

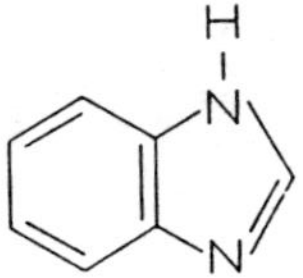

Carbendazim

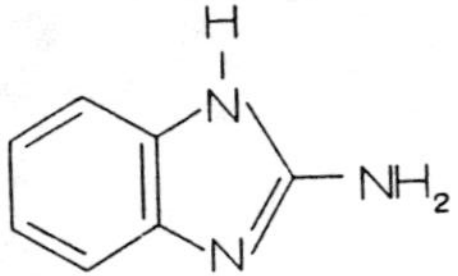

Carbofuran

Carbendazim

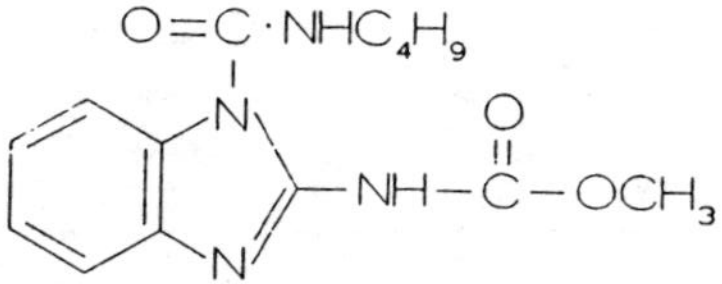

Benomyl

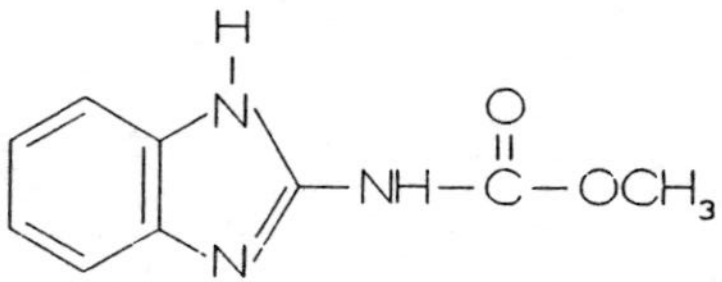

Continued on next page

Pesticide	Structure	Degradation product	Structure
Diphenamid	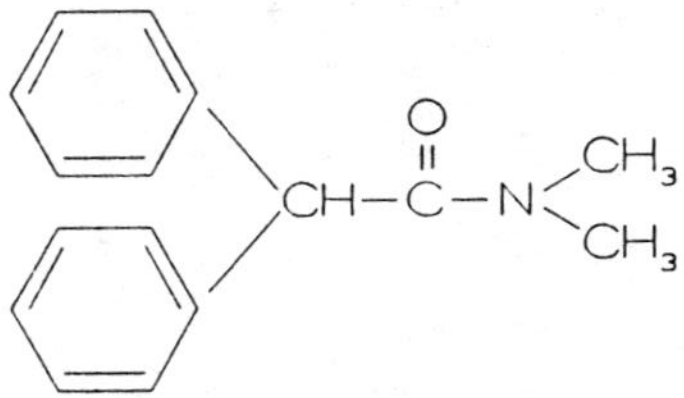	Diphen M-1	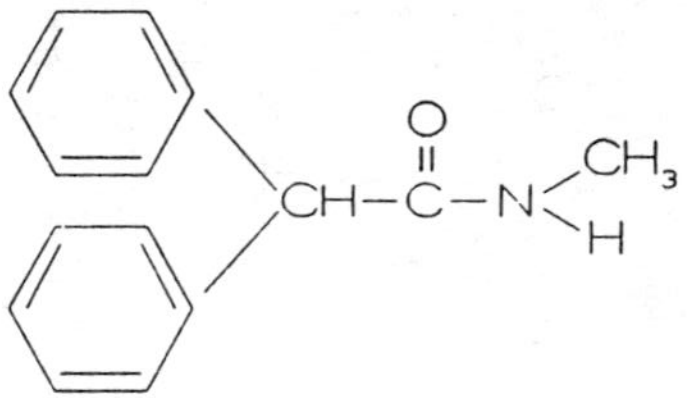
		Diphen M-2	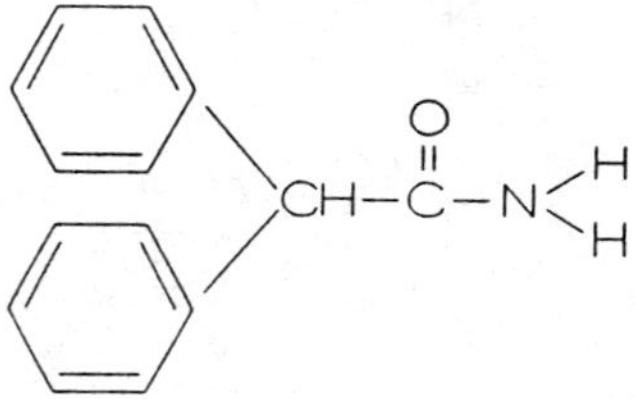
Metham-sodium	$CH_3NH\cdot\overset{S}{\overset{\|}{C}}S-Na^+$	Methyl isothiocynate	$CH_3-N=C=S$

Table II. Degradation products as inducers of pesticide degradation

Inducer/ substrate	Susceptible pesticide	Reference
Inducers of their respective parent compounds		
2,4-dichlorophenol	2,4-D	(8,9)
EPTC sulfoxide	EPTC	(10)
butylate sulfoxide	butylate	(10)
carbofuran phenol	carbofuran	(7)
1-naphthol	carbaryl	(7)
p-nitrophenol	parathion	(6,8,11)
salicylic acid	isofenphos	(8)
2-aminobenzimidazole	carbendazim	(12)
Inducers of degradation of other parent compounds		
butylate sulfoxide	EPTC	(10)
EPTC sulfoxide	butylate	(10)

Hydrolysis Products. Hydrolytic reactions seem to play an important role in the initiation of pesticide metabolism by adapted soil microorganisms (13,14). In several instances, the presence of pesticide hydrolysis products has resulted in induction of a pesticide degrading population of soil microorganisms (6,7). In one of the first studies on the biodegradation of pesticides, Newman and Thomas (9) observed decreased persistence of 2,4-D in soils pretreated with its hydrolysis metabolite, 2,4-dichlorophenol. In that study, pretreatment of soils with closely related compounds such as phenoxyacetic acid and 2-chlorophenol, however, had no effect on the persistence of 2,4-D. Steenson and Walker (15) reported the ability of an *Achromobacter* strain to use 2,4-dichlorophenol as a substrate. In a recent study in our laboratory, the mineralization rate of 2,4-D increased with the number of pretreatments with 2,4-dichlorophenol (8).

Fungicide carbendazim rapidly degraded in soils previously treated with its metabolite 2-aminobenzimidazole. (12,16) 2-Aminobenzimidazole and carbendazim were equally effective in conditioning the soil to carbendazim. Pretreatment of soils with another potential metabolite, benzimidazole, however, resulted in only a slight reduction in the parent compound's persistence. Benzimidazole rings are generally less susceptible to microbial degradation (17), and the benzimidazole-induced nonenhanced degradation of carbendazim (12) is not clearly understood.

Carbofuran phenol, a hydrolysis product of carbofuran, conditioned soils for enhanced degradation of carbofuran under anaerobic environment (7) but did not have any effect in aerobic conditions (8,18). Although the presence of carbofuran phenol in anaerobic soils accelerated carbofuran hydrolysis, the phenol was not used as an energy source and accumulated in the soil. This finding suggests that besides substrate value, other properties of

degradation products may be involved in the induction process. The
different results obtained from aerobic and anaerobic soils reveal
the importance of environmental factors in the interaction between
pesticides and their degradation products.

Oxidation Products. The sulfoxides of EPTC and butylate induced
enhanced degradation of their parent compounds, and the potential of
the sulfoxides to promote rapid degradation of their parent
compounds was similar to that of the parent compounds themselves
(10). Although butylate sulfoxide conditioned the soils,
pretreating soils with butylate sulfone did not have any effect on
the rate of degradation of the parent compound.

Cross-enhancement. Degradation products can also enhance the
degradation of pesticides other than their precursors. Prior
exposure of soils to butylate sulfoxide accelerated the degradation
of EPTC (10). Similarly, a higher butylate degradation rate was
observed in soils exposed to EPTC sulfoxide. This type of cross-
enhancement is generally limited to structurally similar pesticides.
Location differences were also observed in cross-enhancement tests
for EPTC and butylate degradation in soils pretreated with
sulfoxides of EPTC or butylate, suggesting the role of soil
properties in intensifying the enhancement (10).

Degradation Products as Promoters of Pesticide Persistence

Most of the available information on degradation products indicates
their potential to accelerate the degradation of subsequently
applied pesticides. Contrary to this, some degradation products
could prolong the persistence of pesticides in soil.

Hydrolysis Products. Particularly when applied to the soil
repeatedly, 3,5,6-trichloro-2-pyridinol (TCP), a hydrolysis
metabolite of chlorpyrifos and trichlopyr, slowed the degradation of
chlorpyrifos (8). At 100 ppm, TCP inhibited the degradation of
carbofuran and DOWCO 429X in problem soils but did not have any
effect at 1 and 10 ppm (19). Recent studies have demonstrated the
antimicrobial activity of this metabolite (20,21); the inhibitory
effect of TCP could be a result of its microbial toxicity. The
levels required to affect the microbes, however, are so high that
the effect of a single application of chlorpyrifos or trichlopyr may
not be sufficient to produce a significant effect. In contrast,
under field conditions, pesticides are not homogeneously distributed
through soil but are concentrated in and around the particles of a
granular formulation. This could result in a high concentration of
the pesticide or metabolite in a particular microenvironment.

Cross-retardation. Fonofos, an organophosphorus insecticide,
increased the persistence of EPTC in problem soils (22).
Dietholate, an extender used in the new formulations of EPTC, is
structurally very similar to fonofos. In laboratory toxicity
studies, methyl phenyl sulfone (a metabolite of fonofos) was more
toxic to *Photobacterium phosphoreum*, a bioluminescent bacterium
phylogenetically related to several important soil bacteria, than

fonofos (20). The antimicrobial activity of fonofos (22) could result from the toxicity of its degradation product, methyl phenyl sulfone. The mechanism of the inhibitory effect of fonofos under field conditions has yet to be clearly elucidated, however.

Enhanced Degradation of Degradation Products

Repeated applications of some pesticides can also accelerate the degradation of the products formed (Table III). 2-Aminobenzimidazole rapidly dissipated in soils exposed to its parent, carbendazim, with only 6% remaining 4 days after application (16). In these soils, degradation of 2-aminobenzimidazole was faster than that of the parent. The rate of degradation of benzimidazole, a structurally similar metabolite, however, was not affected by carbendazim history. Carbendazim, also a fungitoxic hydrolysis product of benomyl, degraded more rapidly in soil with a benomyl history than in a soil without one (23). The half-life of carbendazim was reduced from 11 to 4 days in benomyl-treated soils. This reduction has been attributed to the short lag-period observed in these soils.

The degradation of desmethyldiphenamid (diphen M-1), a monodemethylated metabolite of diphenamid, was much faster in soils pretreated with its parent compound than in an untreated soil (24). There was no difference in the degradation rate of Diphen M-2, a bidemethylated product of diphenamid, in soils with and without a diphenamid history. The rate of degradation of diphen M-2 was also much slower than for diphenamid or diphen M-1.

Repeated applications of metham-sodium enhanced its transformation rate to methyl isothiocyanate and subsequent mineralization of the pesticidal metabolite. The half-life of methyl isothiocyanate ranges from 0.5 to 50 days with the shorter half-life in soils previously treated with metham-sodium (3).

Laboratory studies have investigated the fate of degradation products as influenced by their prior treatment (8,12). The purpose of these studies was to confirm the substrate value of the degradation products. Self-enhancement of methyl isothiocyanate (3), 2-aminobenzimidazole (12), and 2,4-dichlorophenol (8) has been previously reported. The microbial adaptation to degradation products is significant, particularly when the pesticidal effect is caused by the degradation product.

Table III. Pesticide degradation products susceptible to enhanced degradation

Susceptible degradation product	Reference
2-aminobenzimidazole	(16)
carbendazim	(22)
methyl isothiocyanate	(3)
desmethyldiphenamid	(24)

Conclusions

A better understanding of the mechanisms of pesticide-degradation product(s) interactions is important for studying the fate and the effects of pesticides in the environment. Synergistic interactions could result in inadequate host protection leading to restricted use or withdrawal of parent compound and thus influence the pesticide class to be used in subsequent applications. The information generated from such studies will help make effective use of existing biodegradable pesticides by promoting an understanding of the specificity and development of enhanced biodegradation.

Acknowledgements

Journal Paper No. J-14292 of the Iowa Agriculture and Home Economics Experiment Station, Ames, Iowa, Project No. 2306.

Literature Cited

1. Coats, J.R. chapter 2 in this volume.
2. Lavelgia, J.; Dahm, P.A. *J. Econ. Entomol.* 1975, *4*, 715-718.
3. Smelt, J.H.; Crum, S.J.; Teunissen, W. *J. Environ. Sci. Health* 1989, *B24*, 437-455.
4. Roeth, F.W. *Rev. Weed Sci.* 1986, *2*, 45-65.
5. Racke, K.D. Coats, J.R. *Enhanced Biodegradation of Pesticides in the Environment*, Am. Chem. Soc.: Washington, D.C., 1990.
6. Ferris, I.G.; Lichtenstein, E.P. *J. Agric. Food Chem.* 1980, *28*, 1011-1019.
7. Rajagopal, B.S.; Panda, S.; Sethunathan, N. *Bull. Environ. Contam. Toxicol.* 1986, *36*, 827-832.
8. Somasundaram, L.; Coats, J.R.; Racke, K.D. *J. Environ. Sci. Health* 1989, *B24*, 457-478.
9. Newman, A.S.; Thomas, J.R. *Proc. Soil Sci. Soc. Am.* 1949, *14*, 160-164.
10. Bean, B.W.; Roeth, F.W.; Martin, A.R.; Wilson, R.G. *Weed Sci.* 1988, *36*, 70-77.
11. Sudhakar-Barik; Wahid, P.A.; Ramakrishna, C.; Sethunathan, N. *J. Agric. Food Chem.* 1979, *27*, 1391-92.
12. Yarden, O.; Salomon, R.; Katan, J.; Aharonson, N. *Can. J. Microbiol.* 1990, *36*, 15-23.
13. Sethunathan, N.; Pathak, M.D. *J. Agric. Food Chem.* 1972, *20*, 586-589.
14. Racke, K.D.; Coats, J.R. *J. Agric. Food Chem.* 1987, *35*, 94-99.
15. Steenson, T.I.; Walker, N. *J. Gen. Microbiol.* 1958, *18*, 692-697.
16. Aharonson, N.; Katan, J.; Avidov, E.; Yarden, O. In *Enhanced Biodegradation of Pesticides in the Environment*; Racke, K.D.; Coats, J.R.; Eds.; American Chemical Society, 1990; pp 113-127.
17. Sisler, H.D. In *Biodegradation of Pesticides*; Matsumara, F.; Krishnamurti, C.R. Eds.; Plenum Press: New York, NY, 1982; pp 131-151.
18. Harris, C.R.; Chapman, R.A.; Harris, C.; Tu, C.M. *J. Environ. Sci. Health* 1984, *B19*, 1-11.
19. Chapman, R.A.; Harris, C.R. In *Enhanced Biodegradation of*

Pesticides in the Environment; Racke, K.D.; Coats, J.R.; Eds.; Am. Chem. Soc.: Washington, D.C. **1990**; pp 82-96.

20. Somasundaram, L.; Coats, J.R.; Racke, K.D.; Stahr. H.M. *Bull. Environ. Contam. Toxicol.* **1990**, *44*, 254-259.

21. Felsot, A.; Pedersen, W. **1991**, chapter in this volume.

22. Roeth, F.W.; Wilson, R.G.; Martin, A.R.; Shea, P.J. In *Enhanced Biodegradation of Pesticides in the Environment*; Racke, K.D.; Coats, J.R.; Eds.; Am. Chem. Soc.: Washington, D.C., **1990**; pp 23-36.

23. Yarden, O.; Katan, J.; Aharonson, N.; Ben-Yephet, Y. *Phytopathology* **1985**, *75*, 763-767.

24. Avidov, E.; Aharonson, N.; Katan, J. *Weed Sci.* **1990**, *38*, 186-193.

RECEIVED December 12, 1990

Chapter 13

Pesticidal Activity of Degradation Products

A. S. Felsot[1] and W. L. Pedersen[2]

[1]Center for Economic Entomology, Illinois Natural History Survey,
Champaign, IL 61820
[2]Department of Plant Pathology, University of Illinois,
Urbana, IL 61801

Pesticides may be degraded into compounds that also have pesticidal activity against target and/or untarget pests. Relatively few
studies have documented the level of pesticidal activity; however,
they indicate the products may be more, less, or similar in activity
to the parent pesticide. The prolonged efficacy of many systemic
insecticides against foliar feeding insects is dependant upon absorption from soil by plants and metabolism into water soluble but
equally toxic anticholinesterase agents. In the soil, the more water
soluble transformation products of organophosphorodithioate insecticides are less toxic to insect larvae than the parent compounds.
However, when these compounds are topically applied, toxicity is
similar. Hydrolysis of the insecticide chlorpyrifos to 3,5,6-trichloro-
2-pyridinol results in total loss of insecticidal activity, but the
degradation product is bioactive against several fungal pathogens.
This chapter will discuss the importance of degradation products
from insecticides, fungicides, and herbicides.

From a perspective of environmental quality, the ideal pesticide would be
one that persisted only during the portion of the crop's life cycle that is
vulnerable to root-, foliage, and fruit-eating pests. Thereafter, the pesticide
would be completely transformed into innocuous organic compounds by
chemical or biological reactions. Pesticide dissipation in soil and on foliar
surfaces has been commonly described as a first-order kinetic process, and
therefore the concentration of the active ingredient steadily declines
immediately following application and throughout the growing season. The
biological activity of nearly all currently registered pesticides against insects,
weeds, and, diseases does not persist more than several months into the
growing season; however, pesticide residues can be easily found at "trace

concentrations" (i.e., levels below 0.1 ppm) in the environment or harvested crop long after the growing season. Many times these residues are actually transformation products of the originally applied pesticide; acute toxicity against target or nontarget pests is hardly ever associated with these compounds, but potential long-term effects (e.g., carcinogenicity, ecosystem disruptions) has long been a concern.

For many currently registered products, pesticidal activity depends solely on the amount of parent material present in the soil or on the crop at the time of pest infestation. For some products, however, pesticidal activity results from transformation of the parent pesticide into acutely toxic compounds. The transformation may occur by abiotic or biotic mechanisms. Transformation products may be produced directly in the soil, where they can directly kill pests; alternatively, the parent pesticide may be absorbed by plant roots and translocated through the xylem to tissues that metabolize the compound into other toxicants. In some cases, a pesticide applied to control one kind of pest may exert biological activity against a completely different pest by virtue of its transformation product.

The importance of pesticide transformation products to efficacy against insects, plant diseases, and weeds will be discussed through a review of the literature and personal observations. The discussion will include the pesticidal activity of pesticide transformations in soil and plants against both targeted and untargeted pests.

Insecticidal Activity of Insecticide Degradation Products

Early recognition of the importance of insecticide degradation or trans- formation products to biological activity probably came with the observed phytotoxicity to apples of lead arsenate ($PbHAsO_4$) sprays. It was believed that injury to fruit and foliage was caused by soluble arsenic in the spray mixtures or in residues from spray mixtures (1). Several studies showed that the ratio of lead (Pb) to arsenious acid (As_2O_3) on apples after spraying increased upon weathering of the pesticide (2,3). The reaction was exacer- bated by the addition of lime to lead arsenate but the transformation did not occur with the use of lead arsenate alone, with the use of additives, such as Bordeaux mixture and aluminum sulfate, or after subsequent use of oil sprays.

In the case of synthetic organic insecticides, early studies with insects had shown that the anticholinesterase activity of phosphorothioate or phosphorodithioate compounds like parathion (O,O-diethyl O-p-nitrophenyl phosphorothioate) or malathion (S-[1,2-bis(ethoxycarbonyl)ethyl] O,O-di- methyl phosphorodithioate), respectively, was low compared to the phos- phate and thiolophosphate transformation products known as oxons (i.e., paraoxon and malaoxon, respectively, (4,5) (Table I). Mammalian studies

Table I. Cholinesterase (ChE) Inhibition and Toxicity of Organophosphorus Insecticides and Metabolites

Compound	House Fly Head ChE Inhibition I_{50} (moles x 10^{-6})	LD_{50} $\mu g/fly$	Reference
malathion	20		(5)
oxon	0.0046		"
schradan	150000		(9)
oxide	0.34		"
demeton, thiono	220	ND [1]	(10,11)
sulfoxide	3.60	2.0	"
sulfone	0.83	1.2	"
oxon	0.024	0.7	"
oxon sulfoxide	1.10	8.7	"
oxon sulfone	0.12	3.7	"
demeton, thiolo	3.50	ND	"
sulfoxide	1.50	0.8	"
sulfone	0.60	1.2	"
disulfoton	>100		(12)
sulfoxide	70		"
sulfone	3.50		"
oxon	3.50		"
oxon sulfoxide	1.50		"
oxon sulfone	0.60		"
phorate	25	1.5 [2]	(12,13)
sulfoxide	3.70	5.5	"
sulfone	0.04	4.5	"
oxon	0.50	1.1	"
oxon sulfoxide	0.40	5.5	"
oxon sulfone	0.10	1.5	"

[1] ND=not determined
[2] LD_{50} for phorate expressed as mg/kg

showed that parathion was oxidized in vivo to its oxon (*6-7*), and Metcalf and March (*8*) were the first to show that insect tissue carried out a similar transformation. Other studies showed that transformations of organophosphorus (OP) insecticides at sites other than phosphorus also could produce strong anticholinesterase agents. For example, Shradan (octamethylpyrophosphoramide), which was one of the first OP insecticides, was metabolized by cockroach gut to a toxic phosphoramide oxide that was an even stronger cholinesterase inhibitor than the parent compound (*9*). In a variety of cockroach tissues, the thioether linkages of Systox (demeton) was progressively oxidized to toxic thionophosphate and thiolophosphate sulfoxides and sulfones (*10,11*). The phosphorodithioate insecticides disulfoton (O,O-diethyl S-[2-(ethylthio)ethylphosphorodithioate) and phorate (O,O-diethyl S-(ethylthio)methyl phosphorodithioate) were observed to undergo analogous transformations which produced potent inhibitors of acetylcholinesterase (*12,13*) (Table I).

Throughout the 1950's accumulated evidence showed conclusively that environmental degradation of insecticides could produce toxic compounds with biological activity equal to the parent. Early examples of the recognition of the pesticidal activity of degradation products after application to soils or plants were seen with the epoxidation of aldrin to dieldrin (*14,15*) and the oxidation of thioether linkages in demeton (*11*), a mixture of O,O-diethyl S-(and O)-2-[(ethylthio) ethyl] phosphorothioates. The following discussion focuses strictly on the pesticidal activity of environmental transformation products. Two important categories of transformation products can affect the efficacy of pesticides: (1) transformation products in or on plants that can prolong the control of foliar feeding insects; (2) transformation products in soil that can contribute to the control of soil inhabiting pests, including weeds.

Insecticidally Active Transformation Products in Plants. Pesticide residues may occur on or in plant tissue either by direct application to foliage or through root absorption. Residues following spraying can be transformed by enzymatic processes after absorption into the leaf tissue or by abiotic processes on the foliar surfaces. Following spraying on alfalfa, soybeans, and corn aldrin (1,2,3,4,10,10-hexachloro-1,4,4a,5, 8,8a-hexahydro-1,4 -endo-exo-5,8-dimethanonaphthalene) was converted to its epoxide dieldrin, a more residual and equally toxic compound, see Table II (*16*). Heptachlor (1,4,5,6,7,8,8,-heptachloro-3a,4,7,7a-tetrahydro- 4,7-methanoindene) also was oxidized on plant surfaces to heptachlor epoxide (*17*). Gannon and Decker (*17*) suggested that the conversion of heptachlor required sunlight in addition to plant tissue.

Among op compounds, the acaricide and insecticide Delnav (mixture of cis and trans 2,3-p-dioxanedithiol S,S-bis(O, O-diethyl phosphorodithioate) was converted on seedling bean, cotton and tomato plants to anticholinesterase agents more polar and more potent than the parent compound (20). The compounds were hypothesized to be the phosphorothiolate forms of the Delnav isomers. Similar transformations were seen when the pesticide was exposed to sunlight. Parathion is also known to be abiotically oxidized in the presence of dust, ozone, and sunlight to its potent acetyl-

Table II. Contact Toxicity in Potter Spray Tower Test of Chlorinated Cyclodiene Insecticides to Black Cutworm (*Agrotis ipsilon*)

	Average Corrected % Mortality for % Insecticide Solution				
Compound	0.001	0.010	0.100	1.000	Reference
aldrin	0	40	100	100	(18)
dieldrin	0	42	100	100	(18)
heptachlor	0	0	11	100	(19)
h. epoxide	0	0	20	100	(19)

cholinesterase-inhibiting oxon (21). Any transformations to more potent anticholinesterase agents on foliar surfaces should prolong the residual activity of the insecticide, assuming that the uptake of the newly formed compound, which is likely to have a much higher water solubility than the parent, is not altered significantly by the change in physicochemical properties.

After absorption by cotton leaves, acephate (O,S-dimethyl acetylphosphoramidothioate) is transformed partially to the equally toxic compound methamidophos (O, S-dimethyl phosphoramidothioate) (22), which itself is sold as the insecticide Monitor. Both insecticides are poor cholinesterase inhibitors in vitro, which suggests they are further transformed to more biologically active products in vivo (23). The tobacco budworm exhibits nearly equal sensitivity to both acephate and methamidophos (topical LD_{50} of 13 and 24 μg/g, respectively). In contrast, the boll weevil is very tolerant to acephate (LD_{50} of 938 μg/g) but highly susceptible to methamidophos (LD_{50} of 13 μg/g), presumably because this species lacks the ability to deacetylate acephate (22).

Studies during the late 1940's and early 1950's (24,25) showed that parathion could be absorbed by the roots of plants and cause tissues to be toxic to feeding insects. Even the chlorinated cyclodiene insecticides, which were not considered to be systemics, could be absorbed by root tissue like carrots (14) and potatoes (25,26). This early research recognized that a "derivative" of the parent insecticide could have caused toxicity to the feeding insects (24,14). Indications that absorbed insecticide residues were being transformed into other toxic agents came during studies with aldrin in carrots. Concentrations determined by chemical assays were always lower than concentrations derived from bioassays (14); the second toxic agent was determined to be dieldrin.

One of the advantages of soil-applied systemic insecticides is that the pesticide can be distributed throughout the plant making tissues that might not be covered by sprays lethal to feeding insects. Transformation of the parent insecticide within the plant to equally toxic components would prolong the residual bioactivity of the insecticide at the same time it was quickly dissipating from the soil environment. Alkyl thioether derivatives of phosphorothioate and phosphorodithioate insecticides have proven to be good systemics when applied to soil. Using ^{32}P-labeled demeton and auto-radiography, Metcalf et al. (27) were the first to show that the thiono isomer was translocated in plant tissues and transformed to compounds that were highly inhibitory to cholinesterase. In a series of papers by Fukuto, Metcalf, and coworkers (11,28-31) the toxic plant metabolites of demeton were identified as the sulfoxide and sulfone derivatives of the thiono and thiolo isomers. In addition, the oxon derivatives of the thiono isomer were detected. Toxicity of these plant metabolites against house flies are shown in Table I.

The phosphorodithioates disulfoton and phorate undergo oxidations similar to that of demeton in plants grown in soil containing these pesticides (12,13,32,33). Toxicity of the oxidative metabolites of phorate after topical application to houseflies is very similar to the parent (13) (Table 1). Incorporation of metabolites into an artificial diet, however, showed phorate was significantly more toxic to aphids than its metabolites (34). In the latter case, the lower toxicity of the metabolites was attributed to a more rapid hydrolysis of the metabolites than of phorate both in the diet and after ingestion by the insects.

Systemic soil insecticides like disulfoton have been valuable in prolonging control of aphids on brassica crops (33) where sprays only provide control for several weeks and must be repeatedly applied. Brussel sprouts were shown to absorb disulfoton sulfoxide (O,O-diethyl S-[2-(ethyl-sulfinyl)ethyl] phosphorodithioate) and sulfone (O,O-diethyl S-[2-(ethylsulfonyl)ethyl] phosphorodithioate) directly from soil; after four weeks, the oxygen analog of disulfoton sulfone was the only insecticidal residue detected in the plants (33). Sulfur-containing carbamate insecticides also undergo oxidative transformations in plants to produce highly toxic metabolites. Aldicarb (2-methyl-2-[methylthio]propionaldehyde O-[methylcarbamoyl]-oxime) quickly degrades in soil to aldicarb sulfoxide (2-methyl-2-[methylsulfinyl]propionaldehyde 0-[methylcarbamoyl]oxime) and aldicarb sulfone (2-methyl-2-[methylsulfonyl]propionaldehyde 0-[methylcarbamoyl]-oxime), but it is also quickly absorbed by plants where its toxicity is prolonged by transformation to the highly toxic aldicarb sulfoxide (35,36, and references cited therein). Aldicarb sulfoxide may be further oxidized in plants to the sulfone, but the sulfoxide is much more toxic to boll weevils, aphids and mites (35,36).

Croneton, [2-(ethylthio)methyl]phenyl N-methylcarbamate, was taken up from soil by citrus trees and metabolized within four days to croneton sulfoxide, [2-(ethylsulfenyl)methyl]phenyl N-methylcarbamate. Croneton sulfone, [2-(ethylsulfonyl)methyl]phenyl N-methylcarbamate, appeared in citrus leaves more slowly (37). Residual efficacy of croneton against aphids was probably due to the presence of the sulfoxide and sulfone, which were the predominant metabolites in the leaves and were nearly as toxic as the parent to *Aphis citricola*.

Insecticidally Active Transformation Products in Soil. In early studies, the residual bioactivity of these soil applied compounds was tested using the fruit fly, *Drosophila melanogaster* (15). Edwards et al. (15) were the first to recognize that degradation of the parent insecticide in the soil, like in the plant, could result in lower or higher toxicity. The transformation of aldrin to dieldrin and heptachlor to heptachlor epoxide was representative of the latter situation (15,38,39). In contrast to plants where the insect would be feeding directly on insecticide residues, the toxicity in soil was observed to vary according to the amount of organic matter (OM) in soil (15,40), Table III.

The chlorinated cyclodiene insecticides were recognized to be excellent soil insecticides for control of root feeding insects like the corn rootworm. Although aldrin was considered to be nonpersistent in the soil, dieldrin residues were very persistent and could induce toxicity for years after the original application (41). Corn rootworms quickly developed resistance to aldrin and heptachlor, perhaps because of application of these insecticides by broadcasting (rather than banding) and their extremely long chemical persistence. Despite the suspension and lack of use of the chlorinated cyclodienes, corn rootworm beetles have continued to show resistance to

Table III. Toxicity of Aldrin and Dieldrin in Soil to the Common Field Cricket (*Gryllus pennsylvanicus*), (*40*)

Compound	LD$_{50}$ (μg insecticide/g soil)	
	Sandy Loam (1.5% OM)	Muck (64.6% OM)
aldrin	0.22	5.07
dieldrin	0.55	15.66

aldrin. Ball (*42*) suggested that the lack of return to susceptibility in corn rootworm populations in Nebraska was probably associated with the continuing presence of high levels of dieldrin in soil. Transformation of phosphorodithioate and oxime carbamate insecticides in soils to toxic oxidative metabolites has been extensively studied. The main difference between soils and plant tissues in transformation pathways is that the oxon metabolites are very minor products in soil but constitute significant concentrations in plants (*33,43*). Furthermore, small amounts of phorate sulfoxide are reduced to the parent phorate under certain conditions (*44*).

Phorate and terbufos, which are used extensively for control of corn rootworm are readily oxidized to sulfoxide and sulfone forms in soil. The transformation is partly due to microbial activity and partly due to chemical oxidation (*45,46*). Harris and Chapman (*47,48*) have developed concentration-response curves for these compounds and their metabolites using the common field cricket in a sand and muck soil (Table IV). Although the oxidative metabolites are nearly as toxic as the parent when tested by topical application to crickets, the parent is significantly more toxic in soil. Similarly, oxidative metabolites of isofenphos (1-methylethyl 2-[[ethoxy [(1-methyl-ethyl)amino] phosphinothioyl]-oxy] benzoate) and terbufos have been shown to be less toxic to corn rootworm larvae in soil than the parent (*49*); furthermore, the toxic response is highly dependent on the soil type (Table V).

Because the degradation of phorate in soil is extremely rapid and large amounts of phorate sulfone eventually accumulate, the residual bioactivity of phorate may be due to a synergistic or additive effect of the parent and the metabolites (*47,51*). Because terbufos residues in soil are more persistent than phorate and the formation of terbufos sulfone from terbufos sulfoxide is much slower and less significant than the corresponding transformation of phorate sulfoxide, the residual bioactivity of terbufos may be due primarily to the terbufos residue (*48*).

Table IV. Toxicity to the Common Field Cricket of Phorate, Terbufos, and Oxidative Metabolites by Topical Spray Application and Soil Contact [Adapted from Harris and Chapman (*47*) and Chapman and Harris (*48*)]

	LC_{50}		
	Direct Spray	Soil Treatment (ppm)	
Compound	% solution	sand	muck
phorate	0.018	0.40	11.70
sulfoxide	0.006	7.23	170.00
sulfone	0.010	3.22	79.50
terbufos	0.009	0.08	1.34
sulfoxide	0.003	2.82	58.00
sulfone	0.003	1.03	31.60

Table V. Effect of Soil on Toxicity of Isofenphos, Terbufos, and Oxidative Metabolites to Southern Corn Rootworm (*49*)

	LC_{50} (ppm)		
Compound	clay loam 1.8% OM	silty clay loam 1.9% OM	silt loam 3.5% OM
isofenphos [1]	-	0.59	0.94
oxon	-	5.03	9.92
terbufos	0.09	-	0.19
sulfoxide	0.62	-	6.77

[1] LD_{50} as by direct topical application was 5.7 μg/g insect (*50*).

Aldicarb has been extensively tested for the control of various nematode species. It is thought to work by direct uptake from soil rather than through systemic activity (*52,53*). Several studies have shown that aldicarb sulfoxide and sulfone are less toxic to nematodes than the parent compound (*53,54*). The toxic action of aldicarb may be due more to the rapid uptake of parent compound and quick conversion in vivo to the sulfoxide rather than the direct uptake of the sulfoxide. Batterby et al. (*55*) have shown that control of *Heterodera schachtii* in the field occurred early in the season by commercial granular formulations that give a quick release of aldicarb; repeated incremental applications did not provide control. Like terbufos, therefore, the oxidative metabolites of aldicarb may not contribute to nematicidal activity in soil as much as one would predict from their anticholinesterase activities.

The comparatively lower toxicities of the sulfoxide and sulfone derivatives to insects in soil compared to toxicities measured after direct contact by topical application can be explained by their higher water solubilities compared to the parent. The direct correlation between soil organic matter and pesticide sorption and the inverse correlation between pesticide water solubility and sorption have been well established (*56,57*). Uptake of an insecticide from soil by corn rootworm larvae is also inversely correlated with organic matter content but positively correlated with desorbability and toxicity (*58*). The inverse correlations between soil organic matter and toxicity (*15,59*) or uptake suggest that among a series of analogous compounds of similar contact toxicity, those that are more intensely adsorbed will exhibit comparatively higher toxicities than those that are more weakly adsorbed because the insect cuticle is acting like an organic phase in the soil. Since water solubility greatly influences sorption, soil insecticides of comparatively lower water solubility are more toxic than insecticides of higher water solubility; indeed, a significant negative correlation exists between water solubility and toxicity (*60*).

Fungicidal Activity of Fungicide Degradation Products

Compared with insecticides, few studies on the pesticidal activity of degradation products of fungicides have been published. However, several examples indicate that the degradation products of fungicides may have pesticidal activity. Like most pesticides, fungicide degradation is affected by temperature, water level, soil type, pH, microbial populations, etc. Clemons and Sisler (*61*) reported the fungicide benomyl [1-(butylcarbamoyl)-2-benzimidazole carbamic acid, methyl ester] separated into two components on silica gel thin-layer chromatograms in laboratory analysis. The second product, was identified as carbendazim [2-bendimidazole carbamic acid, methyl ester]. When compared to benomyl, carbendazim was found to be equally toxic to *Neurospora crassa* and *Rhizoctonia solani*, but 30 times less toxic to *Saccromyces pastorianus*. Helwig (*62*) isolated microorganisms from garden soil that were capable of degrading the fungicide to non-fungistatic compounds in vitro. Recently, Yarden et.al. (*63*) studied the degradation of benomyl and carbendazim in disinfected and fungicide treated soils. When soils were disinfected with methyl bromide or solarization or sterilized by autoclaving, the persistence of carbendazim, which was added after treatment, was enhanced. This demonstrated the role of soil microbes in the degradation of cardendazim. In addition, the fungicides thiram and fentin acetate also inhibited degradation of carbendazim. Since thiram and fentin acetate have limited effect on bacterial populations, the authors indicated fungal populations, specifically *Alternaria alternata* and *Bipolaris tetramera*, were involved in the degradation process. In soil with a benomyl history, they also found carbendazim was degraded more rapidly than in soil with no history of benomyl treatment. Therefore, benomyl, which is unstable in aqueous solution, rapidly breaks down to carbendazim, which is adsorbed to organic matter or clay minerals (*64*) and is further degraded by microorganism into nontoxic compounds (*63-65*).

A different type of degradation process is involved with the fungicide fosetyl aluminum. It is a monoethyl phosphonate fungicide and is used for the control of *Phytophthora* sp. It may be applied either as a foliar spray or as a trunk injection (tree crops) for the control of root rots. When fosetyl aluminum enters the plant, the monoethyl phosphonate is converted to phosphonic acid, H_2PO_3, which is ionized to dianion phosphonate, HPO_3^{-2}. Quimette and Coffee (*66*) demonstrated that the application of diethyl and dimethyl phosphonate could result in high levels of phosphonate being present in plant tissues. The same authors (*67*) recently compared three alkyl-substituted phosphonate compounds (methyl, dimethyl, and diethyl phosphonate), potassium phosphonate, and potassium hypophosphonate for their fungicidal activity against 24 isolates of nine species of *Phytophthora*. They found potassium phosphonate was the most effective compound in vitro. However, the ability to reduce stem rots on peppers and *Persea indica*, caused by *Phytophthora capsici* and *P. citricola*, respectively, for the three alkyl-substituted phosphonate were similar to potassium phosphonate in vivo. Potassium hypophosphonate had low antifungal activity in vitro and was ineffective in controlling Phytophthora diseases in vivo. Quimette and Coffee (*67*) concluded, "While the mode of action of phosphonate remains unknown, phosphonate inhibition of Phytophthora might be explained with a specific biochemical process, as is true with some other systemic fungicides such as metalaxyl." In contrast with benomyl, which is rapidly degraded to carbendazin in an aqueous solution and further degraded by microbes in the soil, fosetyl aluminum (a monoethyl phosphonate) is apparently converted to the active compound in the plant.

Pesticidal Activity of Insecticides on Fungi

Backman and Hammond (68) and Hammond, et. al. (69) demonstrated that the insecticide, chlorpyrifos, suppressed the fungal pathogen, *Sclerotium rolfsii* in peanuts. They also reported a synergism for fungicidal activity between chlorpyrifos and the inert ingredients in the emulsifiable concentrate formulation. However, the emulsifiable concentrate formulation was no longer approved for use on peanuts and the granular formulation did not contain the same inert ingredients. Csinos (70) studied the effect of the parent compound, chlorpyrifos, two degradation products, a stabilizer, and a clay carrier on *Sclerotium rolfsii*. Like previous researchers, Csinos found chlorpyrifos suppressed growth of the *Sclerotium rolfsii* in vitro. However, one of the degradation products, 3,5,6-trichloro-2-pyridinol, was more effective than the parent compound in suppressing radial growth, reducing sclerotia formation, and reducing sclerotia germination in vitro. Neither the inert ingredients or the clay carrier were fungicidal in this study. Csinos speculated that the inert ingredients in the emulsifiable concentrate formulation may have "acted as a catalyst for the hydrolysis of chlorpyrifos to 3,5,6-trichloro-2-pyridinol and thus provide antifungal activity" (70). Therefore the parent compound may be an effective insecticide and have little fungicidal activity, but one or more of the degradation products can have beneficial activity against another pest.

Since 1986, we (Pedersen, W.L. *Plant Dis.*, in press) have worked on the effect of the soil insecticide, chlorpyrifos, on soil-borne pathogens of corn, *Zea mays* L. Initially, we evaluated the parent compound and the breakdown product, 3,5,6-trichloro-2-pyridinol, against several plant pathogens in vitro. Both compounds reduced radial growth of *Fusarium graminarium, Diplodia maydis, Colletotrichum graminicola*, and to a lesser extent *Fusarium moniliforme*. Neither chlorpyrifos nor 3,5,6-trichloro-2-pyridinol affected the growth of several species of *Pythium*, or several bacterial pathogens. However, the control of a pathogen in vitro does not always transfer to the control of a pathogen in vivo. Therefore, we evaluated the granular formulation of chlorpyrifos to control soil-borne pathogens in field experiments. At planting, we inoculated the plots by adding oat seeds infested with *Fusarium graminarium, Diplodia maydis, Colletotrichum graminicola*, and *Fusarium moniliforme* with the corn kernels.

The plots were then treated with or without the granular formulation of chlorpyrifos. We evaluated plant growth parameters, and grain yields for three corn hybrids. All studies were done on fields previously cropped to soybeans to reduce the effect of insect control on corn planted in fields previously cropped to corn. We observed increases in both root mass, leaf area, and grain yield for one of the corn hybrids, while the other two hybrids had inconsistent results. Additional research involving fifty corn hybrids at several locations demonstrated a definite relationship between certain corn hybrids and chlorpyrifos. That is, some hybrids consistently have an increase in yield when treated with chlorpyrifos, while others apparently never or rarely are affected by the application of chlorpyrifos. We also have evaluated forty corn inbreds and found similar results. In summary, it appears that corn hybrids that are susceptible to root pathogens or have rather poor root systems, tend to respond to the application of chlorpyrifos, while inbreds and hybrids with well developed root systems do not respond. Based on these results, we felt that chlorpyrifos or 3,5,6-trichloro-2-pyridinol was acting as a fungicide under field conditions.

Herbicidal Activity of Herbicide Degradation Products

Herbicides are applied to agricultural crops either before or after emergence of the target species. When applied before emergence, many herbicides are incorporated into the soil with a tillage implement. Efficacy of these herbicides depends on a sufficiently long persistence following application and during the early part of the growing season. Many factors affect herbicide degradation including: soil type, soil moisture, tillage practice, previous history of herbicide application, and microbial populations in the soil (71-75). Like fungicides, very few studies have demonstrated that herbicidal degradation products have pesticidal activity against weeds.

Thiocarbamate herbicides are rapidly degraded by soil microbes (72). Casida, et al. showed the herbicidal activity of six thiocarbamates actually increased after conversion to thiocarbamate sulfoxides. For example, the level of control of three broadleaf weeds by butylate (s-ethyl diisobutyl thiocarbamate) increased from 23% to 68% after conversion to butylate sulfoxide. Similarly, sulfoxidation of S-ethyl di-N,N-propylthiocarbamate (EPTC) also increased the control of three broadleaf weeds from 58% to 90% Neither of the two thiocarbamate sulfoxide forms injured corn plants. The conversion from thiocarbamates to thiocarbamate sulfoxide has been shown to occur in corn plants (76) for EPTC and in soil for butylate (Thomas, M.V. and Gray, R.A., unpublished, cited in 76).

Tuxhorn et al. (77) also studied the persistence and activity of the thiocarbamate herbicide, butylate, in soils previously untreated or treated with butylate, EPTC, or vernolate (s-propyl dipropylthiocarbamate). They found butylate degraded faster in soils previously treated with butylate than with EPTC, vernolate, or no herbicide. They also found higher herbicidal activity from a bioassay using oats as an indicator plant than from gas-liquid chromatography measurements. When degradation was compared in autoclaved and nonautoclaved soils, degradation was faster in the nonautoclaved soils, but herbicidal activity also was higher in the nonautoclaved soils. The authors concluded, "These results suggest that microbial action increased the herbicidal activity of butylate and that an active metabolite was formed during butylate degradation. Butylate sulfoxide is a likely candidate." They also concluded, "Further research is necessary to establish if this sulfoxide is present in the soil, and the persistence of butylate sulfoxide compared to butylate in soils where enhanced degradation occurs."

Acknowledgments This publication was supported in part by the Illinois Natural History Survey and the Illinois Agricultural Experiment Station, College of Agriculture, University of Illinois.

Literature Cited
1. Fahey, J. E.; Rusk, H. W. *J. Econ. Entomol.*, **1939**, 32, 319-322.
2. Pearce, G. W.; Avens, A. W. *J. Econ. Entomol.*, **1938**, 31, 594-7.
3. Pearce, G. W.; Avens, A. W. *J. Econ. Entomol.*, **1940**, 33, 918-920.
4. Spencer, E. Y.; O'Brien, R. D. *Ann. Rev. Entomol.*, **1957**, 2, 261-278.
5. O'Brien, R. D. *J. Econ. Entomol.*, **1957**, 50, 159-164.
6. Diggle, W. M.; Gage, J. C. *Nature*, **1951**, 168, 998.
7. Gage, J. C. *Biochem. J.*, **1953**, 54, 426-30.
8. Metcalf, R. L.; March, R. B. *Ann. Entomol. Soc. Amer.*, **1953**, 46, 63-74.
9. Casida, J. E.; Chapman, R. K., Stahmann, M. A., Allen, T. C. *J. Econ. Entomol.*, **1954**, 47:64-71.
10. March, R. B.; Metcalf, R. L., Fukuto, T. R., Maxon, M. G. *J. Econ. Entomol.*, **1955**, 48, 355-363.
11. Fukuto, T. R.; Metcalf, R. L., March, R. B., Maxon, M. G. *J. Econ. Entomol.*, **1955**, 48, 347-354.
12. Metcalf, R. L.; T. R. Fukuto, March, R. B. *J. Econ. Entomol.*, **1957**, 50, 435-439.
13. Bowman, J. S.; Casida, J. E. *J. Econ. Entomol.*, **1958**, 51, 838-843.
14. Glasser, R. F.; Blenk, R. G., Dewey, J. E., Hilton, B. D., Weiden, M. H. J. *J. Econ. Entomol.*, **1958**, 51, 337-341.
15. Edwards, C. A.; Beck, S. D., Lichtenstein, E. P. *J. Econ. Entomol.*, **1957**, 50, 622-626.

16. Gannon, N.; Decker, G.C. *J. Econ. Entomol.*, **1958**, 51, 8-11.
17. Gannon, N.; Decker, G. C. *J. Econ. Entomol.*, **1958**, 51, 3-7.
18. Harris, C. R.; Manson, G. F., Mazurek, J. H. *J. Econ. Entomol.*, **1962**, 55, 777-780.
19. Harris, C. R.; Sans, W. W. *J. Econ. Entomol.*, **1972**, 65, 336-341.
20. Casida, J. E.; Ahmed, M. K. *J. Econ. Entomol.*, **1959**, 52, 111-116.
21. Spencer, W. F.; Adams, J D., Hess, R. E., Shoup, J. D., and Spear, R. C. *J. Agric. Food Chem.*, **1980**, 28, 366-71.
22. Bull, D. L. *J. Agric. Food Chem.*, **1979**, 27, 268-72.
23. Suksayretrup, P.; Plapp, F. W., Jr. *J. Agric. Food Chem.*, **1977**, 25, 481-485
24. Questel, D. D.; Connin, R. V. *J. Econ. Entomol.*, **1947**, 40, 914-5.
25. Starnes, O. *J. Econ. Entomol.*, **1950**, 43, 338-342.
26. Terriere, L. C.; Ingalsbe, D. W. *J. Econ. Entomol.*, **1953**, 46, 751-3.
27. Metcalf, R. L.; March, R. B., Fukuto, T. R., Maxon, M. *J. Econ. Entomol.*, **1954**, 47, 1055.
28. Fukuto, T. R.; Wolf, J. P. III, Metcalf, R. L., March. R. B. *J. Econ. Entomol.*, **1956**, 49, 147-151.
29. Fukuto, T. R.; Wolf, J. Pl. III, Metcalf, March, R. B. *J. Econ. Entomol.*, **1957**, 50, 399-.
30. Metcalf, R. L.; March, R. B., Fukuto, T. R., Maxon, M. G. *J. Econ. Entomol.*, **1955**, 48, 364-69.
31. Metcalf, R. L.; Fukuto, T. R., March, R. B., Stafford, E. M. *J. Econ. Entomol.*, **1956**, 49, 738-741.
32. Metcalf, R. L.; Reynolds, H. T., Winton, M., Fukuto, T. R. *J. Econ. Entomol.*, **1959**, 52, 435-439.
33. Suett, D. *Proc. 1977 British Crop Protection Conference--Pests and Diseases*, **1977**, pp 427-434.
34. Ho, S. H.; Galley, D. *J. Pestic. Sci.*, **1982**, 13, 183-188.
35. Coppedge, J. R.; Lindquist, D. A., Bull, D. L., Dorough, H. W. *J. Agric. Food Chem.*, **1967**, 15, 902-910.
36. Sites, R. W.; Cone, W. W. *J. Econ. Entomol.*, **1989**, 82, 1237-1244.
37. Aharonson, N.; Neubauer, I., Ishaaya, I., Raccah, B. *J. Agric. Food Chem.*, **1979**, 27, 265-268.
38. Gannon, N.; Bigger, J. H. *J. Econ. Entomol.*, **1958**, 51, 1-2.
39. Lichtenstein, E. P.; Schulz, K. R. *J. Econ. Entomol.*, **1959**, 52, 118-123.
40. Harris, C. R. *J. Econ. Entomol.*, **1969**, 62, 1437-1441.
41. Chiang, H. C.; Raros, R. S. *J. Econ. Entomol.*, **1968**, 61, 1204-1208.
42. Ball, H. J. *J. Environ. Sci. Health*, **1983**, B18, 735-744.
43. Laveglia, J.; Dahm, P. A. *Ann. Rev. Entomol.*, **1977**, 22, 483-513.
44. Walter-Echols, G.; Lichtenstein, E. P. *J. Econ. Entomol.*, **1977**, 70, 505-509.
45. Getzin, L. W.; Chapman, R. K. *J. Econ. Entomol.*, **1960**, 53, 47-51.

46. Chapman, R. A.; Tu, C. M., Harris, C. R., Dubois, D. *J. Econ. Entomol.*, **1982**, 75, 955-960.
47. Harris, C. R.; Chapman, R. A. *Can. Entomologist*, **1980**, 112, 641-653.
48. Chapman, R. A.; Harris, C. R. *J. Econ. Entomol.*, **1980**, 73, 536-543.
49. Felsot, A. *Proc 37th Ill. Custom Spray Operators Training School*, **1985**, pp 134-138.
50. Hsin, C.-Y.; Coats, J. R. (Abstr.) *No. Cent. Br. Entomol. Soc. Am.*, **1982**, no. 200.
51. Dewey, J. E.; Parker, B. L. *J. Econ. Entomol.*, **1965**, 58, 491-497.
52. Hague, N. G. M.; Pain, B. F. *Pestic. Sci.*, **1973**, 4, 459-465.
53. Batterby, S. *Nematologica*, **1979**, 25, 377-384.
54. Nelmes, A. J. *J. Nematology*, **1970**, 2, 223-227.
55. Batterby, S.; Le Patourel, G. N. J., Wright, D. J. *Ann. Appl. Biol.*, **1980**, 95, 105-113.
56. Felsot, A.; Dahm, P. A. *J. Agric. Food Chem.*, **1979**, 27, 557-563.
57. Chiou, C. T.; Peters, L J., Freed, V. H. *Science*, **1979**, 206, 831-832.
58. Felsot, A. S.; Lew, A. *J. Econ. Entomol.*, **1989**, 82, 389-395.
59. Harris, C. R. *J. Econ. Entomol.*, **1966**, 59, 1221-1225.
60. Harris, C. R.; Bowman, B. T. *J. Econ. Entomol.*, **1981**, 74, 210-212.
61. Clemons, G.P.; Sisler, H.D. *Phytopathology*, **1969**, 59, 706-706.
62. Helwig, A. *Soil Biol. Biochem.*, **1972**. 4, 377-378.
63. Yarden, O.; Katan, J., Aharonson, N., Ben-Yephet, Y. *Phytopathology*, **1985**, 75, 763-767.
64. Helwig, A. *Pestic Sci.*, **1977**, 8, 71-78.
65. Baude, F.J.; Pease, H.L., Holt, R.F. *J. Agric. Food Chem.* **1974**, 22, 413-418.
66. Quimette, D.G.; Coffee, M.D. *Phytopathology*, **1988**, 78, 1150-1155.
67. Quimette, D.G.; Coffee, M.D. *Phytopathology*, **1989**, 79, 761-767.
68. Backman, P.A.; Hammond, J.M. *Peanut Sci.*, **1981**, 8, 129-130.
69. Hammond, J.M.; Backman, P.A., Bass, M.H. *Proc. J. Am. Peanut Res. Educ. Assoc.*, **1979**. 11, 44.
70. Csinos, A.S. *Plant Dis.*, **1985**, 69, 254-256.
71. Mills, J.A.; Witt, W.M. *Weed Sci.* **1989**. 37, 353-359.
72. Tal, A.; Rubin, B., Katan, J., Aharonson, N. *Pestic Sci.* **1989**. 25, 343-353.
73. Choi, J.S.; Fermanian, T.W., Wehner, D.J., Spomer, L.A. *Agron. J.* **1988**. 80: 108-113.
74. Marty, J.L.; Khafif, T., Vega, D., Bastide, J. *Soil Biol. Biochem.* **1986**. 18, 649-653.
75. Gray, R.A.; Joo, G.K. *Weed Sci.* **1985**. 33, 698-702.
76. Casida, J.E.; Gray, R.A., Tilles, H. *Science.* **1974**. 184, 573-574.
77. Tuxhorn, G.L.; Roeth, F.W., Martin, A.R., Wilson, R.G. *Weed Sci.* **1986**. 34, 961-965.

RECEIVED December 6, 1990

Chapter 14

Phytotoxicity of Pesticide Degradation Products

Stephen O. Duke, Thomas B. Moorman, and Charles T. Bryson

Southern Weed Science Laboratory, Agricultural Research Service,
U.S. Department of Agriculture, P.O. Box 350,
Stoneville, MS 38776

Pesticide metabolites and degradation residues can accumulate in soils and plants. Relatively little is known of the potential phytotoxicity of these compounds. High levels of certain pesticide metabolites, particularly herbicides, are phytotoxic. For instance, $\underline{N}$-methyl-$\underline{N}$'-[3-(trifluoromethyl)phenyl]urea, a metabolite of fluometuron, is a weak photosynthesis inhibitor. Crops are not likely to encounter sufficient concentrations of a single weakly phytotoxic pesticide degradation product to cause symptoms in most field situations. Some scientists have speculated that accumulated herbicide degradation products in soils have contributed to a less than anticipated growth of agricultural productivity in cotton and some other crops. However, there is little scientific evidence to validate this and some data indicate that this is not the case. For instance, levels of trifluralin metabolites in soil equivalent to many years of accumulation from high trifluralin use rates have been shown to have no effect on yield of cotton or soybeans. Similar results were obtained with the metabolites of diuron and fluometuron on cotton. In crops genetically engineered to be highly resistant to certain herbicides, metabolite accumulation at high herbicide application rates could be sufficient to cause phytotoxicity in some cases. In the field, there are generally a large number of different degradation products of many different pesticides in the soil. Little is known of possible interactions of these combinations of metabolites on plant health.

The topic of herbicide metabolite phytotoxicity has not been reviewed before and little literature exists on this topic. Much of the literature that exists is "hidden" in papers that emphasize other aspects of herbicides and their degradation products. A non-refereed compilation of the phytotoxicity of thousands of compounds, including some pesticide metabolites is available (1). The pesticide structure-activity and degradation data of many companies is a rich, but usually unavailable, source of this type of information. In this short review we have attempted to provide sufficient coverage of the information that is available to allow the reader to appreciate our limited understanding and knowledge gaps of this topic.

Some herbicides are applied to the plant in a form that is herbicidally inactive at the molecular site of action, but are metabolized to the active form by the target plant or soil microorganisms. Herbicides that must be activated by metabolic activity are termed proherbicides. For instance, the phenoxyalkanoate and aryloxyphenoxy alkanoate herbicides are generally applied as esters that must be hydrolyzed by the plant or microorganisms in order to become active (2). In the case of the aryloxyphenoxy alkanoates, only one of the two enantiomeric forms is herbicidally active. However, in preemergence treatments, soil microorganisms convert the inactive S enantiomer to the active R enantiomer, so that a mixture of the two forms is as herbicidally active as the exclusive R enantiomer (3)(Figure 1). Whether a proherbicide is phytotoxic through design or chance, its metabolically activated form should be viewed as a herbicide rather than as a degradation product. There may be cases of herbicides that have not been identified as proherbicides because the activity of the metabolites have not been examined in an appropriate *in vitro* bioassay. Proherbicides can have several potential advantages over their herbicidal metabolites, including, superior penetration and stability, selectivity based on metabolism to the active form, better translocation, and/or prolonged release of the active form. The active forms of proherbicides have been discussed elsewhere (4-6) and will not be considered in this review.

The phytotoxicity of known metabolites is discussed first, followed by a review and discussion of the possible role of phytotoxic pesticide degradation products on crop yield. The role of pesticide degradation products in the loss of expected yields that have been observed in some crops in the recent past has been a matter of considerable debate, but of little experimentation. Finally, the potential importance of phytotoxicity of herbicide metabolites in crops genetically engineered to be herbicide resistant is discussed.

Phytotoxicity of Specific Pesticide Metabolites

Several factors should be considered in a survey of the phytotoxicity of pesticide metabolites. For instance, the selectivity of many herbicides, such as the sulfonylureas (7), is based on differential metabolism of the herbicide to herbicidally inactive compounds. In these cases, it is highly unlikely that there would be any significant accumulation of phytotoxic metabolites from metabolism of the herbicide by the crop with which it is used. In target species, the herbicide is either not inactivated by metabolic activity or is metabolized to phytotoxic residues. Metabolites may vary between plant species. Just as herbicides can be selective between plant species, metabolites can differ in their phytotoxicity patterns. Microbial and plant metabolism of herbicides can be very different, so that a phytotoxic microbial degradation product might accumulate in soil that would not be produced in the plant. Few studies of the phytotoxicity of pesticide metabolites have considered these contingencies.

Most of the literature describes experiments in which herbicide metabolites of the plant are tested for the same type of herbicidal activity as that possessed by the parent compound. Metabolites might have very different sites of action and/or different species selectivity than the parent compound. Distinct microbial metabolites have seldom been tested for phytotoxicity. Examples of phytotoxic herbicide metabolites are discussed below.

Amitrole (1H-1,2,4-triazol-3-amine) is metabolized to 3-(3-amino-s-triazole-1-yl)-2-aminoproionic acid (3-ATAL) by plants (8). This product is phytotoxic, but much less phytotoxic than the parent compound. Uncharacterized metabolites of amitrole have been reported to be more phytotoxic than the parent compounds (*9, 10*).

Bromoxynil (3,5-dibromo-4-hydroxybenzonitrile) can be converted to 3,5-dibromo-4-hydroxybenzoic acid by a microbial nitrolase. We found this metabolite to be completely ineffective as a photosynthetic inhibitor (unpublished data) and as a growth inhibitor in cotton (see section on herbicide-resistant crops). Interestingly, the 2,6-dibromophenol derivative is a potent growth regulator with auxin-like activity that is probably toxic at sufficiently high concentrations (*5*). Although this compound is not a known metabolite in plants, decarboxylation of the benzoate derivative of bromoxynil is considered a likely route of further degradation in soil (*11*). Decarboxylation of benzoates by soil microbes is very common.

As mentioned above, the basis for selectivity of sufonylurea herbicides is differential metabolism. The hydroxylated metabolites of chlorsulfuron {2-chloro-N-[[(4-methoxy-6-methyl-1,3,5-triazin-2-yl)amino]carbonyl]benzenesulfonamide} formed by plants have reported to be nonphytotoxic (*7*). It is likely that any plant-derived metabolites of sulfonylurea herbicides that accumulate to significant levels are non-phytotoxic because selectivity could not be based on metabolism if phytotoxic metabolites accumulated. Phytotoxic intermediates might only accumulate to insignificant levels in non-target plants.

Three metabolites of diclobenil (2,6-dichlorobenzonitrile) are phytotoxic (*12*). One of the metabolites (2,6-dichlorobenzoic acid) has strong auxin-like activity and the other two (3 and 4-hydroxy-2,6-dichlorobenzonitrile) have stong contact (rapid foliar desiccation) activity. The selectivities of the parent compound, the benzoate, and the hydroxydiclobenils were all different. All three of these metabolites were determined to be as herbicidally active as the parent compound, but only the hydroxydichlobenils had the same mode of action as the parent compound. Chlorosis of the leaf margins of apple trees was attributed to 2,6-dichlorobenzamide, a product of degradation of dichlobenil by soil microbes (*13*). This is an excellent example of the importance of examining metabolites for different types of activity and different selectivity than the parent compound.

Fluometuron {N,N-dimethyl-N'-[3-(trifluoromethyl)phenyl]urea} inhibits photosynthesis by inhibiting photosystem II (PSII). This effect can be rapidly detected by measuring variable fluorescence increases in plant tissues in which PSII is inhibited. The metabolic degradation pathway of fluometuron is shown in Figure 2. We and others have found that N-methyl-N'-[3-(trifluoromethyl)phenyl]-urea (DMFM), the first metabolite of fluometuron , is a weak photosynthesis inhibitor (Figure 3) (*15, 16*). In a leaf disc assay, Rubin and Eshel (*15*) found it to be about two- to five-fold less active than fluometuron as a photosynthesis inhibitor in cotton and redroot pigweed (*Amaranthus retroflexus* L.). However, it was almost as phytotoxic as the parent compound to whole plants (cotton and pigweed) when treated by soil incorporation of the compounds. Pigweed was much more sensitive to both compounds than was cotton. No other principle metabolites of fluometuron had a measureable effect on photosynthesis, however, two metabolites, TFMPU [3-(trifluoromethyl)phenylurea] and TFMA [3-(trifluoromethyl)-aniline], were weakly phytotoxic in both soil and nutrient solution to pigweed and foxtail and to cotton only in nutrient solution. No data have been generated to indicate that DMFM or any other fluometuron metabolites cause phytotoxicity in the field to cotton, a species that is tolerant to fluometuron (*17*). However, no data are available on the potential for phytotoxicity of fluometuron metabolites to species that are not tolerant to the parent compound. DMFM is the principle metabolite of fluometuron in both cotton and cucumber, accumulating to higher levels than the parent compound within 12 and 96 h of application, respectively

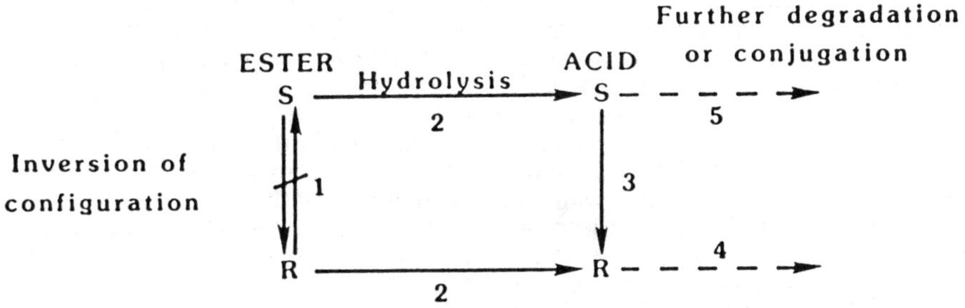

Figure 1. Transformation of enantiomers of fluazifop in different environments. (1) Absent in soil and plant; (2) Rapid in soil and plant; (3) Rapid in soil; absent or very slow in plant; (4) Predominant in soil; significant in plant; (5) Limited in soil; significant in plant. (Adapted from ref. 3; reproduced with permission from ref. 2. Copyright 1988 Marcel Dekker)

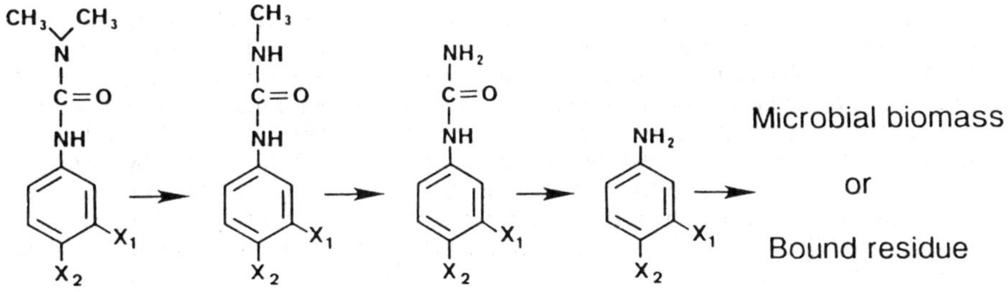

Figure 2. Degradation pathways of the substituted urea herbicides (I) fluometuron (X_1 = CF_3, X_2 = H) or diuron (X_1 = Cl, X_2 = Cl). Metabolites of fluometuron are DMFM (II), TFMPU (III), and TFMA (IV). Diuron metabolites are DCPMU (II), DCPU (III), DCA (IV).

(18). Removal of the second methyl group of fluometuron completely eliminates any effect on photosynthesis (Figure 2) (16) or whole plant growth (19). Metabolites of other PS II inhibitors have been shown to have weak activity. For instance, Matsunaka (20) found 3,4-dichloroaniline (DCA), a plant derivative of propanil [N-3,4-dichlorophenyl)propamide], to be three orders of magnitude less effective as a photosynthesis inhibitor than propanil. DCA is also a metabolite of substituted ureas such as diuron [N'-(3,4-dichlorophenyl)-N,N-dimethylurea].

Although there is no good evidence that the non-selective herbicide glyphosate [N-(phosphonomethyl)glycine] is significantly metabolized by plants, it is metabolized to sarcosine, aminomethylphosphonic acid (AMPA), N-methylphosphinic acid, glycine, N,N-dimethylphosphinic acid, and hydroxymethylphosphonic acid by microorganisms (21). Only AMPA has been reported to significantly inhibit plant growth (22, 23). However, it was a much weaker growth inhibitor than glyphosate in both studies and there was no indication that its phytotoxicity was evoked by inhibition of the shikimate pathway (the mechanism of action of glyphosate). In fact, AMPA stimulated synthesis of anthocyanin, a shikimate pathway product (22). Since glyphosate can be completely metabolized and used as a sole C or P source by soil microbes (24), there is little possibility that AMPA causes any phytotoxicity in field situations.

Methazole [2-(3,4-dichlorophenyl)-4-methyl-1,2,4-oxadiazolidine-3,5-dione] is a proherbicide that is metabolized to the potent PS II inhibitor 1-(3,4-dichloro-phenyl)-3-methylurea (DCPMU) (4, 5, 25). DCPMU is also a metabolite of diuron [N'-(3,4-dichlorophenyl)-N,N-dimethylurea] (Figure 1) and is 10 to 50 % as phytotoxic as diuron (26). Another metabolite, 1-(3,4-dichlorophenyl)urea (DCPU), is about a twentieth as phytotoxic as DCPMU or methazole (27) and 0 to 12 % as phytotoxic as diuron (26).

Hydroxylation of s-triazines to form 2-hydroxy derivatives results in metabolites with no herbicidal activity (28). Monodealkylation of atrazine [6-chloro-N-ethyl-N'-(1-methylethyl)-1,3,5-triazine-2,4-diamine] (formerly 2-chloro-4-(ethylamino)-6-(isoprophylamine)-s-triazine) liberates 2-chloro-4-amino-6-isopropylamino-s-triazine, a compound with only slightly less phytotoxicity than atrazine (29). Monodealkylated triazines generally do not accumulate and they are further degraded to inactive hydroxylated or didealkylated and didealkylated hydroxylated derivatives. However, in some species, such as peas, the monodealkylated product accumulates sufficiently to contribute to the phytotoxicity of atrazine (29). A major nonconjugated soybean metabolite of the as-triazinone herbicide metribuzin [4-amino-6-(1,1-dimethylethyl)-3-(methylthio)-1,2,4-triazin-5(4H)-one], DADK (the deaminated diketo derivative of metribuzin), is not phytotoxic (30).

Trifluralin [2,6-dinitro-N,N'-dipropyl-4-(trifluoromethyl)benzenamine] inhibits growth by inhibiting mitosis at the microtubule level (31). Two metabolites of trifluralin, TR-35M [3-methoxy-2,6-dinitro-N,N'-dipropyl-4-(tri-fluoromethyl)benzenamine] and TR-40 {N-[2,6-dinitro-4-(trifluoromethyl)phenyl]-N-propylpropanamide}, inhibit very weakly (about 1000-fold less than trifluralin) mitosis of goosegrass [Eleucine indica (L.)], a trifluralin-sensitive species (32) (Figure 4). Ten other metabolites had no detectable effect. A trifluralin-resistant biotype of goosegrass was also unaffected by TR-35M and TR-40.

Insecticides can both cause phytotoxicity and be metabolized by plants (33). Whether the insecticide or a metabolic product is responsible for the phytotoxicity has not been determined in most cases. Methyl parathion [O,O-dimethyl-O-(4-nitrophenyl) thiophosphate] is an exception. At high rates methyl and ethyl parathion are phytotoxic to some species, and as early as 1950 the 4-nitrophenol degradation product was suggested as the actual phytotoxic agent

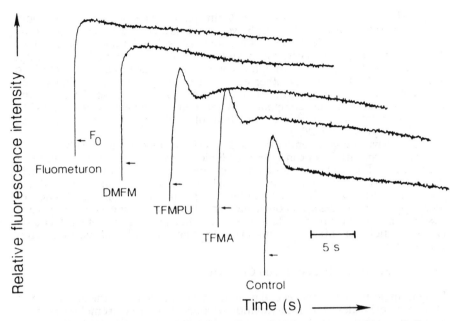

Figure 3. Effects of 10 μM fluometuron and several of its metabolites on variable chlorophyll fluorescence in cotton leaf discs incubated for 2 h on the test solution. Assay procedures were as previously described (*14*).

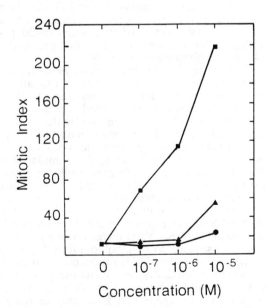

Figure 4. Influence of a 24-h treatment with trifluralin (squares), TR-36M (triangles), or TR-40 (circles) on the mitotic index (cells in division per 1000 cells) of goosegrass root meristem cells (*32*).

(*34*). Methyl parathion degrades to 4-nitrophenol after being sprayed onto plant foliage of cotton and lettuce; however, it is not clear whether degradation occurs on or in the leaf (*35*). Oxidative phosphorylation is inhibited by 4-nitrophenol (*36*). Similarly, the phytotoxicity of fenitrothion [O,O-dimethyl O-(3-methyl-4-nitrophenyl) thiophosphate] to cabbage was attributed to 4-nitrocresol (*37*). Whether other effects of methylparathion on plants, such as inhibition of lipid synthesis (*38*), are due to metabolites of the insecticide or to the parent compound has not been determined. Some insecticides or their metabolites inhibit degradation of herbicides, thereby increasing herbicide phytotoxicity (*20, 39*).

The above examples demonstrate that pesticide metabolites can be phytotoxic by several mechanisms. They can synergize herbicides by interference with herbicide degradation. They can have entirely different mechanisms of action and selectivity than their parent compounds. Soil metabolites can also have different modes of entry into the plant than the parent compound (*e.g.*, soil metabolites of foliar-applied herbicides). Although there are many possible mechanisms by which pesticide metabolites might affect plants, whether they significantly affect crop yields in field situations is unknown or unclear with all pesticide-crop combinations.

Effects of Pesticide Metabolites on Crop Yield

Long-term pesticide use has several potential hazards, including the accumulation of parent herbicides and/or degradation products that are extremely persistent. Pesticide metabolites can be produced in soil by microbial or chemical processes or may be initially formed in plants and later enter the soil in plant residues. The effects of these metabolites, regardless of their source, may be direct or indirect. Metabolites can affect vital plant processes directly (*e.g.*, photosynthesis) or indirectly (*e.g.*, by altering host plant resistance to a pathogen or insect or by affecting the composition of soil microflora). Long-term pesticide use has the potential effect of altering microbial communities in the soil that drive the cycling of plant nutrients, which could result in a crop response. Aside from quantitative effects of degradation products on yields, these compounds may also have effects on food quality.

Assessing the effects of degradation products on yields requires quantitative knowledge of the crop response to a particular dose of toxicant and the levels of metabolites available to the crop under field conditions. Evidence of pesticide metabolite effects on crop yields have been obtained from experiments comparing yield on metabolite-treated soil to that on nontreated soil. Other, more circumstantial evidence can be gathered from experiments that compare the yields of pesticide-treated crops to that of crops grown without pesticides. These types of experiments are advantageous in determining the cumulative effects of both pesticides and metabolites under realistic farm management conditions.

Comparisons of pesticide-treated and untreated crops can be confounded by other factors that affect crop yield, including pest pressure, climatic effects, or other agronomic factors that contribute to yield. The adverse effects of pesticides and their metabolites on crop yields can only be assessed in the absence of losses due to uncontrolled pests. Yield increases due to pest control can be greater than potential losses due to metabolite-induced damage. It is often difficult to separate the effects of degradation products from the parent compound because of their potential similarity in mechanisms of action and their simultaneous presence in the soil. Also, many pesticides (particularly herbicides) have rate-dependent effects on crops that have only marginal tolerance, or rotational crops that may have no tolerance to a pesticide used on a previous crop.

Pesticide degradation products are highly variable in their concentration and persistence in soils. Pesticides that have considerable structural similarity may have similar or common metabolites produced in soils. For example, the weak photosynthesis inhibitor 3,4-dichloroaniline is common metabolite of the photosynthesis-inhibiting herbicides diuron, linuron [N'-(3,4-dichlorophenyl)-N-methoxy-N-methylurea], propanil [N-(3,4-dichlorophenyl)propanamide], and swep (methyl 3,4-dichlorocarbanilate). Many reports indicate that, as pesticides degrade in soil, only small amounts of metabolites appear. If it is assumed that a degradation product has a half-life of twice that of the parent pesticide, the maximum metabolite concentration in soil from a single application would be 50% of the applied amount (*40*).

The available evidence for many different classes of herbicides supports the view that that metabolites generally do not accumulate. For instance, after 20 years of continuous atrazine use, metabolite concentrations ranged from 14 to 296 μg/kg in soil (*41*). Concentrations in the roots and shoots of oats grown in this soil ranged from 73 to 116 μg/kg, but oat growth was not different from growth in soil without atrazine use.

In addition to the formation of extractable metabolites, studies with ^{14}C-labeled pesticides indicate that substantial amounts of the pesticide and/or metabolites are converted to bound residues. Typically, between 20 and 70% of the applied pesticide may become bound residue (*42*). Bound residues include pesticides and degradation products that have become covalently bonded to soil organic matter or compounds that have diffused into the organic or mineral matrix sufficiently to become unextractable by conventional solvent extraction methods. This process occurs fairly quickly and concurrently with the pesticide degradation process.

The unextractable nature of bound residues suggests low bioavailability. However, studies with triazine herbicides showed that strong extraction procedures recovered 10 to 15% of parent pesticide from the bound residue (*43, 44*). In addition to prometryn [N,N'-bis(1-methylethyl)-6-(methylthio)-1,3,5-triazine-2,4-diamine] (20% of added), the metabolites hydroxypropazine [2-hydroxy-4,6-bis(isopropylamino)-s-triazine] (7%) and deisopropylpropetryn [2-methylthio)--amino-6-(isopropylamino)-s-triazine] (11%) were extracted with strong base from the bound residue pool (*43*). Smaller amounts of pesticide and metabolites are released from bound forms by natural processes. Bioavailability, defined as residue uptake by plants, is shown in Table I. These residues are bioavailable in quantities generally below 2% of the applied pesticide. The residues are generally measured as bound ^{14}C or ^{14}C in plants and generally represent both uncharacterized metabolites and parent pesticides. Lee *et al.* (*45*) reported that the uncharacterized bentazon residues released from bound forms (see Table I) had no effect on maize growth and that maize took up approximately 40% of freshly added [^{14}C]bentazon [3-(1-methylethyl)-(1H)-2,1,3-benzothiadiazin-4-(3H)-one 2,2,-dioxide]. All available evidence suggests that while bound residues in soil represent a large pool of parent pesticides and metabolites, only a small fraction of the residues are available to plants.

Plant uptake of pesticide metabolites has been examined in experiments using freshly added metabolites instead of bound residues. Barley uptake of [^{14}C]4-chloroaniline after 20 weeks was only 0.3% of the 1.25 mg/kg added to soil (*50*). Ryegrass treated with a glucoside conjugate of hydroxymonolinuron only took up between 0.19 and 0.23% of the added conjugated metabolite [(*phenyl*-^{14}C)-3-(4-chlorophenyl)-1-(hydroxymethyl)-1-methoxyurea-β-D-glucoside] (*51*). The limited plant uptake of these pesticide metabolites may be related to their relatively strong sorption to soils, their rapid transformation in soil, or poor efficiency of root uptake. Nevertheless, these results tend to confirm the results obtained with the less characterized bound residues discussed previously. Chlorinated anilines taken up by plants appear to become bound principally to the lignin fraction (52).

Table I. Plant Uptake of Pesticide Residues Released from Bound Forms by Natural Processes[a]

Pesticide	Bound[b]	Uptake[c]	Plant
	-(% of applied ^{14}C)-		
Bentazon			
German soil	26	2.3 (34)	Maize
Korean soil	44	1.8 (42)	Maize
Parathion	16	<1	Oats
Methabenzthiazuron			
loessal silt soil	48	0.4	Maize
sandy soil	41	1.0	Maize
Simazine	35	0.08 (20)	Maize

[a]From references 45-49.
[b]Bound residues were characterized as ^{14}C remaining after various incubation periods followed by solvent extraction and includes pesticide and metabolites.
[c]Values in parentheses indicate percentage uptake of freshly added [^{14}C] pesticide.

Because of its high value, cotton is generally treated with higher levels of pesticides than most other crops. Circumstantial evidence has suggested that herbicides or their degradation products could be adversely affecting cotton yields. Trifluralin, linuron, and diuron were introduced in the late 1950's for use in cotton and other crops. The continuous use of these and related herbicides led to the suggestion that these compounds were contributing to the plateau in cotton yields that were reported about this time (Figure 5)(53-55). This plateau occurred during a period of increased genetic potential that should have resulted in substantial yield increases (53). Earlier research had established that dinitroaniline herbicides reduced lateral root growth (56, 57) and potentially could affect nutrient uptake (58, 59). Reddy et al. (60) used the GOSSYM model to simulate cotton yields with different amounts of herbicide-induced root pruning. With no direct experimental data on herbicide or herbicide metabolite effects, they concluded that herbicides were contributing to the cotton yield problem.

Experimental evidence on the effects of these herbicides on cotton yields is mixed. Field-grown cotton treated with trifluralin, fluometuron, and MSMA (sodium methylarsonate) at maximum recommended rates yielded as much as 36% less than weed-free cotton without herbicides (61). Fluometuron applied in continuous cotton monoculture was found in concentrations ranging from 0.1 to 0.8 mg/kg soil 11 months after application in three Arkansas soils, indicating that it has potential to carryover from one season to the next (62). In other field studies, high rates (3.36 kg/ha) of fluometuron reduced cotton yields, but

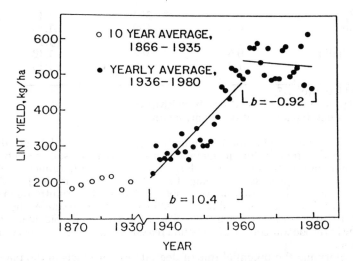

Figure 5. Yield of U.S. cotton from 1866 to 1980. Data from USDA Agricultural Statistic Reporting Service, Crop Reporting Service (*b*= change in lint yield per year). Reproduced with permission from ref. 53. Copyright 1984 Crop Science Society of America, Inc. and American Society of Agronomy, Inc.

trifluralin did not affect yields compared to those obtained from weed-free plots without herbicide (63). In experiments examining long-term trifluralin use, there was no effect on cotton yields (64-66). Even two times the recommended rates of trifluralin or pendimethalin [N-(1-ethylpropyl)-3,4-dimethyl-2,6-dinitrobenzenamine] for 6 years had no significant effect on cotton growth, yields, or fiber quality (Table II).

Table II. Effects of Repeated (1983-88) Dinitroaniline Herbicide Application on Lint Yield and Fiber Quality of Cotton[a]

Treatment	Rate (kg/ha)	Yield (kg/ha)	Lint Value[c] ($/kg)
Control[b]	0	391	1.12
Trifluralin	0.6	439	1.15
	1.2	400	1.14
Pendimethalin	0.6	384	1.12
	1.2	410	1.15

[a]Adapted from ref. 65.
[b]Control plots were kept weed-free by cultivation.
[c]Values reflect impact of cotton quality factors

Dinitroaniline degradation is complex (Figure 6); Golab et al. (67) reported 28 different trifluralin metabolites isolated from soil extracts. The dinitroaniline herbicide trifluralin undergoes N-dealkylation, reduction of nitro substituents, and processes leading to cyclized products (67). Similar patterns of metabolite formation were observed with other dinitroanilines (68, 69). The total solvent-extractable metabolites from dinitroaniline herbicides examined in these studies was less than 10% of the applied herbicide while 20 to 30% of the herbicide was converted to bound residue after 7 to 12 months time since application.

To determine the potential role of degradation products in the plateau of cotton yields in the field, Koskinen et al. (70) examined the effects of 12 trifluralin metabolites on the growth, yield, and fiber quality of cotton in greenhouse experiments. Cotton was grown in soil with metabolites added at rates (0.2 kg/ha), which represented from 14 to > 400 years of accumulation, assuming no metabolite degradation. Seed cotton yields and mean maturity dates were not adversely affected by any of these metabolites. Fiber properties and boll composition were also unaffected, with the exception that one metabolite reduced the lint percentage slightly (71). However, in these greenhouse experiments with well-watered, potted plants, any root pruning effect might not have influenced yield as it might in a drier field environment. Similar concerns over the effects of these metabolites on long-term soybean yields led to similar studies which further established the nonphytotoxic character of these metabolites (72). Four years of trifluralin use with metolachlor [4-amino-6-(1,1-dimethylethyl)-3-(methylthio)-1,2,3,-triazin-5(4H)-one] and fluazifop [(±)-2-[4-[[5-(trifluoromethyl)-2-pyridinyl]-oxy]phenoxy]propanoic acid], along with a standard herbicide treatment regime of dinoseb [2-(1-methylpropyl)-4,6-dinitrophenol], 2,4-DB (2,4-dichlorophenoxy)butanoic acid], and linuron, did not reduce the yield of soybeans below plots with only the standard treatment (73). Trifluralin metabolites are apparently not injurious to soybeans. These conclusions are

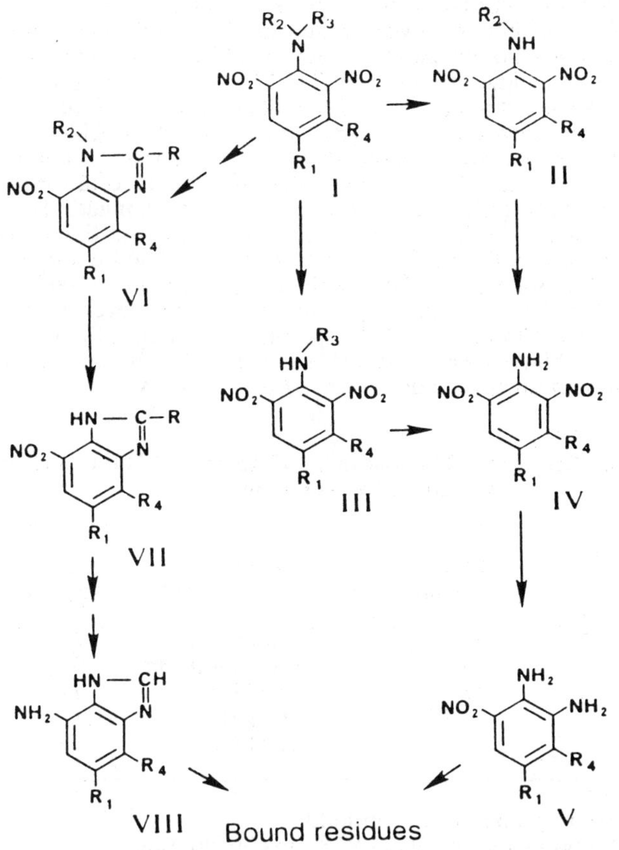

	R_1	R_2	R_3	R_4
Fluchloralin	$-CF_3$	$-CH_2CH_2CH_3$	$-CH_2CH_2Cl$	$-H$
Pendimethalin	$-CH_3$	$-H$	$-CH_2CH_2CH_3$ $\mid$ CH_2CH_3	$-CH_3$
Profluralin	$-CF_3$	$-CH_2CH_2CH_3$	$-CH_2-\triangleleft$	$-H$
Nitralin	$-SO_2CH_3$	$-CH_2CH_2CH_3$	$-CH_2CH_2CH_3$	$-H$
Trifluralin	$-CF_3$	$-CH_2CH_2CH_3$	$-CH_2CH_2CH_3$	$-H$

Figure 6. Principle biotic and abiotic degradation pathways of dinitroaniline herbicides in soils. The most common reactions include dealkylation (See II, III, IV), cyclization (VII), and reduction of the nitro groups (V, VIII).These processes and additional reactions can occur simultaneously or in sequence to produce metabolites not shown here.

consistent with other results showing trifluralin to be 1000 times more active than the major metabolites against a sensitive species, goosegrass (32).

Similar experiments with degradation products of diuron and fluometuron had no effect on cotton growth and yield, with the exception of TFMA (3-[trifluoromethyl] aniline) applied at 1 mg/kg soil (Table III). TFMA reduced seed cotton yield by 11% from that obtained on untreated soil. Potential mechanisms for this effect of TFMA on yield are unknown, but cotton height and the number of blooms were not affected by TFMA. Previous experiments had shown that the toxicity of fluometuron was greater than that of the principal metabolites, but short-term cotton growth was not affected by TFMA (15). However, yield was not examined. The 1 mg/kg soil application rate of TFMA is well above the concentrations that could reasonably be expected in the field. Only 5% of the applied fluometuron was recovered as TFMA from soil during a 72-day incubation, compared with 66% as DMFM and 14% as TFMPU (74). Similar experiments indicate even lower TFMA concentrations in soils 63 days after [^{14}C]fluometuron application (Moorman, unpublished data). Thus, it seems reasonable to conclude that fluometuron and diuron degradation products are not likely to affect cotton yields.

Table III. Effect of Diuron and Fluometuron and Their Metabolites on Cotton Lint Yield in the Greenhouse[a]

Treatment[b] (0.5 mg/kg soil)	Yield (g/plant)	Treatment[b] (1 mg/kg soil)	Yield (g/plant)
Control	27.7	Control	27.2
Diuron	26.1	Fluometuron	30.6[c]
DCPMU	27.9	TFMPU	27.3
DCPU	24.7	DMFM	26.7
DCA	31.0	TFMA	23.9[c]

[a]Adapted from ref. 17.
[b]All treatments were preplant-incorporated.
[c]Significantly different from control at 95 % level of confidence.

No long-term, well-replicated studies of the effects of herbicide use on cotton yield or quality have been documented in which the studies have been carried out on land with no prior history of herbicide use. However, such studies are underway, and preliminary results (after 5 years) have indicated that long term use of trifluralin does not affect cotton yield (66).

In summary, the direct experimental evidence that is available indicates that the principal soil-applied herbicides used for weed control in cotton are not responsible for the yield plateau in cotton. While damage can be induced by higher rates of these herbicides, the accumulation of phytotoxic metabolites affecting cotton growth appears to be most unlikely. The GOSSYM model simulations reported by Reddy et al. (60) appear to greatly overestimate herbicide pruning of roots and the adverse impact of the dinitroaniline herbicides on cotton yield. These simulations were made without input of quantitative data on the effect of herbicides or metabolites on cotton root growth under field conditions. Small losses in yield due to herbicide (not metabolite) damage are probably offset by the increased gain in yield from weed control. There are numerous other potential factors that could influence long-term yield trends, including biological and physical changes in soil due to monoculture, insect management programs, or other farm management practices.

The effects of long-term herbicide use on crops other than cotton have been reported. Fryer et al. (75) examined the effects of MCPA [(4-chloro-2-methyl-phenoxy) acetic acid], simazine (6-chloro-N,N'-diethyl-1,3,5-triazine-2,4-diamine), and linuron on cereals, corn, and carrots, repectively over a 10-year period. Heinonen-Tanski et al. examined the effects of several pesticides on barley yield (76) and carrots (77). Residue levels of the parent herbicides were measured and in most instances were not sufficient to cause accumulation. Linuron was observed to reduce carrot yields in one study (77), but this was correlated with herbicide carryover. Metabolite concentrations were generally 10 % or less of the residual linuron concentration and did not appear to contribute to the phytotoxic response. These studies generally support several conclusions: parameters related to growth or yield (e.g., stand) may be reduced in individual years relative to an untreated crop, but cummulative effects on yield were not observed. Accumulation of phytotoxic residues would result in increasingly large differences between pesticide-treated crop yields and yields of untreated crops. No strong evidence for the existence of metabolites more phytotoxic than the parent herbicide was presented, although the synergistic effects of weakly phytotoxic products with parent herbicides cannot be discounted.

The potential interactions of degradation products with soil microorganisms that affect nutrient cycling, plant pathogens, and other rhizosphere microorganisms have received far less attention. No major effects of trifluralin metabolites were found on microbial decomposition processes in soil (78). However, available evidence indicates that root symbionts, pathogens, and other rhizosphere microorganisms on crops can be affected by pesticides (79, 80). Considering the influence of these microorganisms on crops, potential interactions with pesticides affecting crop growth cannot be completely discounted.

Phytotoxicity in Herbicide-Resistant Crops

Plants can be resistant to herbicides by three processes: 1) rapid metabolism of the herbicide to non-phytotoxic derivatives; 2) insensitivity at the molecular site of action; and 3) sequestration of the herbicide so as to decrease its contact with the molecular target site. Molecular biologists are pursuing the first two approaches to produce crops that are resistant or highly tolerant to herbicides to which they are normally too sensitive for compatibility. Success with either of these approaches will generate crops which can tolerate exposure to relatively high levels of these herbicides and, one assumes, their metabolites. Since most herbicides are metabolized to some extent in plants, high levels of metabolites could accumulate in resistant or tolerant crops that are sprayed with high rates of herbicides, regardless of the mechanism of resistance. Such metabolites could have similar or different mechanisms of phytotoxicity than the parent herbicide.

For instance, as discussed in the first section of this chapter, bromoxynil might be metabolized to plant growth regulator 2,6-dibromophenol by decarboxylation of the 3,5-dibromo-4-hydroxybenzoate derivative of the herbicide. Genetic engineers have recently produced bromoxynil-resistant cotton by inserting the gene for the nitralase from *Klebsiella ozaenae* which generates 3,5-dibromo-4-hydroxy-benzoate (81). If this metabolite were to be efficiently converted to 2,6-dibromophenol in the cotton plant or in soil and if this metabolite accumulated, hormonal-type damage to cotton might result. This is an unlikely route of metabolism in plants since plants are more likely to form glucose ester, amino acid, or lipid conjugates. We have sprayed 4-week-old, greenhouse-grown, bromoxynil-susceptible cotton (two- and four-true leaf stage) with up to ten-fold-higher than normal bromoxynil use rates of 2,6-dibromo-4-hydroxybenzoate and found no hormonal symptoms nor any other phytotoxicity at any time during 2 weeks after treatment. However, field studies should be made because soil microorganisms might be more likely to generate 2,6-dibromophenol from the benzoate than plants.

Summary and Speculations

Although some pesticide metabolites are phytotoxic, only those of proherbicides have been found to be significantly phytotoxic. Little evidence exists to implicate herbicide metabolites directly in crop yield or quality reduction. Most of the data indicate that pesticide metabolites have little or no direct or indirect effect on crop yields. However, the amount of data available on crop yields is limited, and marginal effects of long-term metabolites that could be due to interactions or expressed only under certain environmental conditions would be difficult to prove. In virtually all crop/pesticide combinations, much more experimental data will be necessary before more extensive conclusions can be made.

Literature Cited

1. Creager, R. A.; Wiswesser, W. J. *A Summary of Results on 31,000 Compounds Evaluated for Herbicidal Activity*, U.S.D.A. Tech. Bull. **1987**, *1721*, 657 pp.
2. Duke, S. O. In *Herbicides - Chemistry, Degradation, and Mode of Action*, Kearney, P. C.; Kaufman, D. D., Eds.; Marcel Dekker, Inc.: New York, NY, 1988, Vol. 3; pp. 71-116.
3. Cartwright, D. *Brighton Crop Protect. Conf. - Weeds* 1989, pp. 707-716.
4. Casida, J. E. In *Pesticide Chemistry - Human Welfare and the Environment. Mode of Action, Metabolism and Toxcology*; Miyamoto, J.; Kearney, P. D., Eds; Pergamon Press: Oxford, 1983, Vol. 3; pp. 239-254.
5. Wain, R. L.; Smith, M. S. In *Herbicides - Physiology, Biochemistry, and Ecology*; Audus, L. J., Ed; 2nd Ed.; Academic Press: London, 1976, Vol. 2; pp. 279-302.
6. Kerr, M. W. In *Herbicides and Plant Metabolism*; Dodge, A. D., Ed; Cambridge Univ. Press: Cambridge, 1989; pp. 199-210.
7. Beyer, E. M.; Duffy, M. J.; Hay, J. V.; Schlueter, D. D. In Herbicides - Chemistry, Degradation, and Mode of Action; Kearney, P. C.; Kaufman, D. D., Eds; Marcel Dekker, Inc.: New York, NY, 1988, Vol. 3; pp. 117-189.
8. Carter, M. C. In *Herbicides - Chemistry, Degradation, and Mode of Action*; Kearney, P. C.; Kaufman, D. D., Eds; Marcel Dekker, Inc.: New York, NY, 1976, Vol. 1; pp. 377-398.
9. Ashton, F. M.; Crafts, A. S. *Mode of Action of Herbicides*; 2nd Ed.; 1981, p. 126.
10. Herrett, R. A.; Bagley, W. P. *J. Agric. Food Chem.* **1964**, *12*, 17-20.
11. Frear, D. S. In *Herbicides - Chemistry, Degradation, and Mode of Action;* Kearney, P. C.; Kaufman, D. D., Eds; Marcel Dekker, Inc.: New York, NY, 1976, Vol. 2; pp. 541-607.
12. Verloop, A.; Nimmo, W. B. *Weed Res.* **1969**, *9*, 357-370.
13. Verloop, A.; Daams, J. *Z. PAKranh. PAPath. PASCHUTZ.* **1970**, *5*, 141-146.
14. Vaughn, K. C.; Duke, S. O. *Physiol. Plant.* **1984**, *62*, 510-520.
15. Rubin, B.; Eshel, Y. *Weed Sci.* **1971**, *19*, 592-594.
16. Goren, R. *Weed Res.* **1969**, *9*, 121-135.
17. Moorman, T. B.; Koskinen, W. C. *Bull. Envrion. Contam. Toxicol.* **1990**, *44*, 26-27.

18. Rogers, R. L.; Funderburk, H. H. *J. Agric. Food Chem.* **1968**, *16*, 434-440.
19. Geissbühler, H.; Martin, H.; Voss, G. In *Herbicides - Chemistry, Degradation, and Mode of Action*; Kearney, P. C.; Kaufman, D. D., Eds; Marcel Dekker, Inc.: New York, NY, 1976, Vol. 1; pp. 209-291.
20. Matsunaka, S. *Res. Rev.* **1969**, *25*, 45-58.
21. Duke, S. O. In *Herbicides - Chemistry, Degradation, and Mode of Action*; Kearney, P. C.; Kaufman, D. D., Eds; Marcel Dekker, Inc.: New York, NY, 1988, Vol. 3; pp. 1-70.
22. Hoagland, R. E. *Weed Sci.* **1980**, *28*, 393-400.
23. Lee, T. T.; Dumas, T.; Jevnikar, J. J. *Pestic. Biochem. Physiol.* **1983**, *20*, 354-359.
24. Kishore, G. M.; Jacob, G. S. *J. Biol. Chem.* **1987**, *262*, 12164-12169.
25. Butts, E. G.; Foy, C. L. *Pestic. Biochem. Physiol.* **1974**, *4*, 44-55.
26. Bowmer, K. H.; Adeney, J. A. *Pestic. Sci.* **1978**, *9*, 342-353.
27. Jones, D. W.; Foy, C. L. *Pestic. Biochem. Physiol.* **1972**, *2*, 8-26.
28. Esser, H. O.; Dupuis, G.; Ebert, E.; Vogel, C.; Marco, G. J. In *Herbicides - Chemistry, Degradation, and Mode of Action*; Kearney, P. C.; Kaufman, D. D., Eds; Marcel Dekker, Inc.: New York, NY, 1976, Vol. 1; pp. 129-208.
29. Shimabukuro, R. H. *J. Agric. Food Chem.* **1967**, *15*, 557-562.
30. Frear, D. S.; Swanson, H. R.; Mansager, E. R. *Pestic. Biochem. Physiol.* **1985**, *23*, 56-65.
31. Vaughn, K. C.; Vaughan, M. A. *Amer. Chem. Soc. Symp. Ser.* **1990**, *421*, 364-75.
32. Vaughn, K. C.; Koskinen, W. C. *Weed Sci.* **1987**, *35*, 36-44.
33. Spencer, E. Y. *Res. Rev.* **1965**, *9*, 153-168.
34. Edwards, F. I., Jr.; Smith, F. F. *J. Econ. Entomol.* **1950**, *43*, 471-473.
35. Youngman, R. R.; Toscano, N. C.; Gaston, L. K. *J. Econ. Entomol.* **1989**, *82*, 1317-1322.
36. Kaufmann, D. D. In *Herbicides - Chemistry, Degradation, and Mode of Action*; Kearney, P. C.; Kaufman, D. D., Eds; Marcel Dekker, Inc.: New York, NY, 1976, Vol. 2; pp. 665-707.
37. Tomizawa, C.; Kobayashi, A. *Noyaku Seisan Gijutsu* **1964**, *11*, 30-32.
38. Swamy, G. S.; Veeresh, A. V. *Pestic. Biochem. Physiol.* **1987**, *28*, 341-348.
39. Chang, F.-Y.; Smith, L. W.; Stephenson, G. R. *J. Agric. Food Chem.* **1971**, *19*, 1183-1186.
40. Smith, A. E. In *Environmental Chemistry of Herbicides*; Grover, R., Ed.; CRC Press, Boca Raton, FL, 1988, Vol. 1; pp. 172-200.
41. Khan, S. U.; Saidak, W. J. *Weed Res.* **1981**, *21*, 9-12.
42. Calderbank, A. *Residue Rev.* **1989**, *108*, 71-103.
43. Khan, S. U. *J. Agric. Food Chem.* **1982**, *30*, 175-179.
44. Capriel, P.; Haisch, A.; Khan, S. U. *J. Agric. Food Chem.* **1985**, *33*, 567-569.
45. Lee, J. K.; Führ, F.; Mittelstaedt, W. *Chemosphere* **1988**, *17*, 441-450.
46. Racke, K. D.; Lichtenstein, E. P. *J. Agric. Food Chem.* **1985**, *33*, 938-943.
47. Kloskowski, R.; Fuhr, F. J. *J. Environ. Sci. Health*, **1987**, *B22*, 623-642.
48. Kloskowski, R.; Fuhr, F. *Chemosphere*, **1983**, *12*, 1557-1574.
49. Führ, F.; Mittelstaedt, W. *J. Agric. Food Chem.* **1980**, *28*, 122-125.
50. Freitag, D.; Scheunert, I.; Klein, V.; Korte, F. *J. Agric. Food Chem.* **1984**, *2*, 203-207.

51. Haque, A.; Schuphan, I.; Ebing, W. *Pesticide Sci.* **1982,** *13,* 219-228.
52. Still, G. C.; Balba, H. M.; Mansager, E. R. *J. Agric. Food Chem.* **1981,** *29,* 39-46.
53. Meredith, W. R., Jr., Bridge R. R. In *Genetic Contributions to Yield Gains of Five Major Crop Plants,* Fehr, W. R., Ed.; Am. Soc. Agron., Madison, WI, 1984; pp. 75-87.
54. Starbird, F. R.; Hazera, J. *Cotton and Weed Outlook and Situation,* Econ. Res. Serv., USDA; Washington, D.C. 1982; CW5-32.
55. Hurst, H. R. *Proc. Beltwide Cotton Production-Mechanization Conf.*; National Cotton Council, Memphis, TN, 1977; pp. 61-63.
56. Anderson, W. P.; Richards, A. B.; Whitworth, J. W. *Weeds* **1967,** *15,* 224-227.
57. Oliver, L. R.; Frans, R. E. *Weed Sci.* **1968,** *16,* 199-203.
58. Cathey, G. W.; Sabbe, W. E. *Agron. J.* **1972,** *64,* 254-255.
59. Bucholtz, D. L.; Lavy, T. L. *Agron J.* **1979,** *71,* 24-26.
60. Reddy, V. R.; Baker, D. N.; Whistle, F. D.; Frye, R. E. *Agron. J.* **1987,** *79,* 42-47.
61. Gaylor, M. J.; Buchanan, G. A.; Gilliland, F. R.; Davis, R. L. *Agron. J.* **1983,** *75,* 903-907.
62. Rogers, C. B.; Talbert, R. E.; Mattice, J. D.; Lavy, T. L.; Frans, R. E. *Weed Sci.* **1985,** *34,* 122-130.
63. Hayes, R. M.; Hoskinson, P. E.; Overton, J. R.; Jeffery, L. S. *Weed Sci.* **1981,** *29,* 120-123.
64. Miller, J. H.; Keeley, P. E.; Carter, C. H.; Thullen, R. J. *Weed Sci.* **1975,** *23,* 211-214.
65. Keeling, J. W.; Abernathy, J. R. *Weed Technol.* **1989,** *3,* 527-530.
66. Bryson, C. T.; Webster, L. *Proc. Beltwide Cotton Prod. Res. Conf.* **1988,** *12,* 384.
67. Golab, T.; Althaus, W. A.; Wooten, H. L. *J. Agric. Food Chem.* **1979,** *27,* 163-179.
68. Golab, T.; Bishop, C. E.; Donoho, A. L.; Manthey, J. A.; Zornes, L. L. *Pestic. Biochem. Physiol.* **1975,** *5,* 196-204.
69. Kearney, P. C.; Plimmer, J. R.; Wheeler, W. B.; Kontson, A. *Pestic. Biochem. Physiol.* **1976,** *6,* 229-238.
70. Koskinen, W. C.; Oliver, J. E.; Kearney, P. C.; McWhorter, C. G. *J. Agric. Food Chem.* **1984,** *32,* 1246-1248.
71. Koskinen, W. C.; Leffler, H. R.; Oliver, J. E.; Kearney, P. C.; McWhorter, C. G. *J. Agric. Food Chem.* **1985,** *33,* 958-961.
72. Koskinen, W. C.; Oliver, J. E.; McWhorter, C. G.; Kearney, P. C. *Weed Sci.* **1986,** *34,* 471-473.
73. Heatherly, L. G.; Elmore, C. D. *Weed Technol.* **1991** - In Press.
74. Bozarth, G. A.; Funderburk, H. H. *Weed Sci.* **1971,** *19,* 691-695.
75. Fryer, J. D.; Smith, P. D.; Hance, R. J. *Weed Res.* **1980,** *20,* 103-110.
76. Heinonen-Tanski, H.; Rosenburg, C.; Siltanen, H.; Kilpi, S.; Simojoki, P. *Pestic. Sci.* **1985,** *16,* 341-348.
77. Heinonen-Tanski, H.; Siltanen, H.; Kilpi, S.; Simojoki, P.; Rosenburg; Makinen, S. *Pestic. Sci.* **1986,** *17,* 135-142.
78. Boyette, K. D.; Moorman, T. B.; Koskinen, W. C. *Biol. Fertil. Soils* **1988,** *6,* 100-105.
79. Rodriguez-Kabana, R.; Curk, E. A. *Annu. Rev. Phytopathol.* **1980,** *18,* 311-332.
80. Moorman, T. B. *J. Prod. Agric.* **1989,** *2,* 14-23.
81. Stalker, D. M.; McBride, K. E.; Malyj, L. D. *Science* **1988,** *242,* 419-423.

RECEIVED November 20, 1990

Chapter 15

Effects of Pesticide Degradation Products on Soil Microflora

N. Dwight Camper

Department of Plant Pathology and Physiology, Clemson University, Clemson, SC 29634-0377

Pesticide degradation products have several effects on soil microflora. These effects include: enhancing or decreasing populations; causing enhanced degradation of the parent compound; inhibiting or stimulating respiration, nitrogen transformations or substrate utilization; inhibiting the growth of one or more specific microbe. Pesticide degradation in soils results from microbial actions such as oxidation, hydrolysis, dehalogenation, conjugation, and reduction. Products formed may be inhibitory or stimulatory to various components of the microbial population, some of which may catalyze the degradation of the parent or the product itself. Significant levels of degradation products in soils and their effects on soil microbes need to be defined.

Pesticides reach the soil either through direct application or indirectly as a result of application methods. Once present in the soil, pesticides are subjected to transformations by chemical or biological processes which affect the environmental fate and behavior of these compounds. A summary of the types of reactions involved is shown in Table I; those most frequently documented for pesticide degradations are hydrolysis and oxidation. Persistence of a pesticide in soil depends on the rate of transformation. Products of these transformations include compounds which have undergone only slight molecular changes, or represent complete mineralization; e.g., carbon dioxide, water, ammonia and halogen ions. Nonenzymatic or chemical reactions rarely lead to appreciable changes in chemical structure; thus, it is the biodegradative or enzymatic reactions that bring about major changes in molecular structure (1). Evidence suggests that microbial populations in the soil are the sources of enzyme-catalyzed degradative reactions (2,3,4).

Microorganisms in the soil convert pesticides either to inorganic compounds, or to different organic compounds. The processes involved

0097–6156/91/0459–0205$06.00/0

Table I. Types of reactions important to pesticide transformations by soil microorganisms

dealkylation	ether cleavage
dealkoxylation	hydrolysis[*]
decarboxylation[*]	oxidation[*]
dehalogenation	methylation
reduction	ß-oxidation

[*]May occur as a result of both chemical and biological processes.

include complete mineralization or cometabolic reactions, the formation of new toxicants, or to the production of compounds that may be incorporated into the soil structure (organic matter). Alteration of pesticide molecules by enzymatic mechanisms may involve incidental metabolism where the pesticide does not serve as an energy source, catabolism where the pesticide does serve as an energy source, or detoxification (possibly a resistance mechanism) (5).

Pesticides exert effects on soil microflora both in pure culture and in soil (3). These effects may be either inhibitory or stimulatory, may affect populations of a single species or of many different species, or may specifically affect a biochemical activity of the soil microorganism(s). Adverse effects of pesticides are observed more frequently at application rates well above those recommended for pest control.

Effects of pesticide degradation products on soil microflora are not well documented, perhaps because emphasis has been on assessing the impact of the parent molecule. At least two reasons can be offered to explain the lack of data: 1. Prior to 1960 data on degradation reactions and products formed were not required for registration; and 2. Detection and analysis of minute amounts of compounds was not possible. Improved analytical instrumentation and the availability of ^{14}C-labeled pesticides enabled researchers to trace minute amounts of parent molecule and products in a soil system (4).

This review will concentrate on the effects of degradation products (metabolites) on soil microflora. Areas reviewed are effects on microbial populations, adaptation phenomena that involve a degradation product as the adaptive agent, and effects on specific microbial activities.

Microbial Populations

Pesticides, when applied at the recommended rates, generally either have no effect on the total culturable indigenous microbial population in the soil, or cause a temporary shift in certain components of the microbial population (1,3,5). Certain species within a population may be inhibited or stimulated; eventually an equilibrium is reestablished. The effect varies depending on the pesticide, time of incubation and various environmental conditions.

Variation of pesticide metabolite effects and the lack of a consistent pattern of effect has been documented. For example terbufos sulfoxide (a degradation product of terbufos { S-[[1,1-(dimethylethyl)thio]methyl]-O,O-diethyl phosphorodithionate} stimulated bacterial numbers (+55%) after two weeks incubation in a loamy sand soil; similar results were obtained on fungal populations (+66%) with terbufos sulfone (6).

Microbial populations in hazardous waste contaminated soils from three different sites were enumerated. They were more active (heterotrophic activity with radiolabeled amino acid mixture), and more abundant and active than populations in adjacent, noncontaminated soil (7). Abundance and activity of bacterial populations varied from site to site, which may be related to nutrient status or the presence of inhibitory chemicals. Phenols are produced as a result of the degradation of a number of different pesticides [2,4-D (2,4-dichlorophenoxyacetic acid), 2,4,5-T (2,4,5-trichlorophenoxyacetic acid), parathion (diethyl-4-nitrophenylphosphorothioate), and others]. In the hazardous waste sites examined, the capacity to degrade phenols varied from site to site, and was not correlated with either numbers of bacteria present, the number of phenol degraders present, or phenol concentration present in the soil.

Parathion degradation in cranberry soils was enhanced by pretreatment with either parathion or p-nitrophenol (a degradation product) (8). The number of aerobic bacteria increased in p-nitrophenol-pretreated soil compared to controls. The presence of p-nitrophenol applied directly, or derived from parathion, apparently resulted in an increase of parathion-degrading microorganisms. Sudhakar-Barik and Wahid (9) showed that parathion-hydrolyzing microorganisms proliferated during the metabolism of p-nitrophenol. Three successive applications of parathion resulted in an increase in the number of microorganisms capable of hydrolyzing the parent molecule from undetectable levels to 1.4×10^7 and 4.3×10^9 /g soil for the second and third successive applications, respectively.

Repeated application of pesticides or their degradation products may alter activities of enzymes or shift the route of a degradative pathway, which may simply be a specific population shift. Repeated application of parathion or its hydrolysis product, p-nitrophenol, to flooded soil resulted in a shift in degradation from a reductive mechanism (nitro group reduction) to one primarily of hydrolysis (10). This shift was a consequence of an increased population of parathion-hydrolyzing microorganisms that used p-nitrophenol as an energy source. Either there were more of these microbes, or less likely microbes with either enhanced levels of the hydrolytic enzyme, or a more active form of the enzyme. A separate study showed that mineralization of parathion increased with the number of successive pretreatments with p-nitrophenol (11). Similar results were obtained when flooded soils were pretreated with the hydrolysis products of carbaryl (1-naphthyl methylcarbamate) and carbofuran (2,3-dihydro-2,2-dimethyl-7-benzofuranylmethyl-carbamate), 1-naphthol and carbofuran phenol, respectively; accelerated hydrolysis of the parent compounds was observed (12). Pretreatment of soil with 2,4-dichlorophenol resulted in a more than

70% increase in the mineralization of 2,4-D; similar results were obtained with hydrolysis products of isofenphos (*11*). In all of the cases cited, the accelerated degradation of the parent compound seems to be a consequence of a microbial population shift as result of exposure to a degradation product.

ß-oxidation is an important metabolic activity of microorganisms and a similar degradation type of reaction may be involved in pesticide metabolism. Compounds such as 2,4-DB (2,4-dichlorophenoxybutyric acid) are converted to 2,4-D, which is then inhibitory (*13*). Sesone (2,4-dichlorophenoxyethyl sulfate) is hydrolyzed to 2,4-dichlorophenoxyethanol, which is further oxidized to 2,4-D. Related compounds such as 2,4,5-trichlorophenoxyethyl sulfate and 2-methyl-4-chlorophenoxyethyl sulfate behave in a similar manner (*14*). Populations of bacteria were reduced by 2,4-D in soil receiving repeated applications of the herbicide (*15*). In other soils, total microbial biomass levels in soils receiving 35 years of annual treatments were not different from levels in untreated soils (*16*).

Adaptation of Microflora to Degradation Products

A characteristic of microbial populations is the ability to adapt to the presence of an increasing number and quantity of chemicals in the environment. Such adaptation is the measurable ability of an organism to show insensitivity or reduced sensitivity to chemicals that might otherwise be toxic. Possible mechanisms involved are: enrichment, mutation, induction, or selective inhibition of one or more members of the microbial community. Adaptation, as it is used in the literature reviewed, generally does not suggest a mechanism of adaptation; therefore, "adaptation" as used in pesticide degradation studies may refer only to microbial population changes, or to a true genetic change resulting in more rapid degradation for example. The pesticide may also be acting as an inducer of a specific enzyme, which would result in an adapted microbe. The case of penicillin-resistance in bacteria through induction of penicillinase is well known (*17*). Such microbial adaptation has been observed with pesticides. For example, the second addition of BHC (1,2,3,4,5,6-hexachlorocyclohexane) to a submerged rice field resulted in a higher BHC-degrading activity (*18*). Similar responses to pesticide degradation products have been observed. Rouchaud et al. (*19*) obtained results indicating that repeated application of chlorfenvinphos [diethyl-1-(2,4-dichlorophenyl)-2-chlorovinyl phosphate] in a cauliflower monoculture situation resulted in an adapted microbial population and higher rates of degradation of the insecticide. Adaptation of microbial populations to parent pesticides has been demonstrated and therefore, adaptation to degradation products might be expected which could result in increased degradation rates of parent molecules.

The growth of an <u>Achromobacter</u> strain on peptone agar with either 2,4-dichlorophenol or 4-chloro-2-methylphenol (degradation products of 2,4-D and MCPA [(4-chloro-2-methylphenoxy)acetic acid], respectively), resulted in

cultures which were adapted to oxidize 2,4-D and MCPA (20). A strain of
Flavobacterium peregrinum, when grown with either MCPA or 2-chloro-4-
methylphenoxyacetic acid, could degrade 2,4-D and 4-chloro-2-methylphenol,
but not MCPA. The inability to degrade MCPA may indicate the inability to
use this compound as a carbon or energy source, perhaps because of the
presence of peptone in the growth medium.

The phenomenon of accelerated pesticide degradation induced by prior
pesticide treatment is referred to as enhanced degradation (21). One
explanation of this phenomenon assumes adaptation of a specific component
of the microbial community, which results in a competitive advantage and
subsequent proliferation. This adaptation phenomena may result from
interactions between the soil, pesticide, microbes and environmental
conditions (11). The adapted population(s) can more rapidly degrade the
pesticide upon the next application. Enhanced degradation has been
demonstrated in chemostats, soil slurries, and soils with herbicides,
insecticides, fungicides, and other xenobiotics. The loss of efficacy of a
particular pesticide may be related to its degradation and the presence of a
degradation product, which then serves as the inducing agent.

The effect of pretreatment of soils with EPTC (S-ethyl
dipropylcarbamothioate) and butylate [S-ethyl bis(2-
methylpropyl)carbamothioate] metabolites on subsequent EPTC and butylate
degradation was reported by Bean et al (22). Butylate sulfone did not
enhance the rate of EPTC or butylate degradation. However, EPTC and
butylate sulfoxides induced enhanced degradation of both parent compounds.
In a Clay Center type soil, cross-enhancement for EPTC and butylate
degradation was less when pretreated with the sulfoxides than with either
parent compound. In a Scottsbluff type soil, cross-enhancement was similar
for pretreatment with either the parent molecule or the sulfoxide derivatives.

A comprehensive study of eight pesticides and their hydrolysis products
in a clay loam soil was reported by Somasundaram et al (23). Results showed
that 2,4-dichlorophenol, p-nitrophenol and salicylic acid predisposed the soil
for enhanced degradation of 2,4-D, parathion and isofenphos {1-methyl-
2[[ethoxy[9(1-methylethyl)amino]-phosphinothioyl]oxy]benzoate}, respectively.
The persistence of chlorpyrifos (O,O-diethyl O-3,5,6-trichloro-2-pyridyl
phosphorothioate) was increased by prior treatment of the soil with its
degradation product, 3,5,6-trichloro-2-pyridinol. Repeated application of
carbofuran phenol, 2-isopropyl-4-methyl-6-hydroxypyrimidine, methyl phenyl
sulfone, thiophenol, isopropyl salicylate, 2,4,5-trichlorophenol {hydrolysis
products of carbofuran, diazinon [O,O-diethyl O-(2-isopropyl-6-methyl-4-
pyrimidinyl)phosphorothioate], fonofos (ethyl S-phenyl
ethylphosphonothiolothionate), isofenphos, and 2,4,5-T, respectively)} had no
effect on degradation rates of their parent molecules.

Microbial Activities

Metabolic activities of soil microorganisms are many and varied. These activities include: nitrogen transformations and fixation; respiratory activities as evidenced by CO_2 evolution and O_2 uptake; substrate utilization (cellulose and protein decomposition, and glucose metabolism); and various enzymatic activities. Biological and chemical degradation affect the persistence of pesticides in soil, and thus affect the duration of effects on microorganisms (24). Pesticide metabolites may be present in larger quantities than the parent compound, and relatively less toxic. However, they may accumulate to potentially toxic levels.

Respiration. Respiration, indicated by oxygen uptake, is an indicator of the activity of aerobic microorganisms involved in substrate decomposition in the soil.   Degradation   products   of   DDT   [1,1,1-trichloro-2,2-bis(p-chlorophenyl)ether], PCPA (p-chlorophenylacetic acid) and DDA [2,2-bis(p-chlorophenyl)acetic acid] were inhibitory to oxygen consumption, but only at 100 μg/g soil (25). The nonpolar metabolites and the parent compound were not inhibitory. Phorate [diethyl S-(ethylthiomethyl)-phosphorothiolothioate] and terbufos, and their sulfones and sulfoxides, did not inhibit oxygen uptake by a loamy sand soil (6). More oxygen was consumed when the soil was treated with the sulfone and sulfoxide derivatives as compared to the controls. Populations of bacteria and fungi were greater in soils treated with the sulfoxide and sulfone derivatives, respectively; thus, enhanced populations may account for stimulated oxygen consumption reported.

Nitrogen Transformations.   One of the degradation products of several phenylamides is 3,4-dichloroaniline. This metabolite inhibited the oxidation of ammonia-nitrogen to nitrite-nitrogen, but was inactive against organisms that oxidize nitrite-nitrogen to nitrate-nitrogen (26). High concentrations of two   other   degradation   products,   3,4-dichlorophenylurea   and   3,3',4,4'-tetrachloroazobenzene, were required to inhibit nitrification.   Another metabolite, 3-(3,4-dichlorophenyl)-1-methylurea, inhibited the oxidation of nitrite-nitrogen to nitrate-nitrogen by Nitrobacter spp., resulting in accumulation of nitrite-nitrogen in soil from ammonia-nitrogen oxidation. High concentrations of the sulfoxide and sulfone metabolites of terbufos and phorate decreased nitrification, but less that the parent molecules (6).

A study of benomyl [methyl-1-(butylcarbamoyl)-2-benzimidazole] and its hydrolysis products MBC (methyl-2-benzimidazole carbamate) and AB (2-aminobenzimidazole) on nitrification in a flooded soil showed no effect except at 1000μg/ml for MBC and at 100 and 1000μg/ml for AB (27). Both products were less toxic than the parent compound. AB was toxic to Nitrosomas sp. and N. agilis, but only at 100 μg/ml. Benomyl was more toxic to both organisms than AB; MBC had no effect at 10 and 100 μg/ml.

Studies with 2,4-D showed that there were no long-term effects on N mineralization in treated soil (16,28).

Rhizobia and Azotobacter are important in the nitrogen cycle in the soil.

Their association with the roots of legume crops also positions them to interact with pesticides and their metabolites in the soil. Carbofuran, carbaryl or their metabolites at normal and a 10x rate had no effect on Rhizobium sp. growth (*29*). Carbofuran and 3-hydroxycarbofuran at 0.5 and 5 μg/ml did not affect the growth of Azotobacter chroococcum in a nitrogen-containing medium; 3-ketocarbofuran (at 5, but not 0.5 μg/ml) delayed A. chroococcum growth, which then was comparable to the control. Carbaryl or 1-naphthol had no effect on growth of the bacterium. When grown in a nitrogen-free medium carbofuran and both metabolites inhibited the growth of A. chroococcum; the metabolites were more inhibitory than the parent compound. Carbaryl and 1-naphthol were both inhibitory to growth. The different results obtained with the two media were interpreted as an effect of the pesticides and their metabolites on the nitrogen-fixing systems of A. chroococcum; e.g., whether nitrogenase was active or not. Carbofuran is converted to 3-hydroxcarbofuran, then to 3-ketocarbofuran. The latter was the most toxic of the three compounds tested.

Substrate Utilization. Microorganisms are important in the decomposition of many substances in the soil; i.e., cellulose, proteins, other types of organic and inorganic compounds. The effect of trifluralin [2,6-dinitro-N,N-dipropyl-4-(trifluromethyl)-benzenamine] and twelve of its metabolites on the decomposition of glucose, cellulose and protein was evaluated in a Dundee silt loam soil (*30*). Decomposition was measured as $^{14}CO_2$ production from ^{14}C-labeled substrates mixed with the soil and incubated in biometer flasks. Trifluralin increased the rate of metabolism of glucose; five of its metabolites inhibited glucose utilization, but in a transitory manner only (Table II). Only one metabolite, TR-9, inhibited glucose utilization over the 30-day observation period. Neither trifluralin or any of its metabolites affected the decomposition of protein. However, seven of the metabolites, when applied at rates exceeding those expected in the soil, increased the rate of cellulose decomposition (Table II). The parent compound had no effect on cellulose mineralization. Considering the potential concentration of metabolites in the soil at normal application rates of the parent compound, and the type and duration of the effects of these compounds, it was concluded that trifluralin metabolites would have minimal effects on the various decomposition processes in the soil. The degradation product of 2,4-DB, 2,4-D, has minimal effects on cellulose or plant residue decomposition, except at very high concentrations (*31,32*).

Toxicity. Pesticide metabolites may be toxic to soil microbes and hence, have an adverse effect on induction or inhibition of microbial degradation by the effective organisms. Oxygen-limited cultures of Klebsiella pneumoniae reduced 4-methylsulfinyl phenol (100 to 125 μg/ml) to 4-methylthiophenol, which at 200 μg/ml was toxic to the bacterium (*33*). Thus, accumulation of the thiophenol could result in inhibition of the degradation of the

TABLE II. Effect of trifluralin and metabolites on mineral-ization of ^{14}C-glucose and ^{14}C-cellulose in soil. Values are shown as a percentage of mineralization in control soils (stimulation, S or inhibition, I). TR-28 and TR-32 were compared with benzene controls. Adapted from Boyette et al. 1988 (30)

| Compound (0.5 mg/ml soil) | ^{14}CO$_2$ evolution from: | |
	^{14}C-glucose (+N) Maximum % S or I	^{14}C-cellulose Maximum % S
trifluralin	5 (I)	-
TR-2*	-	18
TR-3	2 (S)	23
TR-6	17 (I)	36
TR-9	26 (I)	-
TR-13	10 (I)	22
TR-15	-	-
TR-17	2 (S)	-
TR-21	-	17
TR-28	-	-
TR-32	12 (I)	47
TR-36	-	-
TR-40	10(I)	18
Benzene cont.	-	-
Methanol cont.	-	-

*TR-2: 2,6-dinitro-N-propyl-4-(trifluoromethyl)benzenenamine; TR-3: 2,6-dinitro-4-(trifluoromethyl)-benzenamine;TR-6:3-nitro-5-(trifluoromethyl)-1,2-benzenediamine; TR-9: 5-(trifluoromethyl)-1,2,3-benzenetriamine; TR-13: 2-ethyl-7-nitro-1-propyl-5-(trifluoromethyl)-1H-benzimidazole;TR-15:2-ethyl-4-nitro-6-(trifluoromethyl)-1H-benzimidazole;   TR-17:   7-nitro-1-propyl-5-(trifluoromethyl)-1H-benzimidazole;   TR-21:   4-(dipropylamino)-3,5-dinitrobenzoic   acid;   TR-28:   2,2'-azoxybis-(6-nitro-N-propyl-4-(trifluoromethyl)benzenamine);   TR-32:   2,2'-azobis-(6-nitro-N-propyl-4-(trifluoromethyl)benzenamine);TR-36:3-methoxy-2,6-dinitro-N,N-dipropyl-4-(trifluoromethyl)benzenamine;   TR-40:   N-(2,6-dinitro-4-(trifluoromethyl)phenyl)-N-propyl-proanamide.

sulfinylphenol and possibly the parent compound, fensulfothion {O,O-diethyl O-[4-(methylsulfinyl)-phenyl]phosphorothionate}. Somasundaram et al (34) used the Beckman Microtox system to assess the toxicity of several pesticide metabolites to (Photobacterium phosphoreum) as a model system. Metabolites of chloropyrifos -- 3,5,6-trichloro-2-pyridinol, 2,4-D -- 2,4-dichlorophenol, 2,4,5-T -- 2,4,5-trichlorophenol, and isofenphos -- isopropyl salicylate, were more toxic than their respective parent molecules. A degradation product of carbaryl, 1-naphthol, was toxic to <u>Pseudomonas</u> sp., <u>Nitrosomas</u> sp., and <u>Nitrobacter</u> sp. (*10*), but does not seem to be used as an energy source for microorganisms (*34*). The decreased mineralization and increased persistence of chlorpyrifos resulting from repeated soil application of its degradation product, 3,5,6-trichloro-2-pyridinol, may be because of its toxicity to the effective degraders. The growth of <u>Pseudomonas</u> sp. was inhibited by 2,4-dichlorophenol at 25 mg/l, whereas the parent compound (2,4-D) had no effect at concentrations up to 2000 mg/l (*35*). The relatively low toxicity of salicylic acid and 2-isopropyl-4-methyl-6-hydroxylpyrmidine (degradation products of isofenphos and diazinon, respectively) may be related to their rapid metabolism to nontoxic compounds and also result in enhanced degradation of the parent compounds (*34,36*).

Racke et al. (*37*) reported an interesting study where the hydrolysis product of chlorpyrifos, 3,5,6-trichloro-2-pyridinol (TCP), apparently accumulated to toxic levels in certain soils treated with the parent compound. TCP was not detected in some treated soils; rather, only the parent compound was present. Direct treatment of fresh soil samples with 5 mg/l TCP resulted in its rapid mineralization only in those soils in which TCP had not accumulated to toxic levels after chlorpyrifos treatment. Microbial adaptation as a result of insecticide treatment may be the reason for the rapid degradation of TCP. TCP at 50 mg/l inhibited its mineralization in all soils tested indicating the potential for microbial toxicity.

Fungitoxicity of methoxychlor [1,1,1-trichloro-2,2-bis-(4-methoxyphenyl)ethane], fenitrothion (dimethyl-3-methyl-4-nitrophenyl phosphorothionate) and their metabolites were evaluated against <u>Mortierella</u> <u>pusilla</u>, <u>Mortierella</u> <u>isabellina</u>, <u>Trichoderma</u> <u>virdie</u> and <u>Saprolegnia</u> <u>parasitica</u> (*38*). Concentrations tested were 1.6 to 400 μg/ml except for <u>S</u>. <u>parasitica</u> which was more sensitive and the concentration range tested was 0.2 to 440 μg/ml. Hydroxychlor [2,2-bis-(p-hydroxyphenyl)-1,1,1-trichloroethane], S-methylfenitrothion and 3-methyl-4-nitrophenol were the most toxic compounds tested and were more toxic than the parent molecules. In all cases the more toxic compounds were also more water soluble.

Conclusions

Segments of the microbial community may increase under conditions where a enzymes system provides a detoxification or cometabolic mechanism, which would reduce the potential inhibitory level of a metabolite or parent pesticide molecule. Increased microbial populations may also be attributed to utilization of the metabolite as an energy source. Either scenario would give a competitive advantage to certain segments of the microbial community.

Adaptation, a characteristic of microbial populations, enables some components of the community to survive exposure to potentially toxic chemicals. Frequently, the pesticide metabolite serves as the inducing agent and results in a more rapid degradation of either the metabolite or the parent pesticide, or both. The end result can be reduced efficacy of the pesticide upon repeated applications in a monoculture situation. However, the mechanism of the adaptation is unknown.

Pesticide metabolites include some potentially toxic compounds; e.g., phenols, anilines and organic acids. These compounds can inhibit specific microbial enzymes, interfere with energy (ATP) generation, or other microbial activities such as nitrogen transformations. These activities contribute to many attributes of the soil and adverse effects of pesticide metabolites could alter one or more of these attributes; e.g., soil fertility and pesticide efficacy. These effects singly or in combination could affect agricultural productivity.

Two central questions remain: 1. What constitutes a significant level of a pesticide degradation product in the soil, and 2. What constitutes a significant effect of these chemicals on soil microflora? A significant level of degradation products has not been defined to date; however, a level equivalent to 10 percent of the original concentration of the parent compound may be justification for further research (perhaps first suggested by Blinn in 1973, 39). Further research certainly should include determination of product identity and concentration in the soil, and persistence. Significant effects on soil microflora will be more difficult to define. Atlas et al (40) suggest that the most relevant effects to be evaluated are essential cycling processes in the soil; e.g., N and S cycling. They suggest that evaluation of cycling processes may be more realistic than an evaluation of population shifts. A stress on one component of the microbial community may be compensated for by increased activity of another. Johnen and Drew (41) suggest that if microbial populations in the field require 60 days or more to recover from pesticide treatment to pretreatment levels, the effect should be considered of major importance and further tests be conducted. Concerns about environmental quality will most likely dictate that the status of pesticide degradation products in the soil be examined more extensively in the future.

Acknowledgments

Technical Contribution No. 3091 of the South Carolina Agricultural Experiment Station, Clemson University. Assistance and reviews provided by Dr. M. J. B. Paynter, Professor of Microbiology, Clemson University are gratefully acknowledged.

Literature Cited

1. Alexander,M. *Science* 1981, *211*, 132-138.
2. Hill,I.R.; Wright, S.J.L. In *Pesticide Microbiology*; Hill,I.R.; Wright,S.J.L., Eds.; Academic Press: NY, 1978; pp 79-136.
3. Grossbard,E. In *Herbicides: Physiology, Biochemistry, Ecology*; Audus,L.J., Ed.; Academic Press: NY, 1976; pp 99-147.

4. Smith,A.E. In *Environmental Chemistry of Herbicides*; Grover,R., Ed.,; CRC Press: Boca Raton, FL, 1988; pp 171-200.
5. Matsumura, F.; Benezet, H.J. In *Pesticide Microbiology*; Hill,I.R.; Wright,S.J.L., Eds.; Academic Press: NY, 1978; pp 623-667.
6. Tu, C.M. *Bull. Environ. Contam. Toxicol.* **1980**, *24*, 13-19.
7. Dean-Ross,D. *Bull. Environ. Contam. Toxicol.* **1989**, *43*, 511-517.
8. Ferris,I.G.; Lichtenstein,E.P. *J. Agric. Food Chem.*, **1980**, *28*, 1011-1019.
9. Sudhakar-Barik; Wahid,P.A.; Ramakrishna,C.; Sethunathan, N. *J. Agric. Food Chem.* **1979**, *27*, 1391-1392.
10. Ramarkishna,C.; Sethunathan,N. *J. Appl. Bacteriol.* **1983**, *54*, 191-195.
11. Somasundaram,L; Coats,J.R. Enhanced Biodegradation of Pesticides in the Environment; Racke,K.D.; Coats,J.R., Eds.; ACS Symposium Series, No 426; Am. Chem. Soc.: Washington, DC, 1990; pp 128-140.
12. Rajagopal,B.S.; Panda,S; Sethunathan,N. *Bull. Environ. Contam. Toxicol.* **1986**, *36*, 827-832.
13. Gutenmann,W.H.; Loos,M.A.; Alexander,M.; Lisk,D.J. *Soil Sci. Soc. Am. Proc.* **1964**, *28*, 205.
14. Metcalf, R.L. In *Pesticides In The Environment*; White-Stevens,R., Ed.; Marcel Dekker Inc.: New York, NY, 1971; Vol 1; pp 1-144.
15. Breazeale,F.B.; Camper, N.D. *Appl. Microbiol.* **1970**, *19*, 379-380.
16. Biederbeck,V.O.; Cambell,C.A.; Smith,A.E. *J. Environ. Qual.* **1989**, *16*, 257-262.
17. Pollack,M.R. In *Drug Resistance in Microorganisms*; Wolstenholme, G.E.W.; O'Connor,C.M., Eds.; Little, Brown and Co.: Boston, MS, 1957; pp 78-95.
18. Raghu,K; MacRae,I.C. *Science* **1966**, *154*, 263-264.
19. Rouchaud,J.; Metsue,M.; Van de Steene,F.; Pelerents,C.; Gillet,J.; Benoit,F.; Ceustermans,N.; Vanparys,L. *Bull. Environ. Contam. Toxicol.* **1989**, *42*, 409-416.
20. Steenson,T.I.; Walker,N. *J. Gen. Microbiol.* **1958**, *18*, 692-697.
21. Roeth,F.W. *Reviews of Weed Science*; Weed Sci Soc of Am: Champaign, IL, **1986**; Vol 2, pp 45-65.
22. Bean,B.W.; Roeth,F.W.; Martin,A.R.; Wilson,R.G. *Weed Sci.*, **1988**, *36*, 70-77.
23. Somasundaram,L.; Coats,J.R.; Racke,K.D. *J. Environ. Sci. Health* **1989**, *B24*, 457-478.
24. Moorman,T.B. *J. Prod. Agric.* **1989**, *2*, 14-23.
25. Subba-Rao, R.V.; Alexander, M. *Bull. Environ. Contam. Toxicol.* **1980**, *25*, 215-220.
26. Corke,C.T.; Thompson,F.R. *Can. J. Microbiol.* **1970**, *16*, 567-571.
27. Ramakrishna, C.; Gowda, T.K.S.; Sethunathan, N. *Bull. Environ. Contam. Toxicol.* **1979**, *21*, 328-333.
28. Grossbard,E. *Weed Res.* **1971**, *11*, 263-265.
29. Kale,S.P.; Murthy,N.B.; Raghu,K. *Bull. Environ. Contam. Toxicol.* **1989**, *42*, 769-772.

30. Boyette,K.D.; Moorman,T.B.; Koskinen,W.C. *Biol. Fertil. Soils* **1988**, *6*, 100-105.
31. Fletcher,K.; Freedman,B. *Can. J. For. Res.* **1986**, *16*, 6-9.
32. Gottschalk,M.R.; Shure,D.J. *Ecol.* **1979**, *60*, 143-151.
33. MacRae, I.C.; Cameron, A.J. *Appl. Environ. Microbiol.* **1985**, *49*, 236-237.
34. Somasundaram,L.; Coats,J.R.; Racke,K.D.; Stahr,H.M. *Bull. Environ. Contam. Toxicol.* **1990**, *44*, 254-259.
35. Tyler,J.E.; Finn,R.K. *Appl. Microbiol.* **1974**, *28*, 181-184.
36. Sethunathan,N.;Pathak,M.D. *J. Agric. Food Chem.* **1972**, *20*, 586-589.
37. Racke,K.D.; Coats,J.R.; Titus,K.R. *J. Environ. Sci. Health.* **1988**, *B26*, 527-539.
38. Baarschers, W.H.; Bhaarath, A.I.; Hazenberg, M.; Todd, J.E. *Can. J. Bot.* **1980**, *58*, 426-431.
39. Blinn, R.C. *ACS Astr.* **1973**, No. 33, 166th ACS Meeting, Chicago, IL.
40. Atlas, R.M.; Pramer, D.; Bartha, R. *Soil Biol. Biochem.* **1978**, *10*, 231-239.
41. Johnen, B.G.; Drew, E.A. *Soil Sci.* **1977**, *123*, 319-324.

RECEIVED November 8, 1990

Chapter 16

Pesticide Transformation Products in Surface Waters

Effects on Aquatic Biota

Kristin E. Day

Rivers Research Branch, National Water Research Institute,
Department of the Environment, Burlington,
Ontario L7R 4A6, Canada

This paper reviews the scientific literature for information on the effects of transformation and/or breakdown products of pesticides on aquatic biota. The results indicate that such compounds can be less, more or similar in toxicity when compared to the parent chemical. In many cases, the toxicity of transformation products is reduced, particularly for herbicides, but this is dependent upon the group of organism tested i.e., plant or animal. In addition, several metabolites of the organophosphate and carbamate insecticides were more toxic or at least similar in toxicity to the parent compound, especially to fish. Factors which must be considered when evaluating the hazards of transformation products of pesticides in aquatic ecosystems include a) the rate at which the compounds appear and disappear b) concentrations of residues in the field c) time of exposure for aquatic biota and d) compartmentalization of transformation products in the ecosystem. Areas of concern where future research is indicated include the required toxicity testing of transformation products for new chemicals prior to registration and the interactive effects of the parent compound, its transformation products and any formulation adjuvants on the toxic response of non-target organisms.

The direct and indirect contamination of surface waters by pesticides is known to occur via aerial drift, watershed runoff and accidental spillage during the widespread use of these chemicals in both agriculture and forestry. In addition, many pesticides are applied directly to aquatic ecosystems to control noxious biting insects and aquatic weeds. Most pesticides, with the exception of very persistent compounds, undergo transformations either chemically (e.g., isomerisation, hydrolysis, photolysis, etc.) or biologically (metabolism by organisms) soon after application. The rate of transformation is dependent upon a variety of factors including the physicochemical properties of the

0097–6156/91/0459–0217$07.25/0
Published 1991 American Chemical Society

pesticide, temperature, moisture, pH, and light in the surrounding environment, presence and abundance of organic matter and micro- and macroflora and fauna of the ecosystem.

Numerous papers have been published on the effects of various pesticides on non-target aquatic organisms at concentrations simulating the contamination of surface waters at field application rates. Most toxicity studies, however, do not take into account the possible transformation of pesticides into compounds of equal or greater toxicity than their precursors, but instead presume that transformation of a chemical results in compounds that are less persistent and less toxic. With the exception of a few studies on fish (1-4), daphnids (5,6), and algae (7,8), most studies have not examined the effects of degradation and/or tranformation products of pesticides on aquatic biota. It is the purpose of this paper to review the available information on the toxic effects of pesticide transformation products to biota in aquatic ecosystems and to suggest future considerations for research in this area of ecotoxicology.

Toxicity of the Transformation Products of Insecticides to Aquatic Biota

Organochlorines. Although the chorinated hydrocarbons are known to be fairly persistent in the environment, they can be converted under natural conditions to even more stable and sometimes more toxic residues than the parent compounds (9). Amongst insecticides, p,p'-DDT (1,1,1-trichloro-2,2-bis (p-chlorophenyl) ethane) and its metabolites, p,p'-DDE (1,1-dichloro-2,2-bis (p-chlorophenyl) ethylene) and p,p'-TDE (1,1-dichloro-2,2-bis (p-chlorophenyl) ethane) are the most thoroughly studied compounds for their comparative effects on aquatic biota. p,p'-DDT has been shown to be toxic to many aquatic organisms at various concentrations (for a review, see reference 10) but its transformation products can have similar, lower or greater toxicity than the parent compound depending on the species of organism tested and the duration of the toxicity test. For example, Sanders and Cope (11) found that p,p'-DDT is 100-fold more toxic to stonefly larvae (Pteronarcys spp.) than p,p'-TDE. In contrast, Kouyoumjian and Uglow (12) found that for the planarian worm Polycelis felina, p,p'-TDE was the most toxic and p,p'-DDT the least toxic, with p,p'-DDE showing intermediate toxicity in acute studies. At sublethal concentrations, both p,p'-DDT and p,p'-TDE were shown to reduce the righting time of animals turned onto their backs and this was presumed to be due to an effect on the nervous system.

Studies with fish have shown that the toxicity of p,p'-DDT is 5 to 10X more than p,p'-TDE and p,p'-DDE in the same test system (13-15). In studies on sublethal toxicity, Peterson (16) monitored the selection of temperature by juvenile Atlantic salmon (Salmo salar) previously exposed to p,p'-DDT or its metabolites and found that low concentrations produced no effect on temperature selection but as concentrations of chemicals increased, the temperature selected by the fish also increased. Fish were the most sensitive, in this respect, to p,p'-DDE and showed decreasing sensitivity to o,p'-DDT, p,p'-TDE and p,p'-DDT. Conversely, Gardner (13) found that p,p'-DDE did not produce any temperature preference for the same species of fish and at concentrations similar to p,p'-DDT or any of its analogues (e.g., methoxychlor).

Some data are also available for the effects of *p,p'*-DDT and its metabolites on algae. Luard (*17*) studied the effects of *p,p'*-TDE, *p,p'*-DDE and *p,p'*-DDT on C^{14} uptake by *Scenedesmus quadricauda* and found that concentrations of 0.1 to 1000 µg/L were generally nontoxic; low concentrations of *p,p'*-TDE were stimulatory. *p,p'*-DDT and *p,p'*-DDE have been shown to have similar toxicity towards other algal species (*18,19*), although in other studies, this metabolite (*p,p'*-DDE) was less toxic than the parent compound (*20*).

Endosulfan (1,4,5,6,7,7-hexachloro-5-norbornene-2,3 dimethanol cyclic sulfite) is one of only a few cyclodiene organochlorine pesticides which are still registered for use in North America. This product consists of two isomers, α-endosulfan and ß-endosulfan, which differ both in their toxicities to aquatic organisms and their persistence (*21,22*). Transformation products of both isomers of endosulfan include endosulfan diol, endosulfan ether, endosulfan sulfate, endosulfan α-hydroxy ether and endosulfan lactone (Figure 1), but their presence in aquatic ecosystems depends upon pH and levels of dissolved oxygen in the medium. Residues found in surface waters adjacent to agricultural land have been limited to the oxidation product, endosulfan sulfate, and the hydrolysis product, endosulfan diol (*23*). Aquatic organisms, particularly fish, are highly sensitive to both endosulfan and endosulfan sulfate. For example, the toxicities of both compounds to the guppy, *Lebistes reticulatus*, and the goldfish, *Carassius auratus*, have been shown to be generally within an order of magnitude of each other, with the parent compound being the most toxic (0.8 to 10 µg/L vs. 1.6 to 17.5 µg/L) (*24*). However, the endosulfan diol, as with other transformation products not containing a sulfur group, is several thousand times less toxic than endosulfan to these same species of fish e.g., 1 to 10 mg/kg vs. 0.01 to 0.001 mg/kg respectively. The conversion of the active substances to more polar compounds with reduced uptake by the animal has been given as the explanation for decreased toxicity (*25*).

Aquatic algae (e.g., *Chlorella vulgaris* and *Phormidium* spp.) are not particularly sensitive to endosulfan or its metabolites. Knauf and Schulze (*24*) and Goebels *et al.* (*26*) observed that continuous 5-d bioassays with endosulfan and its metabolites at concentrations of 1 mg/L had no effect on the physiological activity of the algae. Rates of cell division, photosynthesis and biomass production were not affected at levels below 2 mg/L. Endosulfan sulfate, released into the water during metabolism by green algae, did not impair the physiological properties, reproductive rate or photosynthetic rate of algae. Concentrations of endosulfan in waters from agricultural regions of Canada have been reported to be < 1.0 µg/L (*27*) and, therefore, aquatic algae will not be at risk in these surface waters. However, there is a limited safety factor with regard to sensitive species of fish and the concentrations of endosulfan and endosulfan sulfate present in these waters. In addition, since endosulfan and its more polar degradation products appear to be concentrated several thousand times by sediments and suspended solids, the impact of these residues on benthic organism and filter feeders may merit particular attention (*23*).

Batterton *et al.* (*28*) studied the effects of several chlorinated hydrocarbons containing an endomethylene bridge and their transformation products, namely, aldrin, photoaldrin, photodieldrin, metabolites F and G of dieldrin and ketoendrin, on the growth responses of the blue-green algae, *Anacystis nidulans* and *Agmenellum quadruplicatum*. In nature, aldrin is converted to photoaldrin,

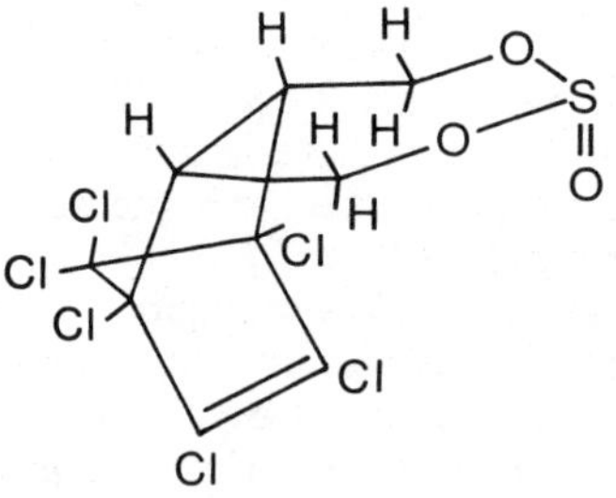

α – endosulfan

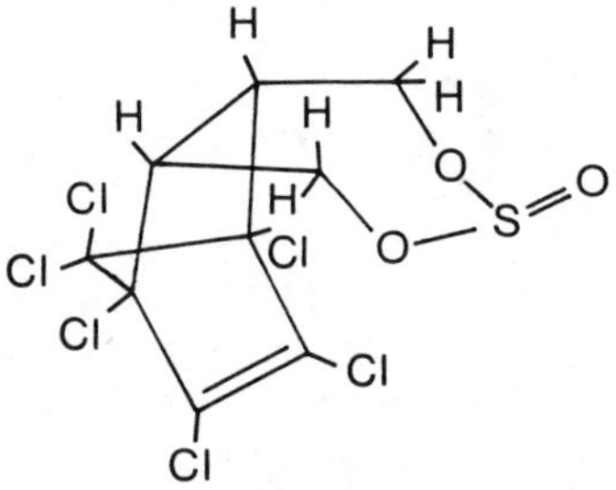

β – endosulfan

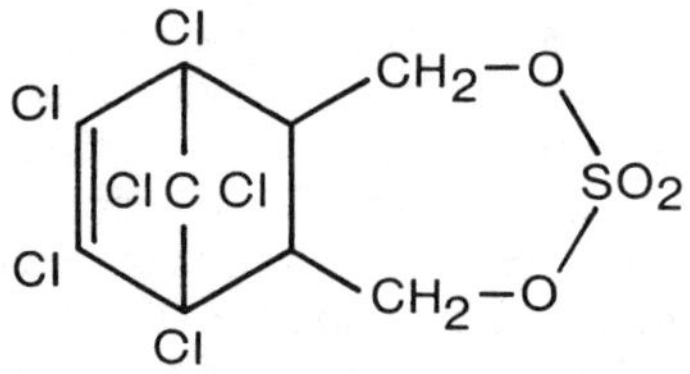

endosulfan sulfate

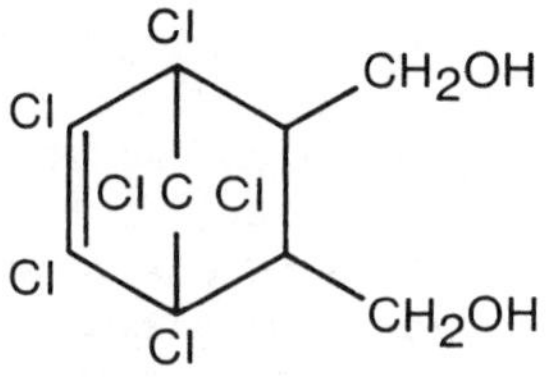

endosulfan diol

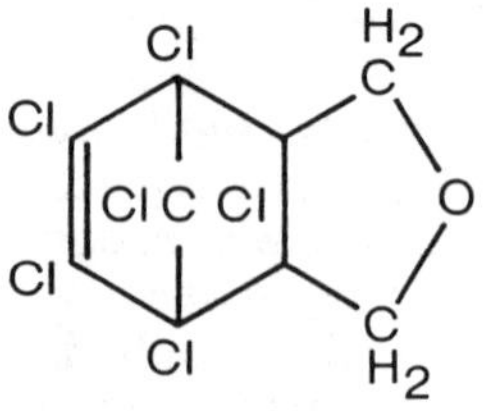

endosulfan ether

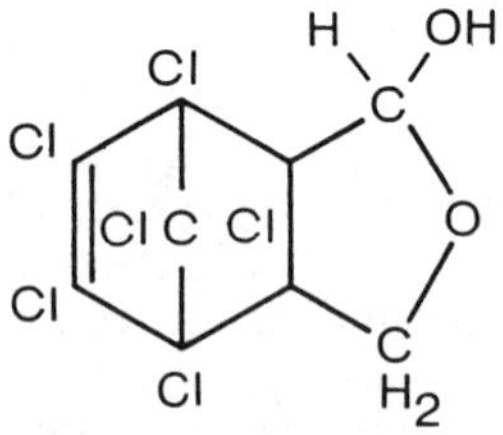

endosulfan α–hydroxy ether

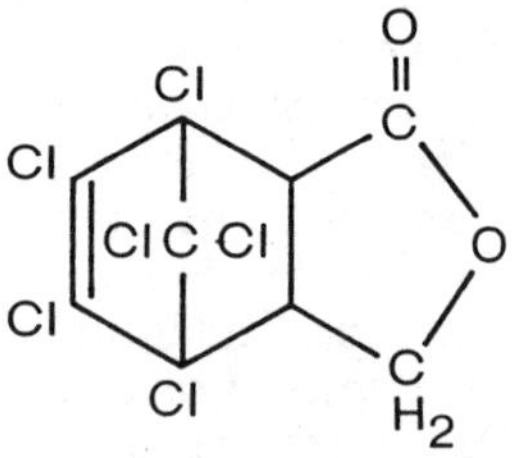

endosulfan lactone

FIGURE 1. Some transformation products of endosulfan.

epoxidized to dieldrin, and dieldrin transformed to aldrin trans-diol which may be the active toxicant (*29*). Results indicated that transformation products of aldrin and dieldrin can be at least as inhibitory to algal growth as the parent compounds, but concentrations used in this study were much higher for both the parent and metabolites than concentrations found in nature. Both blue-green algae were more tolerant of ketoendrin than endrin in this study.

Organophosphates and Carbamates. Organophosphate and carbamate insecticides are diverse and extensively used groups of chemicals which have replaced the more persistent and environmentally damaging organochlorines. Organophosphorus insecticides are normally esters, amides, or thiol derivatives of phosphoric, phosphonic, phosphorothioic or phosphonothioic acids and the carbamates are alkyl and aryl esters of carbamic acid. Thus, these compounds can be expected to be subject to hydrolysis and to be chemically reactive. For example, parathion has been shown to form paraoxon and the *S*-ethyl and *S*-phenyl isomers under the influence of sunlight and ultraviolet radiation (*30*). Limited information is available on the effects of transformation products of organophosphates and carbamates to aquatic biota with the exception of the organophosphates, fenitrothion and malathion, and the carbamates, aminocarb, aldicarb, carbaryl and mexacarbate.

Fenitrothion(*O,O*-dimethyl-*O*-(3-methyl-4-nitrophenyl)-phosphorothioate)is a phosphorothioate compound which requires transformation to fenitrooxon, the oxidative desulfuration metabolite of fenitrothion, for toxicity to organisms. This activation step usually occurs under the influence of mixed function oxidases (*31*). However, the conversion of phosphorothioate insecticides to the more potent phosphoroate analogue can also occur by photo- or chemical oxidation under natural conditions (*32*). The known transformation products of fenitrothion are aminofenitrothion (microbial), carboxyfenitrothion (photolysis), demethylaminofenitrothion (microbial, anaerobic), demethyl-fenitrothion (microbial) and fenitrooxon (oxidation) (*33*) (Figure 2). Miyamoto *et al.* (*1*) studied the relative toxicity of fenitrothion and its various degradation products to the killifish, *Oryzias latipes*, and found that all compounds were less toxic than the parent fenitrothion with the exception of 3-methyl-4-aminophenol. In *Daphnia*, fenitrothion and fenitrooxon were the most toxic among the compounds tested. Hiraoka *et al.*. (*4*) exposed the same species of fish to several dilutions of a fenitrothion emulsion placed in natural sunlight for 47 d and compared the results to a control group of fish exposed to an untreated fenitrothion emulsion. The effects of the treated solution on the hatching rate of fertilized eggs and the rate of survival of larvae were greater than those of the untreated solution, e.g., percent hatch of eggs was 46 to 64% in the treated solution vs. 90 to 94% hatching success in the untreated solution. In acute tests on adult fish, the number of survivors at 24 and 48 h in the untreated solution was 60 to 70% whereas none survived at a concentration of 4 mg/L. The authors concluded that exposure of fenitrothion to sunlight results in degradation and/or transformation to products which are more toxic than the parent compound; unfortunately, the authors did not identify these products or measure their concentrations.

Moody *et al.* (*34*) studied the fate of fenitrothion in stream water following aerial spraying and found traces of demethylamino-fenitrothion, *S*-

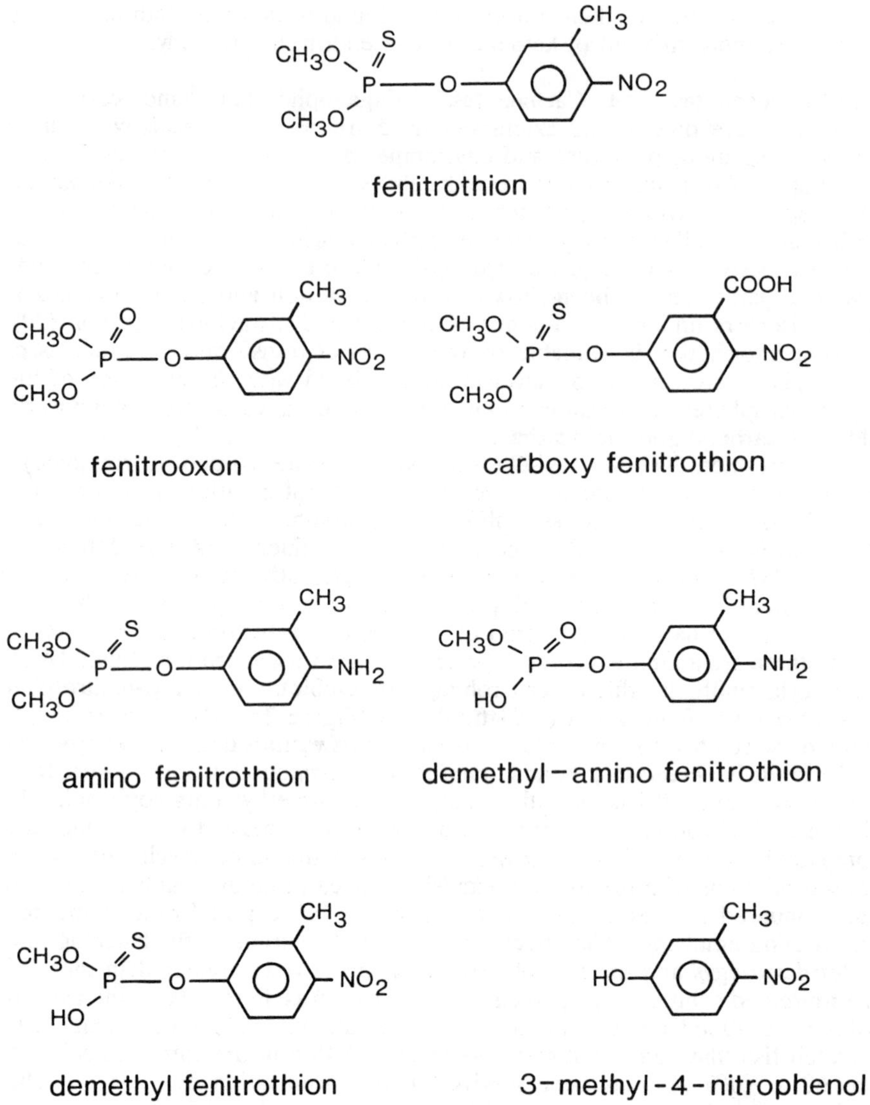

FIGURE 2. Some transformation products of fenitrothion.

methylfenitrothion and aminofenitrothion up to 100 h following application. Ohmae *et al..* (*35*) studied the environmental behaviour of fenitrothion and its decomposition products after operational aerial application and found that the initial concentration of fenitrothion in water (38 μg/L) immediately after application was rapidly reduced, although small amounts of the parent compound (0.02 μg/L) and 3-methyl-4-nitrophenol were detected after 49 days. 3-Methyl-4-nitrophenol has been shown to be an inhibitor of the enzyme, ribonucleotide reductase, which is a key regulatory enzyme in DNA synthesis in mammals (*36*) and therefore it is possible that sublethal concentrations of this chemical in aquatic ecosystems could have detrimental effects on biota. 3-Methyl-4-aminophenol has not been detected in field studies but low concentrations of this chemical could cause toxic effects based on the results of Miyamoto *et al.* (*1*). Technical fenitrothion is also known to contain an impurity, *S*-methyl fenitrothion, which is significantly more toxic than fenitrothion (*37*) but this impurity would have been present in both solutions in the study by Hiraoka *et al.* (*4*) and is not likely the cause of the observed toxicity.

The organophosphate, malathion (*O,O*-dimethyl phosphorodithioate of diethyl mercaptosuccinate) undergoes hydrolysis in aqueous solution to several products dependent upon the pH of the medium (*38*). In basic solutions the primary products are diethyl fumarate and dimethyl phophorodithioic acid. In acid solutions the products are dimethyl phophorothionic acid and 2-mercaptodiethyl succinate. Bender (*39*) found that the basic hydrolysis product, diethyl fumarate was more toxic to the fathead minnow, *Pimephales promelas*, in short term toxicity bioassays (96-h LD_{50}'s) and continuous exposure tests of 14 d duration.

Aminocarb (4-dimethylamino-3-methylphenyl N-methylcarbamate) is a broad spectrum carbamate insecticide applied extensively throughout the world but used particularly in forestry in Canada to control the spruce budworm, *Choristoneura fumiferana* (*40*). Aminocarb has been studied by many authors and many transformation products have been identified; for example, AA (4-amino-m-tolyl N-methylcarbamate), AC (4-amino-3-methylphenol), FA (4-formamido-m-tolyl N-methylcarbamate), FC (N-(4-hydroxy-2-methylphenyl)-N-methylformamide), MFA (4-methylformamido-m-tolyl N-methylcarbamate), MAA (4-methylamino-m-tolyl N-methylcarbamate), MAC (3-methyl-4-(methylamino)phenyl-N-methylcarbamate) have been identified (Figure 3) as well as phenol, methylamine and CO_2 (*40,41*). Szeto *et al.* (*2*) determined the toxicity of the oxidative demethylation metabolites (i.e., MFA, MAA, FA and AA) by measuring the inhibition of brain acetylcholinesterase (AChE) in brook trout (*Salvelinus fontinalis*). They found that the toxicity expressed as the *in vitro* molar concentrations at which 50% of the enzyme is inhibited (I_{50}'s) could be ranked as AA (3.62 X 10^{-6}) > MAA (7.92 X 10^{-6}) > aminocarb (1.01 X 10^{-6}) > MFA (4.29 X 10^{-5}) > FA (7.11 X 10^{-5}). These data correlate very well with the LC_{50}'s of aminocarb (5.7 mg/L) and its metabolites (e.g., 1.7 mg/L for AA, 0.35 mg/L for MAA, >15 mg/L for MFA and 15 mg/L for FA) to rainbow trout (*Salmo gairdneri*) determined by Lamb and Roney (*42*). *In vivo* recovery of fish brain (brook trout) AChE activity after transfer of fish from water contaminated with aminocarb or MAA was also studied by Szeto *et al.* (*2*). Mortality was greater and levels of AChE in living fish were lower in animals exposed to MAA compared to those exposed to similar concentrations of

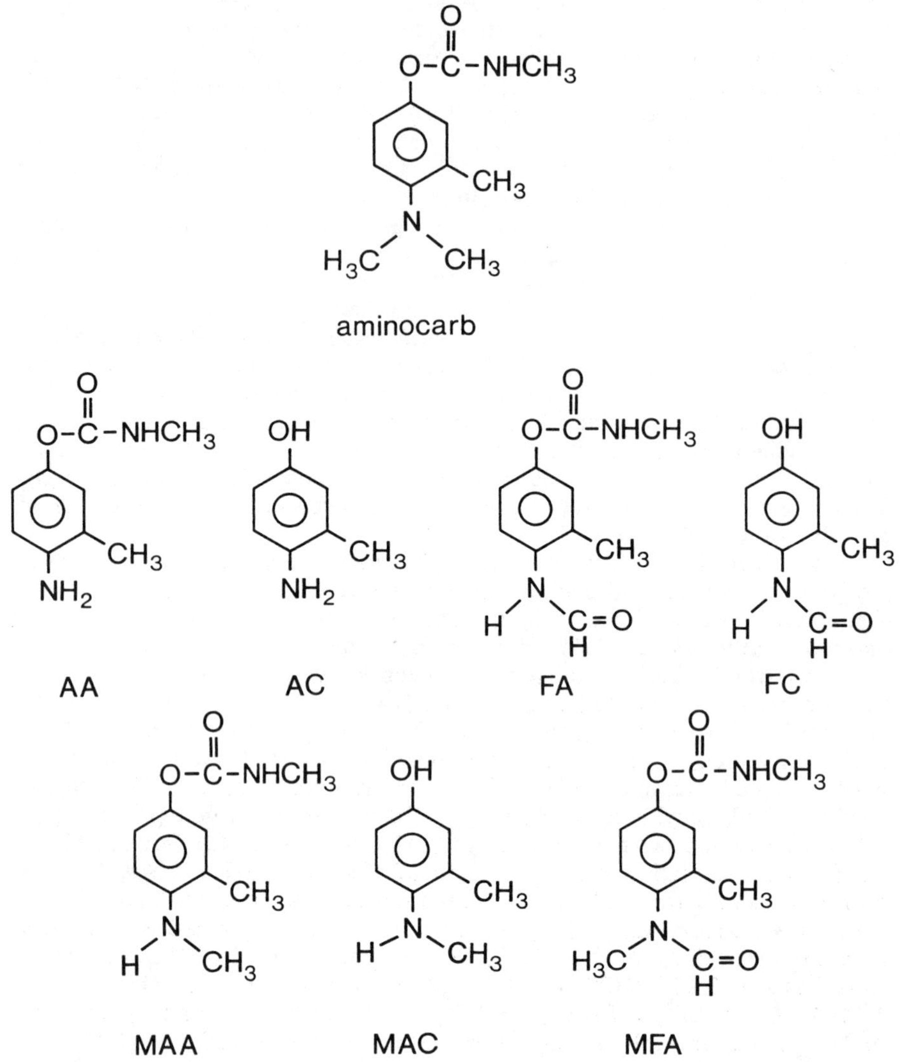

FIGURE 3. Some transformation products of aminocarb.

aminocarb. The authors concluded that according to enzyme inhibition, the metabolites AA and MAA were more potent than the parent compound and MFA and FA were less potent.

Monitoring studies suggest that concentrations of aminocarb in natural woodland waters rarely exceed 10 μg/L although considerably higher concentrations have occasionally been cited (i.e., 25 to 53 μg/L) (43). These residues are within the range of the lowest concentrations of aminocarb and metabolites which have been shown to cause effects; for example, Szeto et al. (2) observed reductions in fish brain AChE at concentrations of 25 μg/L. Although residues of MAA have been detected in fish tissues after exposure to various concentrations of aminocarb in the laboratory (44) the metabolite has never been detected in natural water after aerial spray although few field studies have analysed for this tranformation product. Ernst et al. (45) found three metabolites (FA, AC and MAC) to persist for 24 days following spray for spruce budworm control in New Brunswick but FC, MAA and AA were not detected.

Aldicarb (2-methyl-2[methylthio]propionaldehyde O-[methyl carbamoyl]-oxime) is a highly water soluble and widely used carbamate insecticide and nematocide. Due to its toxicity to mammals as a potent AChE inhibitor, aldicarb is only applied in granuler formulation to the soil where it is mobilized and released by moisture (46). It then undergoes rapid microbial oxidation to the relatively stable aldicarb sulfoxide and followed by slower oxidation to aldicarb sulfone (Figure 4). The degradation and transport of this compound in water, especially groundwater, has been studied widely due to recent findings of the parent compound and its metabolites in drinking water in parts of Canada and the United States (47,48). Foran et al. (5) studied the acute toxicity of aldicarb, aldicarb sulfoxide and aldicarb sulfone to the cladoceran, *Daphnia laevis*, and found that aldicarb sulfoxide was similar in toxicity to the parent compound (e.g., range 43 to 65 μg/L) for both adults and juveniles. However, aldicarb sulfone was almost an order of magnitude less toxic than aldicarb and aldicarb sulfoxide (369 μg/L vs. 51 for adults; 556 μg/L vs. 65 for juveniles). Similar results were obtained for bluegill sunfish (49); for example, static 72-h LC_{50}'s were approximately 100 μg/L for aldicarb, 400 μg/L for aldicarb sulfoxide and over 1000 μg/L for aldicarb sulfone. These data indicate that aldicarb and the first oxidative metabolite, aldicarb sulfoxide, are equal in toxicity and, therefore, microbial oxidation does not lessen the toxic impact of aldicarb contamination of surface waters; however, further degradation of aldicarb sulfoxide as well as aldicarb sulfone, to corresponding oximes and nitriles is thought to occur fairly rapidly with additional degradation to aldehydes, acids and alcohols, none of which are toxicologically significant (47).

In 96-h bioassay experiments with the monoalkyl carbamate, mexcarbate (4-dimethylamino-3,5-xylylmethylcarbamate), Macek et al. (49) found that the hydrolysis product, 4-amino-3,5-xylenol, was 70X more lethal to bluegills (*Lepomis macrochirus*) than the parent compound but the intermediate breakdown product (4-dimethylamino-3,5-xylenol) between these two chemicals was not as toxic.

Stewart et al.(50) found that the primary breakdown product of carbaryl (1-naphthyl N-methyl carbamate), 1-naphthol, was more toxic to molluscs and three species of marine fish than the parent compound in 24- and 48-h laboratory experiments. These results were confirmed by the studies of Butler et al. (51)

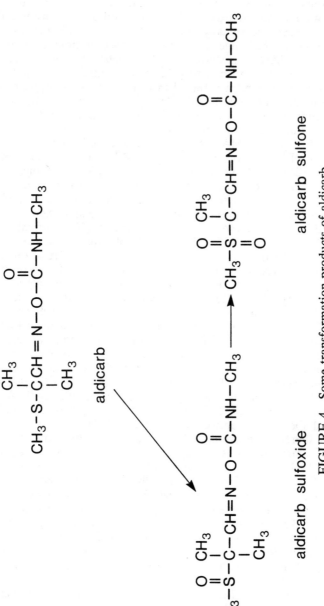

FIGURE 4. Some transformation products of aldicarb.

and Murty (*52*) for juvenile cockleclam (*Clinocardium nuttalli*) and the fish (*Mystus cavasius*).

Synthetic Pyrethroids. The synthetic pyrethroids are a class of lipophilic insecticides which have been marketed for agricultural uses for approximately 10 years. This group of chemicals can be manufactured as a mixture of complex molecules with one or more optically active centers (e.g., permethrin, cypermethrin and fenvalerate) or as single isomers (e.g., deltamethrin). The environmental fate and effects of these compounds have been described by various authors, and they are known as insecticides which are highly toxic under laboratory conditions to fish and other aquatic organisms (*53*) and are very easily degraded in the natural environment (*54*). There are three main chemical reactions involved in this degradation i.e., isomerisation, hydrolysis and oxidation. Isomerisation usually involves the cyclopropane ring and is initiated by sunlight, but the process may be affected by the presence of pigments or humic substances (*55*). Hydrolysis occurs at the ester bond and results in a breaking of the parent molecule into two fragments, the acid and alcohol moieties. Oxidation may occur at a variety of sites.

The stereochemical structure of pyrethroid insecticides greatly influences their toxicity to aquatic organisms (*56*). *Cis* 1*R*, α*S*-deltamethrin is the only enantiomer present in the product registered for agricultrual use (*55*). However, this parent compound (isomer 1) has been shown to be converted to three other isomers in natural water exposed to sunlight (i.e., *cis* 1*S*, α*S* (isomer 2'), *trans* 1*R*, α*S* (isomer 3), and *trans* 1*S*, α*R* (isomer 4')) (*57*) (Figure 5). Four other isomers (1', 2, 3' and 4) are known to exist but have not been found in treated water. Day and Maguire (*6*) found that of these isomers, only isomers 1,2,3 and 4 were toxic to juvenile *Daphnia magna*, with the parent compound being approximately 10X more toxic. These results indicate therefore that isomerization of isomer-1 to isomer-3 in natural water is only a partial detoxification step as far as some aquatic organisms are concerned, although isomer-3 has not been reported in ponds oversprayed with deltamethrin (*58,59*). The isomer pair (3+3') has, however, been found on pasture forage and litter (*60*) and alfalfa (*61*).

There have been several studies on the toxicities of the products of ester hydrolysis of pyrethroids to aquatic organisms (*6,7,53*). The major degradation products of these insecticides are considerably more polar than the parent molecules and all have been shown to be much less toxic to fish and invertebrates than the parent compounds (Table I).

In contrast, Stratton and Corke (*7*) found that two to five of the degradation products of permethrin were significantly more toxic towards algae and cyanobacteria than the parent compound. For example, permethrin is relatively nontoxic towards phototrophic microorganisms, with EC_{50} values for growth and photosynthesis being > 10 and > 100 mg/L, respectively; however, 3-phenoxybenzaldehyde (PBald) and 3-phenoxybenzyl alcohol (PBalc), followed by benzoic acid, 3-hydroxybenzoic acid and 3-phenoxybenzoic acid had values ranging from 2 to 6 mg/L for growth and 30 to 70 mg/L for photosynthesis. The cyanobacteria were more sensitive than green algae and the authors attributed this to the basic cellular organizational differences between these organisms (procaryotic vs. eucaryotic cells). However, the concentrations used

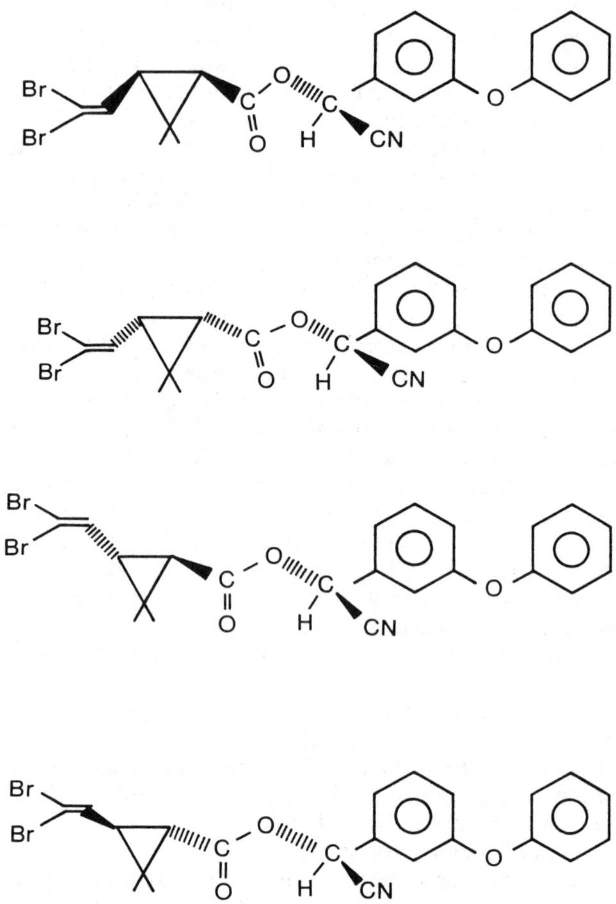

1 - isomer
1R, 3R, αS

2′- isomer
1S, 3S, αS

3 - isomer
1R, 3S, αS

4′ - isomer
1S, 3R, αS

FIGURE 5. Toxic isomers of deltamethrin.

TABLE I. ACUTE TOXICITY OF PYRETHROID METABOLITES

Chemical	*Daphnia magna* 48 h EC$_{50}$ μg/L	Fish 96 h LC$_{50}$ μg/L	Reference
3-(2,2-dichlorovinyl)-2,3-dimethyl cyclopropane carboxylic acid (DCVA)[a]	130,000	3,000	53
3-(2,2-dibromovinyl)-2,2-dimethylcyclopropane carboxylic acid (DBCA)[b]	> 50	-	6
(1RS)-cis-3-(chloro-3,3,3-trifluoroprop-1-enyl)-2,2-dimethyl cyclopropane carboxylic acid[c]	100,000	> 16,000	53
3-phenoxybenzyl alcohol	10,000	3,000-7,000	53
(PBalc)	> 25	-	6
3-phenoxybenzoic acid	85,000	13,000-36,000	53
(PBacid)	> 25	_	6
3-phenoxybenzaldehyde (PBald)	> 25	-	6
2-(4-chloropheynl)-3-methyl butyric acid	-	> 10,000	53

[a] degradation product of permethrin
[b] degradation product of deltamethrin
[c] degradation product of lambda-cyhalothrin

in this study were much higher than concentrations expected in natural waters contaminated with pyrethroids (53). Therefore, it is the toxicity of the parent pyrethroid molecule (or its active isomers) rather than its degradation products which are of potential concern in aquatic ecosystems.

Toxicity of the Transformation Products of Herbicides to Aquatic Biota

Relatively fewer data exist on the effects of transformation products of herbicides on aquatic biota compared to those available for insecticides.

Triazine herbicides are a group of heterocyclic nitrogen compounds which have been largely responsible for the substantial increases in corn yields in North America observed over the last 25 years (62). Of these herbicides, atrazine (2-chloro-4-ethylamino-6-isopropylamino-1,3,5-triazine) is the most heavily applied agricultural pesticide in North America and it is used to control broadleaf and grassy weeds in corn and sorghum (63).

Atrazine has been detected in lakes and streams at levels ranging from 0.1 to 30 μg/L with peak concentrations up to 1000 μg/L known to occur in surface runoff from agricultural fields adjacent to bodies of water during times of application (64). Many studies have been conducted on the effects of atrazine on various species of aquatic flora under controlled conditions and these studies have found that at concentrations of 1 to 5 μg/L and exposure periods of 5 min. to 7 weeks, adverse effects on photosynthesis, growth and oxygen evolution of aquatic plants have occurred. Higher concentrations have altered species composition, reduced carbon uptake and reduced reproduction (63).

The half-life of atrazine in aquatic environments has been shown to range from 3.2 days to 7 to 8 months (63). The major route of degradation is thought to be hydrolysis to hydroxyatrazine (2-hydroxy-4-ethylamino-6-isopropylamino-1,3,5-triazine) although N-dealkylation through removal of the ethyl- or the isopropyl group (Figure 6) has also been shown to occur (65). Few studies have measured the concentrations of these metabolites in the natural environment and only one study to date has examined the effects of these degradation products of atrazine on aquatic biota. Stratton (8) found that atrazine was 4 to 10 times more effective than its transformation products in producing reductions in growth, inhibition of photosynthesis, and acetylene-reducing ability in two species of green algae, *Chlorella pyrenoidosa* and *Scenedesmus quadricauda*, and three species of cyanobacteria, *Anabaena*. For example, atrazine reduced growth by 50% at 0.03 to 5.0 mg/L and inhibited photosynthesis by 50% at 0.1 to 0.5 mg/L. Comparable values for deethylated atrazine (2-chloro-4-amino-6-isopropylamino-1,3,5-triazine) were 1.0 to 8.5 mg/L for growth reduction and 0.7 to 4.8 mg/L for photosynthetic inhibition. For deisopropylated atrazine (2-chloro-4-ethylamino-6-amino-1,3,5-triazine) these values were 2.5 to >10 mg/L and 3.6 to 9.3 for the same physiological functions. Hydroxyatrazine and diaminoatrazine (2-chloro-4,6-diamino,1,3,5-triazine) were nontoxic to most cultures tested. Acetylene reduction with cyanobacteria was found to be insensitive to all of the test compounds with the exception of atrazine which had an EC_{50} of 55 mg/L to *Anabaena inaequalis*. This study concluded that atrazine degradation products would not normally be present in the aquatic environment at levels inhibitory to algae and cyanobacteria.

Several other transformation products of herbicides have been studied for

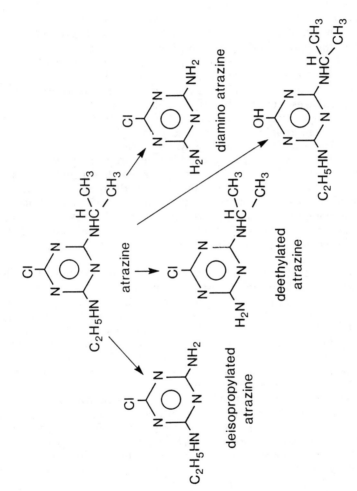

FIGURE 6. Some transformation products of atrazine.

their effects on photosynthetic microorganisms. For example, 3,4-dichloroaniline, a major degradation product of the amide herbicide, propanil (3',4'-dichloropropionanilide) has been shown to be up to 10X less inhibitory than the parent compound towards phototrophic organisms (66). In addition, this metabolite had no effect on algal populations in experimental enclosures treated with this chemical (67).

The effects of the carbanilate, chlorpropham (isopropyl m-chlorocarbanilate) and 3-chloroaniline, a metabolite, on populations of the cyanobacterium, *Anacyctis midulans* and the alga, *Chlamydomonas reinhardii* were monitored in large scale batch cultures by Maule and Wright (68). 3-Chloroaniline was less inhibitory than the parent herbicide, chlorpropham; for example, 6.1 to 9.5 μg chlorpropham/mL inhibited growth rates and final yield of *A. nidulans* vs. 5.4 to 19.9 μg 3-chloroaniline/mL for the same parameters. Similar results were obtained for *C. reinhardii*. In addition, chlorpropham caused marked morphological changes such as increase in cell size, the formation of multi-layered cell envelopes, etc. at concentrations which still permitted substantial growth whereas 3-chloroaniline produced no such changes.

Triclopyr (3,5,6-trichloro-2-pyridinyloxyacetic acid) (Figure 7) is the active ingredient in several relatively new herbicides which are formulated as either an amine (Garlon 3A) or an ester (Garlon 4). These herbicides are used in selective post-emergent control of woody plants and broadleaf vegetation in forest site preparation and conifer release programs. In such applications, these chemicals can potentially reach surface water through drift or inadvertent overspray of aquatic areas and this raises concerns about the potential hazards of the active ingredient or its transformation products to aquatic life, particularly fish.

Wan *et al.* (3) evaluated the acute toxicity of both formulated products, technical triclopyr and several transformation products to juvenile Pacific salmonids and found that the order of increasing toxicity as LC_{50} was: Garlon 3A (347 ± 44 mg/L), triclopyr (7.9 ± 0.7 mg/L), pyridine (3.7 ± 0.8 mg/L), pyridinol (2.1 ± 0.2 mg/L), Garlon 4 (2.0 ± 0.2 mg/L) and triclopyr ester (0.7 ± 0.2 mg/L). Both triclopyr and the amine formulation, Garlon 3A, were less toxic than both the triclopyr ester and the major transformation products; the metabolic product, pyridinol, was as toxic as the formulated triclopyr product, Garlon 4, which contains the triclopyr ester.

The triclopyr ester (2-butoxyethyl 3,5,6-trichloro-2-pyridinyloxyacetate) has a short half-life and undergoes hydrolysis in water and soil to the active acid, triclopyr, within 6 to 24 h (69). The acid is resistant to further hydrolysis but susceptible to photolysis and transformation by microorganisms to 3,5,6-trichloro-2-pyridinol and CO_2. Absorption by fish and metabolism to the acid form of the chemical and excretion of the acid back into the water is also a known route of transformation (3). Under field conditions, the concentration of Garlon 3A in a stream unintentionally oversprayed during an aerial operation would not likely exceed a level greater than 10 mg/L in 15 cm water even at the highest recommended rate of application (i.e., 10 kg active ingredient/ha). Therefore, the potential for this formulation or its transformation products to cause fish kills is small when it is used under prescribed conditions. However, the use of the lower recommended rate of Garlon 4, (i.e., 2.4 kg active ingredient/ha), has some potential to generate toxic concentrations (approximately 4 mg/L in

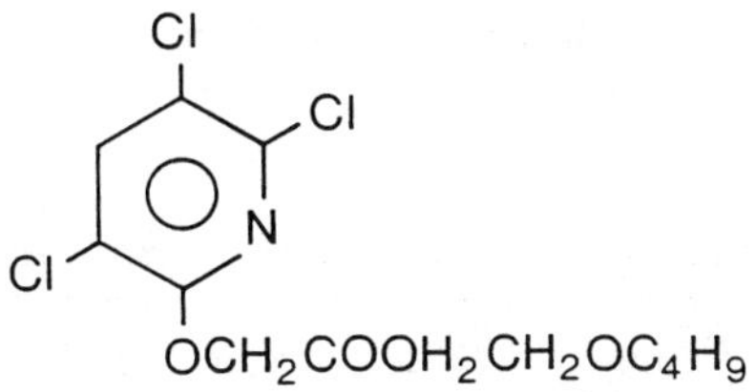

Triclopyr ester

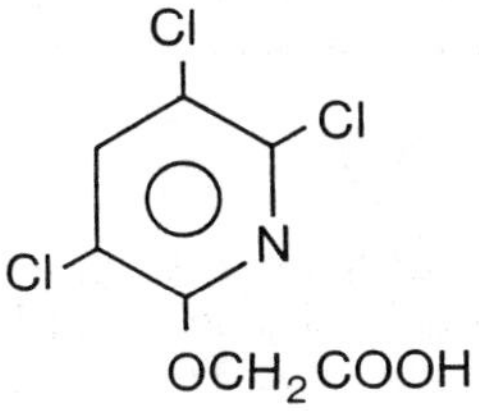

Triclopyr

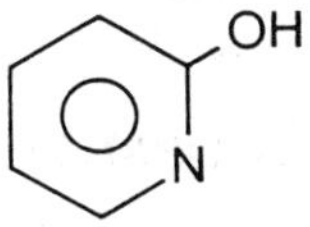

Pyridinol

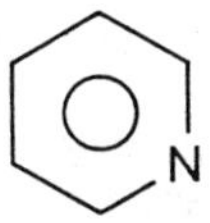

Pyridine

FIGURE 7. Some transformation products of triclopyr.

15 cm water) if the residues of the parent compound or its degradation products are not rapidly diluted and flushed out of the aquatic system.

The phenoxy alkanoic acid herbicides such as 2,4-D (2,4-dichlorophenoxy acetic acid) are also formulated as esters or amine salts to improve their solubility in oil or water respectively. The ester formulations (e.g., butoxyethanol ester (BEE) and propylene glycol butyl ether ester (PGBEE)) have been shown to rapidly convert to the acid in water; for example, approximately 21 and 38% hydrolysis of BEE and PGBEE, respectively, to 2,4-D acid occurred within a 3 h 50% water volume replacement time in continuous flow bioassays with chinook salmon (*Oncorhynchus tshawytscha*) and steelhead-rainbow trout (*Salmo gairdneri*) (*70*).

The acute toxicities of the ester formulations are approximately 100X more toxic than the acid formulation (*71*) presumably due to the quicker absorption of nonpolar (esters) rather than polar (acid) compounds via passive transport through the gill membranes of fish. Thus the rapid conversion of the ester results in the reduced toxicity of 2,4-D to fish although the toxicity of other transformation products to fish and other aquatic organisms has not been tested.

Other Pesticides

Tributyltin compounds (e.g., bis (tributyltin oxide) and tributyltin fluoride) are biocides used in organotin antifouling paint formulations which are classified as pesticides under the Pest Control Products Act in Canada and FIFRA in the U.S.A. They are known to be extremely toxic to aquatic organisms and several countries have restricted their use (*72*). Tributyltin degrades by successive debutylation to dibutyltin, monobutyltin and finally to inorganic tin. Studies on the acute toxicity of these transformation products to aquatic invertebrates and algae have shown that tributyltin is the most toxic chemical with toxicity decreasing with successive degradation (Table II).

TABLE II. TOXICITY OF TRANSFORMATION PRODUCTS OF TRIBUTYLTIN COMPOUNDS

SPECIES	TEST	COMPOUND	CONCENTRATION	REFERENCE
Daphnia magna	24 h EC_{50}	Tributlytin	0.013 mg/L	73
		Dibutlytin	49.0 mg/L	"
		Monobutyltin	0.9 mg/L	"
Ankistrodesmus falcatus	IC_{50}	Tributlytin	0.02 mg/L	74
		Dibutyltin	6.8 mg/L	"
		Monobutyltin	25.0 mg/L	"
Skeletonema costatum	EC_{50}	Tributyltin	0.36 µg/L	75
		Dibutyltin	40.0 µg/L	"
S. costatum	LC_{50}	Tributyltin	11.5 µg/L	75
		Dibutyltin	>500.0 µg/L	"

Very little information on the toxicity of other transformation products of pesticides, particularly fungicides, to aquatic biota is available in the open literature.

Discussion

The transformation and/or degradation of pesticides in the natural environment results in chemicals which have the potential to be deleterious to non-target organisms. A review of the scientific literature for information on the effects of these transformation products on aquatic biota has indicated that such compounds can be less, more or similar in toxicity when compared to the parent chemical (Table III). In many cases, the toxicity of these products is less than the toxicity of the parent compound, particularly with herbicides, although this is somewhat dependent upon the group of organism tested i.e., plant or animal and the specific metabolite. Several metabolites of organophosphate and carbamate insecticides can be more toxic than the parent compound or similar in toxicity e.g., fenitrooxon and 3-methyl-4-aminophenol of fenitrothion; AA and MAA of aminocarb; 1-naphthol of carbaryl. More information on a wider variety of pesticides, particularly organophosphates, carbamates and synthetic pyrethroids would be necessary to conclude that the transformation products of most pesticides are less toxic than the parent compound.

There are several factors which must be considered when evaluating the hazards of transformation products of pesticides in aquatic ecosystems. The breakdown and/or conversion of pesticides to their respective metabolites in the natural environment is a continuous process which occurs from the time of application (or before) and beyond. The nature of the compounds formed, the rate at which they appear and disappear, their concentrations and their compartmentalization into various parts of the ecosystem are dependent upon the physical, biological and chemical properties of the system involved. These complicating factors make it difficult to estimate the actual concentrations and the length of time to which organisms are exposed to any given metabolite in the environment. Some metabolites may not be biologically available to organisms or are present for such a limited duration (e.g., minutes to hours) that there is no time for a toxic reaction and response to occur. On the other hand, the uptake, storage and/or eventual metabolism of pesticides by biota and/or the microbial release of pesticides bound to particulate organic matter over time may result in a continuous input of low levels of toxic metabolites into the environment. Few studies have examined the sublethal, long-term effects of persistent transformation products on biota in the aquatic environment.

Much of the available information on the toxicity of transformation products to specific organisms is based on laboratory toxicity data where conditions are very different from those in the field. For example, the presence of natural sunlight produces radiation of various wavelengths which cannot be simulated in the laboratory and which can cause photodegradation and transformation. In addition, most laboratory tests are static and concentrations of chemicals may not diminish as rapidly as they would under field conditions especially in the lotic (running water) environment. The concentrations of transformation products used in laboratory toxicity tests often exceed by several orders of magnitude the actual concentrations of chemicals which occur following

TABLE III. RELATIVE TOXICITY OF PESTICIDE TRANSFORMATION PRODUCTS TO THE PARENT COMPOUND

CHEMICAL & METABOLITE	SENSITIVITY		
	LESS	EQUAL	MORE
Insecticides			
DDT			
TDE	ab		ab
DDE	c	ac	a
Endosulfan			
endosulfan sulfate			a
endosulfan diol	a	c	
Aldrin			
photodieldrin	c		
Endrin			
ketoendrin	c		
Fenitrothion			
fenitrooxon			b
aminofenitrothion	a		
carboxyfenitrothion	a		
demethylfenitrothion	a		
3-methyl-4-nitrophenol	a		
3-methyl-4-aminophenol			a
Malathion			
diethyl fumarate			a
dimethyl phophorodithioic acid	a		
Aminocarb			
AA			a
MAA			a
MFA	a		
FA	a		
Aldicarb			
aldicarb sulfoxide		ab	
aldicarb sulfone	ab		
Mexcarbate			
4-amino 3,5-xylenol			a
4-dimethylamino-3,5-xylenol	a		

Continued on next page

TABLE III. (cont.)

CHEMICAL & METABOLITE	SENSITIVITY		
	LESS	EQUAL	MORE
Insecticides			
Carbaryl			
1-naphthol			ab
Deltamethrin			
isomer 2'	ab		
isomer 3	ab		
isomer 4'	ab		
PBald	ab		c
PBacid	ab		c
PBalc	ab		c
Herbicides			
Atrazine			
deethylated	c		
deisopropylated	c		
diaminoatrazine	c		
hydroxyatrazine	c		
Propanil			
3,4-dichloroaniline	c		
Chlorpropham			
3-chloroaniline	c		
Triclopyr			
pyridinol		a	
pyridine	a		
Biocides			
Tributyltin			
dibutyltin	bc		
monobutyltin	bc		

a = fish; b = invertebrates; c = algae

application. Realistic concentrations of residues of pesticide transformation products may be below the limits of detection. In addition, few field studies monitor for residues of transformation products under field conditions.

An area that requires further research is an evaluation of the interactive effects of the parent compound, its various transformation products and other formulation adjuvants on the toxic response of the non-target organism. No contaminant is ever present alone in the environment but is always in association with other organic and inorganic chemicals including its own transformation products. Stratton (8) found that when atrazine and its dealkylated breakdown products were combined and tested against the blue-green alga, A. inaequalis, synergistic, antagonistic and additive interaction responses were observed depending upon the test system employed. For example, whenever atrazine was present in a mixture with deisopropylated- or deethylated-atrazine, antagonism occurred using the photosynthetic response as a parameter i.e., the apparent inhibitory effects of the individual toxicants were reduced. In contrast, a synergistic interaction was recorded for culture growth i.e., toxicity was enhanced. Combinations of deethylated- and deisopropylated-atrazine interacted antagonistically towards photosynthesis and additively towards the growth yield. Additive interactions occur when the overall toxicity of the mixture is greater than that of the component compounds. Bender (39) found that a pronounced synergistic effect was demonstrated between malathion and its two basic hydrolysis products, diethyl fumarate and dimethyl phophorodithioic acid, when fathead minnows (P. promelas) were exposed continuously for 14 d to these compounds.

The interactive effects of pesticides, their transformation products, and any other chemicals (toxic or non-toxic) could lead to situations in the natural environment where degradation products of low individual toxicity still pose a serious threat to non-target organisms when in combination. Such situations could easily arise in the field when pesticides are used over a prolonged time period (days) or when several applications are applied during the growing season

The determination of the toxicity of pesticide transformation products to aquatic biota is a difficult task due to the uncertainties of concentrations, duration of exposure and availability; however, for new pesticides being considered for registration under the Pest Control Products Act in Canada, it has been suggested that toxicity tests be conducted on any transformation product which is present at concentrations > 10% of the applied pesticide or accumulates over the course of laboratory transformation and/or field dissipation studies (Commercial Chemicals Branch, Environment Canada, personal communication). The toxicity of such transformation products would initially be tested at a concentration comparable to that which results from an application at the maximum label-recommended rate to provide a worst-case scenario. If no toxicity is observed, no further testing would be required; however, if toxicity is observed, a number of lower concentrations would be tested to generate a dose-response curve. Such extensive testing of toxic transformation products of new chemicals coming on the market would ensure that pesticides with very toxic breakdown products or those chemicals which convert to more toxic isomers would be detected.

Literature Cited

1. Miyamoto, J.; Mikami, N.; Kihara, K.; Takimoto, Y.; Kohda, H.; Suzuki, H. *J.Pesticide Sci.* **1978**, *3*, 35.
2. Szeto, S.Y.; Sundaram, K.M.S; Feng, J. *J. Environ. Sci. Health* **1985**, *B20*, 559.
3. Wan, M.T.; Moul, D.J.; Watts, R.G. *Bull. Environ. Contam. Toxicol.* **1987**, *39*, 721.
4. Hiraoka, Y.; Tanaka, J.; Okuda, H. *Bull. Environ. Contam. Toxicol.* **1990**, *44*, 210.
5. Foran, J.A.; Germuska, P.J.; Delfino, J.J. *Bull. Environ. Contam. Toxicol.* **1985**, *35*, 546.
6. Day, K.E.; Maguire, R.J. *Environ. Toxicol. Chem.* **1990**, *9*, 1297.
7. Stratton, G.W.; Corke, C.T. *Environ. Poll.* **1982**, *29*, 71.
8. Stratton, G.W. *Arch. Environ. Contam. Toxicol.* **1984**, *13*, 35.
9. Fuhremann, T.W.; Lichtenstein, E.P. *J.Agric. Food Chem.* **1980**, *28*, 446.
10. WHO Task Group. *DDT and its derivatives - Environmental Aspects*; Environmental Health Criteria 83; World Health Organization; Finland; **1989**, 98 pp.
11. Sanders, H.O.; Cope, O.B. *Limnol. Oceanog.* **1968**, *13*, 112.
12. Kouyoumjian, H.H.; Uglow, R.F. *Environ. Pollut.* **1974**, *7*, 103.
13. Gardner, D.R. *Pest. Biochem. Physiol.* **1973**, *2*, 437.
14. Mayer, F.L.; Ellersieck, M.R. *Manual of acute toxicity: interpretation and data base for 410 chemicals and 66 species of freshwater animals*; US Fish and Wildlife Ser. Publ. No. 160; Washington, DC; **1986**, 274 pp.
15. Mayer, F.L. *Acute toxicity handbook of chemicals to estuarine organisms*; National Technical Information Ser.; NTIS PB87-188686; Washington, DC; **1987**, 274 pp.
16. Peterson, R.H. *J.Fish Res. Board Can.* **1973**, *30*, 1091.
17. Luard, E.J. *Phycologia* **1973**, *12*, 29.
18. Bowes, G.W.; Gee, R.W. *J. Bioenergetics*, **1971**, *2*, 47.
19. Mosser, J.L.; Teng, T.; Walther, W.G.; Wurster, C.F. *Bull. Environ. Contam. Toxicol.* **1974**, *12*, 665.
20. Butler, G.L. *Residue Rev.* **1977**, *66*, 19.
21. Priyamvada Devi, A.; Rato, D.M.R.; Tilak, K.S.; Murty, A.S. *Bull. Environ. Contam. Toxicol.* **1981**, *27*, 239.
22. Cotham, W.E.; Bidleman, T.F. *J. Agric. Food Chem.* **1989**, *37*, 824.
23. National Research Council Canada. *Endosulfan: Its Effects on Environmental Quality*; NRCC No. 14098; Ottawa, Canada, **1975**, 100 pp.
24. Knauf, W.; Schulze, E.F. **1973**. Cited in National Research Council Canada, *Endosulfan: Its Effects on Environmental Quality*; NRCC No. 14098; Ottawa, Canada, 1975, 100 pp.
25. Geike, F. *Z. angew. Entomol.* **1970**, *65*, 98.
26. Goebel, H.; Gorbach, S.; Knauf, W.; Rimpau, R.H.; Huttenbach, H. *Residue Rev.* **1982**, *83*, 1.
27. Wan, M.T. *J. Environ. Sci. Health.* **1989**, *B24*, 183.
28. Batterton, J.C.; Boush, G.M.; Matsumura, F. *Bull. Environ. Contam. Toxicol.* **1971**, *6*, 589.
29. Patil, K.C.; Matsumura, F.; Boush, G.M. *Environ. Sci. Technol.* **1972**, *6*, 629.
30. Dauterman, W.C. *WHO Bull.*, **1971**, *44*, 133.

31. Roberts, J.R.; Greenhalgh, R.; Marshall, W.K. *Fenitrothion: The Long-Term Effects of its Use in Forest Ecosystems*; NRCC No. 16073; Ottawa, Canada, **1977**, 628 pp.
32. Weinberger, P.; Greenhalgh, R.; Sher, D.; Ouellette, M. *Bull. Environ. Contam. Toxicol.* **1982**, *28*, 484.
33. Fairchild, W.L.; Ernst, W.R.; Mallet, V.N. In *Environmental Effects of Fenitrothion Use in Forestry*; Ernst, W.R.; Pearce, P.A.; Pollock, T.L. (Eds.); Department of the Environment, Ottawa, Canada, **1989**; pp. 109-166.
34. Moody, R.P.; Greenhalgh, R.; Lockhart, L.; Weinberger, P. *Bull. Environ. Contam. Toxicol.* **1978**, *19*, 8.
35. Ohmae, T.M.; Uno, T.; Okada, T.; Onji, Y.; Terada, I.; Tanigawa, K. *J. Pestic. Sci.* **1981**, *6*, 437.
36. Wright, J.A.; Hermonat, M.W.; Hards, R.G. *Bull. Environ. Contam. Toxicol.* **1982**, *28*, 480.
37. Kovacicova, J.; Batora, V.; Truchlik, S. *Pestic. Sci.* **1973**, *4*, 759.
38. Muhlman, V.R.; Schrader, G. *Z. Naturf.* **1957**, *12*, 196.
39. Bender, M.E. *Wat. Res.* **1969**, *3*, 571.
40. National Research Council of Canada *Aminocarb: The Effects of its Use on the Forest and the Human Environment*; NRCC No. 18979; Ottawa, Canada, **1982**, 253 pp.
41. Leger, D.A.; Mallet, V.N. *J. Agri. Food Chem.* **1988**, *36*, 185.
42. Lamb, D.W.; Roney, D.J. Mobay Chem. Corp. Report 44208 **1975**, cited in Szeto *et al.*, 1985.
43. Coady, L.W. Canada Dept. Environ. Surveillance Report EPS-5-AR-78-1; **1978**, Ottawa, Canada
44. Szeto, S.Y.; Holmes, S.B. *J. Environ. Sci. Health.* **1980**, *B17*, 51.
45. Ernst, W.R.; Julien, G.; Doe, K.; Parker, R. Environmental Protection Service EPS-5-R-81-3, **1981**, 51 pp.
46. Miles, C.J.; Delfino, J.J. *J. Agri. Food Chem.* **1985**, *35*, 455.
47. Baron, R.L.; Merriam, T.L. *Rev. Environ. Contam. Toxicol.* **1988**, *105*, 1.
48. Moye, H.A.; Miles, C.J. *Rev. Environ. Cont. Toxicol.*, **1988**, *105*, 99.
49. Clarkson, V.A. Union Carbide Agricultural Products Co. Inc., Report File No. 10493, **1968**. Cited in Baron, R.L.; Merriam, T.L. *Rev. Environ. Contam. Toxicol.*, **1988**, *105*, 1.
50. Stewart, N.E.; Millemann, R.E.; Breese, W.P. *Trans. Am. Fish Soc.* **1967**, *96*, 25.
51. Butler, J.A., Millemann, R.E.; Stewart, N.E. *J. Fish Res. Bd Can.* **1968**, *25*, 1621.
52. Murty, A.S. *Toxicity of Pesticides to Fish*; CRC Press Inc, Boca Raton, Florida, **1986**; p. 27-33.
53. Hill, I.R. *Pestic. Sci.* **1989**, *27*, 429.
54. Demoute, J. *Pestic. Sci.* **1989**, *27*, 385.
55. National Research Council of Canada. *Pyrethroids: Their Effects on Aquatic and Terresrial Ecosystems*, NRCC No. 24376; Ottawa, Canada, **1986**, 303 pp.
56. Bradbury, S.P.; Symonik, D.M.; Coats, J.R.; Atchison, G.J. *Bull. Environ. Contam. Toxicol.* **1987**, *38*, 727.
57. Maguire, R.J. *J. Agri. Food Chem.* **1990**, *38*, 1613.
58. Muir, D.C.G.; Rawn, G.P.; Grift, N.P. *J.Agric. Food Chem.* **1985**, *33*, 603.

59. Maguire, R.J.; Carey, J.H.; Hart, J.H.; Tkacz, R.J.; Lee, H. *J. Agric. Food Chem.* **1989**, *37*, 1153.
60. Hill, B.D.; Johnson, D.L. *J. Agric. Food Chem.* **1987**, *35*, 373.
61. Hill, B.D.; Inaba, D.J.; Charmetski, W.A. *J. Agric. Food Chem.* **1989**, *37*, 1150.
62. McEwen, F.L.; Stephenson, G.R. *The Use and Significance of Pesticides in the Environment*; John Wiley & Sons, New York, **1979**; 538 pp.
63. Eisler, R. *Atrazine Hazards to Fish, Wildlife, and Invertebrates: A Synoptic Review*, U.S. Fish Wildl. Serv. Biol. Rep. *85*, 53 pp.
64. DeNoyelles, F.; Kettle, W.D.; Sinn, D.E. *Ecology* **1982**, *63*, 1285.
65. Jones, T.W.; Kemp, W.M.; Stevenson, J.C.; Means, J.C. *J. Environ. Qual.*, **1982**, *11*, 632.
66. Wright, S.J.L.; Stainthorpe, A.F.; Downs, J.D. *Acta Phytopathol. hung*, **1977**, *12*, 51.
67. Kuiper, J.; Hanstveit, A.O. *Ecotoxicol. Environ. Safety* **1984**, *8*, 34.
68. Maule, A.; Wright, S.J.L. *J. App. Bacteriol.* **1984**, *57*, 369.
69. McCall, P.J.; Gavit, P.D. *Environ. Toxicol. Chem.* **1986**, *5*, 879.
70. Finlayson, B.J.; Verrue, K.M. *Arch. Environ. Contam. Toxicol.* **1985**, *14*, 153.
71. Dodson, J.J.; Mayfield, C.I. *Trans. Amer. Fish. Soc.* **1979**, *108*, 632.
72. Maguire, R.J. *Appl. Organometal Chem.* **1987**, *1*, 475.
73. Vighi, M.; Calamari, D. *Chemosphere* **1985**, *14*, 1925.
74. Wong, P.T.S.; Chau, Y.K.; Kramar, O.; Bengert, G.A. *Can. J. Fish. Aquat. Sci.* **1982**, *39*, 483.
75. Walsh, G.E.; McLaughlan, L.L.; Lores, E.M.; Louie, M.K.; Deans, C.H. *Chemosphere* **1985**, *14*, 383.

RECEIVED November 8, 1990

Chapter 17

Toxicological Significance of Bound Residues in Livestock and Crops

L. J. Lawrence and M. R. McLean

Pharmacology and Toxicology Research Laboratory,
3945 Simpson Lane, Richmond, KY 40475

The difficulty in establishing the chemical nature and in evaluating the
toxicological significance of bound drug and pesticide residues in
edible livestock and crop tissues presents a dilemma for both the
scientist and the regulatory agencies, i.e., the EPA and the FDA.
Bioavailability studies are one approach to assessing of the
toxicological significance of bound residues. The bioavailability to
rats of three agrochemicals and the bound residues which result from
their use in cattle, swine and strawberry fruit are presented. Febantel,
an anthelmintic used in the beef cattle industry, is 85.7% bioavailable
whereas bound residues of febantel are only 0.3% bioavailable.
Results from similar studies with olaquindox, an antibiotic and growth
promoter used in swine, and chloropicrin, a fumigant used in the
production of strawberries, are also presented.

In the strictest sense the toxicological significance of any chemical, natural or
synthetic, is based on the establishment of a dose-response relationship, the most
fundamental and pervasive principle in toxicology (1). For most compounds,
especially the newer drugs and pesticides, this relationship is correlative in nature
since, in general, there is a degree of uncertainty about the true identity of the toxic
agent(s), the actual molecular site of the toxic event, or both. These uncertainties
have their basis in biological considerations which, for convenience and clarity, are
divided here into two components. The first consideration is that the administered
chemical is subject to absorption, distribution, biotransformation and elimination
which, of course, depend upon the physiochemical characteristics of the molecule as
well as the biochemical and physiological characteristics of the individual organism.
These processes, generally referred to as toxicokinetic parameters, play a crucial role
in determining both the identity and the concentration of the potentially toxic agent(s).
Secondly, the biological response to any chemical varies depending upon both the
presence or the absence of a susceptable molecular target site as well as the
biochemical characteristics of the molecular target (2). In addition, there may be
more than one site of action, which further complicates the situation. In practice, the
objective is to identify a measurable end-point that is correlated with the toxic event.
    Furthermore, each of the factors, i.e., the dose, the toxicokinetic parameters
and the target (organism), are intimately related to each other. Therefore it is

0097–6156/91/0459–0242$06.00/0

inappropriate from a toxicological viewpoint, to isolate one variable from the other when assessing the toxicity of a compound (*3*).

Toxicological Significance of Bound Residues. Weber (*4*) has defined bound residues as those residues (parent or degradation product) which are covalently bound to biological macromolecules; are not extractable from the macromolecules by exhaustive extraction, denaturation, or solubilization techniques; and are not the result of endogenous incorporation. Therefore, it follows that establishing the chemical nature of bound residues is not easily accomplished. The matrix of bound residues can be both ill-defined and difficult to formulate into the traditional "neat" dose for the purposes of toxicity testing. These problems as well as the practical limitations in obtaining sufficient quantities of the bound residue limits the determination of a dose-response relationship in bioassays and laboratory animals. However, one can quantify the absorption of the bound residues from the gastrointestinal tract of laboratory animals and, in combination with other data, be in a position to make inferences about the toxicological significance of the bound residue. This approach has merit since a response can only be expressed when the potential toxicant, in sufficient concentration, reaches its site of action.

The technique of determining bioavailability as proposed by Gallo-Torres (*5*) is the most advanced and comprehensive method for quantifying the intestinal absorption of bound residues, or any radiolabelled compound. For an informative overview of the characteristics of the gastrointestinal tract as they relate to the bioavailability method the reader is referred to the work published by Huber *et al.* (*6*)

Classification of Residues

In many cases drug and pesticide residues in edible livestock and crop tissues are for the most part extractable using standard residue analysis techniques. This is generally accomplished by using extracting solvents, singly or in combination, at a reasonable temperature, i. e., homogenization at room temperature, refluxing or Soxhlet extraction. Commonly used are polar and nonpolar solvents, buffer or salt solutions at various pHs, enzyme treatments or mild acids and bases. Although the solubility characteristics of the parent drug or pesticide serve as guidelines, multiple extraction conditions are necessary since, in addition to covalent binding, an unextractable residue may result from ionic interactions, hydrogen bonding, van der Waal's forces or hydrophobic bonding (*6*).

The extractable residue typically consists of the parent compound, the phase I metabolites (*e. g.*, dehalogenated, desulfurated, hydroxylated, hydrolyzed, oxidized or reduced products) and the Phase II conjugates (*e. g.*, to an amino acid, glutathione, glucose, sulfate or glucuronic acid) (*7*). The parent and the Phase I products are usually extractable with organic solvents. The conjugates are soluble in polar solvents and, in fact, are frequently referred to as "water-soluble metabolites" (*8*). In most cases the quantification and the identification of the extractable residue(s) is a routinely successful endeavor however this does not, of course, constitute a toxicological evaluation of the extracted residue (*9*). Indeed, since the extractable metabolites usually retain chemical properties similar to the parent drug or pesticide, structure-activity relationships can be employed for this purpose. In addition, bioassay techniques (Ames test, acute and chronic toxicity studies, etc.) are available, although there is no one battery of tests that can fully address the potential toxicity of metabolites (*9*). Furthermore, the practical limitations of time, expense and technical knowledge forbid a thorough toxicological evaluation of every extractable metabolite/residue resulting from every drug and pesticide.

In some cases residues are found to be distributed into practically every class of biomolecule, *e. g.*, carbohydrates, proteins, nucleic acids, lipids, lignin, polyketides and terpenes. This may come about by the degradation of part or all of the drug or pesticide to carbon dioxide or to products identical to normal enzyme substrates. Obviously these "residues" are not a health hazard to humans, yet for the analyst the diversity of the compounds makes accurate quantitation and identification difficult if not impossible. The problem is further compounded when a large portion of the endogenously incorporated residue is in the unextractable fraction. It follows then that the correct positioning of the radiolabel within the molecule is critically important to limit these complications.

In every case the initial work focuses on developing extraction techniques for the solubilization of the residue. Of course, there are many cell organelles and macromolecules which are insoluble under these conditions yet may contain some form of covalently bound residue. In some instances rigorous extracting conditions (e. g., high temperature distillation (10) or strong mineral acids and bases) may be successfully employed to release the bound residues. Nevertheless, the usefulness of these techniques must be evaluated on a case by case basis since the degradation products are almost certainly not representative of the residue *in situ*. In fact, the value of identifying the extracted product(s) and more importantly evaluating their toxicological significance is probably lost.

In summary, the residue chemist typically categorizes drug and pesticide residues as (11): 1) free metabolites, 2) conjugated metabolites, 3) natural products and 4) bound.

Genesis of Bound Residues

As is known to every toxicologist and pesticide chemist, Phase I enzymes introduce into xenobiotics, which are typically lipophilic, a reactive functional group (7, 11-13) During Phase II metabolism small endogenous polar molecules are conjugated to the reactive group, such as a hydroxide moiety, thereby increasing the water solubility of the xenobiotic (7, 13) In principle and practice these reactions serve to both detoxify and to promote the elimination, via the bile or urine, of the xenobiotic from the body. Nevertheless, it is also the case that the toxicity of many xenobiotics results from their bioactivation to electrophilic metabolites (Phase I) which covalently bind to nucleophilic centers of macromolecules, primarily proteins and nucleic acids (14-16). Metabolically derived electrophiles include epoxides, free radicals, various carbocations and nitrinium ions. Of course electrophilic moieties may also be found in the parent molecule; aldehydes, polarized double bonds, alkyl halides and alkyl sulfates to mention a few. Nucleophilic sites in proteins include the thiol group of cysteine, the sulfur atom of methionine, the primary amino groups of arginine and lysine and the secondary amino group of histidine. Within the nucleic acids are the amino groups of purines, the oxygen atoms of purines and pyrimidines, and the phosphate oxygen. At face value one might suspect that the number of possible reactions would be bewildering but, because of thermodynamic considerations, these reactions tend to be selective. In general terms, "hard and soft" electrophiles react respectively with "hard and soft" nucleophiles. Coles (14) has tabulated a number of electrophiles and nucleophiles in accordance with their degree of hardness. No doubt, the knowledge obtained in this increasingly active research area could prove to be beneficial for both the synthetic design of drug and pesticides as well as the identification of bound residues.

The reactions between drug and pesticide electrophiles, parent or metabolites, and the nucleophilic centers in macromolecules explains how bound residues, as defined by Weber (4), can be produced. For the treated animal such reactions may be the progenitor of carcinogenicity, necrogenicity, teratogenicity or allergenicity. No doubt the popularization of these observations have fueled, and rightly so, the growing concern over the safety of drug and pesticide residues in foods. There are however many instances in which bound residues have not been associated with any biological effect in the food animal (*17*), i. e., the binding of reactive metabolites to macromolecules does not *a priori* result in toxicity. In short, the potential for harm to humans due to the reactions of drug and pesticides and their metabolites is difficult to predict.

In contrast to animals, plants lack specific organs for the elimination of xenobiotics. Pesticides are generally metabolized to water soluble conjugates that remain within the cell or the plant sap (*13*). Additionally, many xenobiotics are incorporated as insoluble terminal residues within lignin or other cell wall polymers. The latter, of course, are bound residues since routine extracting solvents do not solubilize cell wall polymers. Lignin, for example, is a major binding site for aromatic and heteroaromatic rings derived from pesticide molecules (*18, 19*).

At issue however is the production of reactive, second generation metabolites when the bound residues are consumed by humans. On the other hand, it is frequently proposed that the dangers of reactive metabolites to humans are abolished as soon as they are covalently bound to animal or plant macromolecules (*17*); however, there is a paucity of data to support this contention.

The Regulatory Perspective

Under the authority of the Federal Insecticide, Fungicide and Rodenticide Act (FIFRA) the Environmental Protection Agency (EPA) has the responsibility for registering and reregistering pesticides. Complimenting FIFRA is the Food, Drug and Cosmetic Act which charges the Food and Drug Administration (FDA) to establish the safety tolerances of drug residues in food products. Although the residue requirements of both agencies serve the same purpose, i. e., ensuring the safety of humans, domesticated plants and animals and the environment, the guidelines differ somewhat depending upon their respective legislated missions.

EPA. In simple terms, the registration and reregistration process for a pesticide includes the collection of residue data in representative plants and animals (*20*). Animal metabolism studies are required if use of the pesticide could result in residues in the edible tissues of livestock, in milk or in eggs. The magnitude and nature of the residue must be determined in two representative animal species: the laying hen and the lactating goat or dairy cow. Any metabolite that represents greater than 10% of the terminal radioactivity may be subjected to further characterization. This includes the free metabolites, conjugated metabolites, bound residues and endogenously incorporated products.

Plant metabolism studies are generally required for three different rotational crops (root, leafy vegetable, and small grain). Again, the magnitude and nature of the residues must be reported and the metabolites accounting for 10% or more of the terminal radioactivity must be identified. In some instances the uncharacterized bound residue fraction can be excluded if the fraction is < 0.1 ppm (parent compound equivalents) and the Toxicology Branch of the EPA has not expressed a concern.

FDA. The FDA requires information on the residue levels in the targeted food commodity for the establishment and enforcement of tolerance limits and withdrawal intervals (*21*). The targeted plant or animal must be dosed both at the use level requested and in the manner proposed by the petitioner. The magnitude and nature of the residues must be reported and, depending upon the toxicological concern, any metabolite greater than 10% of the terminal radioactivity or greater than 0.1 ppm (parent compound equivalents) may have to be structurally identified. Since this is often impossible for bound residues, the FDA has a provision that allows for the deduction from the total toxic residue all or part of the bound residue fraction derived from a noncarcinogen. In particular, that portion (for example 50%) of the bound residue which is not bioavailable is deductable.

Although this is a simplistic outline of the EPA and FDA residue requirements, it is hopefully apparent that the regulatory guidelines complement each other. The major point of division is in the acceptance by the FDA of the bioavailability technique as an alternative to structurally identifying the bound residues, i.e., those > 10% of the terminal radioactivity or > 0.1 ppm (parent equivalents), resulting from noncarcinogens. The EPA does not have this provision. Nevertheless, the aim of the entire process can be summed up in one statement; namely, what dose of residues is safe for the consumer. However, as noted at the beginning of this chapter, this is a difficult question to answer.

Febantel and Olaquindox

Febantel [N-2(N'-N"-bis -methoxycarbonyl-guanidino)-3-methoxyacetamido, 5-phenylthio, benzene] is a pro-benzimidazole that undergoes cyclization in the gut to fenbendazole (22, 23). In addition, febantel is oxidized by beef liver microsomes in vitro to the sulfate followed by ring closure to yield oxfendazole (22), which is also the thio-oxidation product of fenbendazole. Fenbendazole and oxfendazole are the pharmacologically active compounds. As a group the benzimidazoles, and the pro-benzimidazoles, are widely used in the beef cattle industry as broad spectrum anthelmintics (24). The therapeutic efficacy of these compounds is derived from their preferential binding to helminth tubulin, thereby preventing co-polymerization and the formation of microtubules (25). However, febantel has been shown to be embryotoxic in rats (23).

Olaquindox [N-(2-hydroxyethyl)-3-methyl-2-quinoxalinecarbamide-1, 4-dioxide] is a member of the quinoxaline-di-N-oxide class of compounds. These compounds are effective antibiotics and are used for both the acute treatment and prophylaxis of dysentery and other enteropathic diseases of swine (26, 27). Olaquindox is also being proposed for use as a feed additive to enhance the feed conversion and growth rate of swine (28). Olaquindox and many of the other quinoxaline-di-N-oxides have been shown to be mutagenic (29, 30). Although the carcinogenic potential of olaquindox has not been established (31), the quinoxalines as a group are recognized as carcinogenic (30).

Bound Residues in Liver. The oral administration of [14C]febantel to beef cattle and [14C]olaquindox to swine resulted in total residue levels of 201 and 29 ppb (parent equivalents), respectively, in the livers after 7 and 15 days of withdrawal, respectively (Table I). When the [14C]febantel treated beef liver was successively extracted with water followed by ethyl acetate, 51.6 and 7.6 percent, respectively, of the terminal radioactivity was solubilized. The remaining radiocarbon, 39.8%, was classified as bound residues. Similarly, extraction of the [14C]olaquindox treated swine liver with 50% methanol removed 53.2% of the terminal radioactivity leaving 56.6% as bound. Although a number of extracting conditions were attempted (i. e., various combinations of polar and nonpolar solvents, mild solutions of acetic acid, as well as refluxing and Soxhlet extraction) for both the beef and swine liver, those reported here were the most effective. Nevertheless, since the bound residue levels of both [14C]febantel and [14C]olaquindox exceeded the FDA residue requirements, i. e., > 10% of the terminal radioactivity, bioavailability studies were initiated using the Gallo-Torres method (5).

The bioavailability of the bound residues of [14C]febantel and [14C]olaquindox were determined in parallel with the bioavailability of the parent

Table I. Extraction of [^{14}C]Febantel, [^{14}C]Olaquindox and [^{14}C]Chloropicrin From Beef Liver, Swine Liver and Mature Strawberry Fruit, Respectively

Compound/ Tissue	Total Residue (ppb)	Percent of Radiocarbon In:			
		Solvent		Bound	
		A	B	Residue	Wash
[^{14}C]Febantel[1] Beef Liver	201	51.6	7.6	39.8	NA
[^{14}C]Olaquindox[2] Swine Liver	29	53.2	NA	56.6	0.45
[^{14}C]Chloropicrin[3] Strawberry Fruit	8,320	53.7	5.7	39.7	4.80

NA = Not Applicable.

[1]A beef cow was orally dosed with 5 mg [^{14}C]febantel/kg body wt. then sacrificed 7 days later. The entire liver was excised, cubed and stored frozen until processing. Semi-frozen liver was then chopped (Hobart Food Processor) with dry ice and then the carbon dioxide allowed to dissipate at -20° C. Subsamples of the liver were homogenized (Polytron, Brinkman) at room temperature with Water (Solvent A) at a ratio of 5 ml solvent:1 gm liver. This was repeated two additional times at a ratio of 2:1. The extraction was repeated with Ethyl Acetate (Solvent B). Soluble and insoluble residues were separated by centrifugation at 26,890 x g for 20 min at 4° C (Sorvall, RC-2). Total and bound radiocarbon levels were determined by combustion then LSC. Extractable radiocarbon was quantitated by direct LSC.

[2]A swine was orally dosed with 3.25 mg [^{14}C]olaquindox/kg body wt. for 7 consecutive days. After a 15 day withdrawal period the animal was sacrificed. The liver was excised and processed as described above. Subsamples were homogenized at room temperature with 50% Methanol (Solvent A) at a ratio of 4 ml solvent:1 gm liver. Homogenization was repeated two additional times at a ratio of 2 ml solvent:1 gm liver. Soluble and insoluble radiocarbon was separated by centrifugation and then quantitated as described above.

[3]The field site (Watsonville, CA) was fumigated with 3.125 ml [^{14}C]chloropicrin/ft[2] and thereafter commercial methods were utilized (see text). Mature fruit (Muir strain) was harvested as required and the in-life phase was terminated on day 126, post-planting. The combined fruit was minced and then homogenized with 1 N Sodium Bicarbonate (Solvent A) at a ratio of 1 ml solvent:1 gm fruit three times in succession. This was then repeated with Acetonitrile (Solvent B). Soluble and insoluble radiocarbon was separated by centrifugation and then quantitated as described above.

compounds (Table II). [^{14}C]Febantel was readily absorbed from the fortified liver. The urine, bile, liver and GI tract of the rats contained 51.9, 31.4, 0.6 and 1.8 percent, respectively, of the administered radiocarbon. The feces contained 11.5%. In contrast, radiocarbon elimination was almost exclusively via the feces, 99.9%, when the bile duct cannulated rats were given beef liver containing bound residues of [^{14}C]febantel. Only 0.3% of the total dose of bound radiocarbon was absorbed. These findings establish that [^{14}C]febantel *per se* is 85.7% bioavailable but that bound residues of [^{14}C]febantel are only 0.3% bioavailable.

[^{14}C]Olaquindox was also readily absorbed and excreted in the bile and urine of the rats -- 27.1 and 63.8 percent, respectively, of the administered radiocarbon. The feces contained 6.1% of the radiocarbon. Surprisingly, the bile and urine from the rats dosed with the bound residues of [^{14}C]olaquindox contained 5.8 and 14.8 percent, respectively, of the administered radiocarbon. The remainder of the dose, 80.5%, was localized in the feces. In summary, [^{14}C]olaquindox is 90.9% bioavailable whereas bound residues of [^{14}C]olaquindox are about 20.6% bioavailable.

Chloropicrin

Chloropicrin [Trichloronitromethane] is an extremely volatile and toxic fumigant that is widely used as a preplanting soil sterilant, especially in the strawberry industry. Chloropicrin is typically injected six inches into the soil which is then covered for two days with sheets of polyethylene plastic to prevent its loss resulting from volatilization. Planting begins 14 days after soil treatment. Under laboratory conditions the hydrolysis of chloropicrin is negligible; however, simultaneous exposure to light in the blue-UV region, < 300 nm, results in the degradation of chloropicrin to carbon dioxide and both nitric and hydrochloric acids (*32*). In sunlight chloropicrin photohydrolizes with a half-life of about three days and at 254 nm it is completely degraded in six hours. In addition, chloropicrin is very susceptible to nucleophilic attack and is therefore an effective alkylating agent. The mode of action of this highly toxic compound would appear to be due to its photohydrolysis to nitric and hydrochloric acid, its alkylating properties and its narcotic effects, which is typical of small,lipophilic polyhalogenated compounds. Chloropicrin is however metabolized by *Pseudomonas putida* via reductive dehalogenation to nitromethane (*33*). This process is apparently mediated by the enzyme P-450 cam which catalyzes a two electron reduction of the carbon-halogen bond (*33*).

Bound Residues in Strawberry Fruit. Portions of the data reported here originated from a detailed investigation of the disposition of [^{14}C]chloropicrin under commercial agricultural practices. Further information on the soil degradation products and their distribution into strawberries, green beans and beets will be reported elsewhere.

The mature strawberry fruit was found to contain a total of 8.32 ppm (parent equivalents) of [^{14}C]chloropicrin residues (Table I). The most effective extraction method was successive homogenization with 1 N sodium bicarbonate followed by acetonitrile, which removed 53.7% and 5.7%, respectively, of the total terminal radioactivity. The balance of the radioactivity, 39.8%, was classified as bound. In accordance with the EPA residue guidelines, i. e., > 10% of the terminal radioactivity was unextractable, we fractionated the insoluble cell wall materials. The procedure published by Langebartels and Harms (*34*) was followed with the notable exception that we used chlorite oxidation for the delignification of the cell wall material (*35*).

Table II. Bioavailability of [14C]Febantel, [14C]Olaquindox and [14C]Chloropicrin and Bound Residues of [14C]Febantel, [14C]Olaquindox and [14C]Chloropicrin [1]

Excreta/Tissue	Parent Compound[2]			Bound Residues[3]		
	Febantel	Olaquindox	Chloropicrin	Febantel	Olaquindox	Chloropicrin
ABSORBED						
Bile	31.4	27.1	0.8	0.0	5.8	3.5
Urine	51.9	63.8	3.6	0.0	14.8	2.1
Liver	0.6	0.0	3.9	0.3	0.0	12.4
GI Tract	1.8	0.0	4.3	0.0	0.0	7.2
Carcass	0.0	0.0	0.0	0.0	0.0	0.0
SUBTOTAL	85.7	90.9	11.9	0.3	20.6	25.2
UNABSORBED						
Feces	11.5	6.1	0.0	99.9	80.5	8.8
GI Contents	0.0	0.0	44.0	0.0	0.0	50.5
SUBTOTAL	11.5	6.1	44.0	99.9	80.5	59.3
TOTAL RECOVERY	97.2	97.0	55.9	100.2	101.1	84.5

Column header: Percent of Administered Radiocarbon

[1] Sprague-Dawley rats were cannulated and otherwise prepared as described by Gallo-Torres (5) with the exception that sodium taurocholate was not infused during the olaquindox and the chloropicrin bioavailability experiments. Each preparation of parent or bound residue was orally administered to the rats 24 hours after establishing a blood-free flow of bile > 10 ml.

[2] [14C]Febantel, dissolved in 100 μl dimethylformamide, was mixed with control beef liver that had been extracted and otherwise prepared in the manner described below for the bound residues of [14C]febantel (n=1). [14C]Olaquindox was dissolved in water (n=2). [14C]Chloropicrin was dissolved in water (n=1).

[3] Bound residues of [14C]febantel in beef liver and [14C]olaquindox in swine liver were lyopholized to dryness to remove the extracting solvents, stored frozen and then rehydrated just before dosing the rats, n = 3 for both compounds. Bound residues of [14C]chloropicrin in strawberry fruit were air dried, stored frozen and then rehydrated just prior to dosing the rat (n=1).

The fractionation results are presented in Table III. The radioactivity was found to be distributed throughout the cell wall polymers, with the most intractable polymers, i. e., lignin, hemicellulose and cellulose, accounting for about 24.3% of the total terminal radioactivity. It was, of course, of interest to determine the bioavailability of the bound residue.

Since there was very little available material, a single bile duct cannulated rat was given the bound residues. In parallel, a separate bile duct cannulated rat was administered a similar amount of radiocarbon as [^{14}C]chloropicrin. The results are shown in Table II. The total recovery of radiocarbon from the [^{14}C]chloropicrin treated rat was uncharacteristically low, i. e., 55.9%, however this low recovery is attributed to the highly volatile nature of chloropicrin. The majority of the administered radiocarbon was localized in the GI tract contents, i. e, 44.0%. The bile, urine, liver and GI tract contained 0.8, 3.6, 3.9 and 4.3 percent of the administered radiocarbon, respectively. The total recovery of radiocarbon from the rat given the bound [^{14}C]residues was also low. However, considering that only 34,419 dpm was administered, this is not unexpected. Of the radiocarbon that was absorbed, 3.5% was in the bile, 2.1% in the urine, 12.4% in the liver and 7.2% in the GI tract. The majority of the radiocarbon remained in the GI tract contents while the balance was found in the feces, 50.5 and 8.8 percent, respectively.

The results for the bioavailability of [^{14}C]chloropicrin are inconclusive. If anything, they point out the difficulties with using a volatile compound. Nevertheless, at least 25% of the bound residues in strawberries grown in soil fumigated with [^{14}C]chloropicrin are bioavailable. Interestingly, the protein fraction (Table III) contains about 25% of the bound radioactivity.

Discussion

We have reported herein on the bioavailability of bound residues of febantel, olaquindox and chloropicrin. The bioavailability of bound residues of febantel in beef liver was found to be practically zero, which strongly suggests that these residues are of little toxicological significance to the consumer. On the other hand, bound residues of olaquindox in swine liver were found to be about 20% bioavailable. Due to the carcinogenic potential of the quinoxalines, this finding suggests that these residues may be of toxicological concern and should therefore be further characterized. Similarly, a portion of the bound residues of chloropicrin in strawberry fruit were also found to be bioavailable, i. e., in the range of 20 to 30 percent. The chemical characterization of these residues revealed that the radiocarbon had been incorporated into the cell wall macromolecules of the fruit. Since we were unable to structurally identify the residues as natural products, we are not in a position to conclude that these residues are not of toxicological concern. However, it has been well established that chloropicrin can be degraded to carbon dioxide by the action of soil microbes, *Pseudomonas sp.* in particular (*33*), as well as by photohydrolysis (*32*). Given the ubiquitous distribution of the radioactivity within the cell walls as well as the distribution of the radioactivity throughout the plant tissues (data not shown), it is probable that this is an example of the endogenous incorporation of mineralized carbon.

The bioavailability of bound residues resulting from a number of drug and pesticides have been reported in the literature. Unfortunately, the technique of determining the bioavailability of most of these residues has varied. In particular, the bile duct may not have been cannulated (*36-38*) or bile salts were not replaced (*36-39*). The former is essential since, as has been demonstrated here, the bile is a viable excretory route for a portion of the absorbable bound residues. Merely finding the radioactivity in the feces of non-bile duct cannulated animals says very little about the

Table III. Percent Distribution of Bound Residue Radiocarbon Present in Strawberry Fruit Grown to Maturity in Soil Fumigated with [14C]Chloropicrin [1]

Filter-able	Starch	Protein	Pectin	Lignin	Hemi-cellulose	Cellulose	Insoluble	Filter Residue
2.4	0.8	5.7	4.1	9.7	7.0	7.6	1.0	1.9

[1]The bound radioactivity was fractionated in succession according to the procedures of Langebartels and Harms (34) with the notable exception that chlorite oxidation was used for delignification (35). Filterable: One gram of dried residue was washed with 50 mM Potassium Phosphate buffer, pH 7, and the suspension vacuum filtered (1.5 μm, Whatman AH-934). Starch: 0.2 ml alpha-Amylase (Type IA, Sigma) in 50 mM Potassium Phosphate buffer (pH 7) for 20 hr at 30° C. Protein: 300 mg Pronase E (Type XXV, Sigma) in 50 mM Tris-HCl buffer, pH 7.2, for 16 hr at 30° C. Pectin: 50 mM EDTA in 50 mM Sodium Acetate buffer, pH 4.5, for 6 hr at 80° C. Lignin: 40 ml Water to which was added 0.1 ml Glacial Acetic Acid followed by 0.25 gm Sodium Chlorite every hour for a total of 4 hr at 70° C. The filtrate was neutralized with 1 N Sodium Hydroxide then purged with nitrogen to remove chlorine dioxide. Hemicellulose: 24% Potassium Hydroxide for 24 hr at 27° C then neutralized with 6 N Acetic Acid to pH 4.5, shaken for 1 hr, centrifuged and the supernatant decanted. Cellulose: suspend pellet in 72% Sulfuric Acid for 4 hr at ambient temperature then neutralize with 24% Potassium Hydroxide. Insoluble Residue: defined as that residue which is retained on the filter from the cellulose hydrolysis step. Filter Residue: Sum of the residue retained on the filters exclusive of the final insoluble residue. Percent Radiocarbon in Fraction = (Net dpm in fraction - Net dpm filterable) / Net dpm in intact bound residue. Means of two independent fractionations.

residues bioavailability. Bile salt replacement, *e. g.*, sodium taurocholate infusion in the rat, is often times deleted because it is a rather tedious technique requiring the placement , either into the duodenum or forestomach, and subsequent maintenance of a second cannula. As we have shown for olaquindox and chloropicrin, bile salt replacement is not required to demonstrate the bioavailability of bound residues. However, since the physiological well-being of the non-bile salt infused rat is not ensured, it is conceivable that more or less of the bound olaquindox or chloropicrin residues may be bioavailable when strictly adhering to the Gallo-Torres technique (5).

In conclusion, the bioavailability technique of Gallo-Torres is technologically advanced to ensure a physiologically intact animal, or at least one that is a close approximation. In addition, the rat it the most widely used laboratory animal in toxicology; therefore, the bioavailability data is comparable to the existing data base on the toxicology of the drug or pesticide in question. Furthermore, it has been proposed that the bioavailability technique should be a first step in evaluating the magnitude of the toxicological concern of the total residue burden in edible livestock and crop tissues and not just the bound residues. This is particularly true for noncarcinogens.

Acknowledgments

The authors wish to acknowledge John and Stephen Wilhelm of Niklor Chemical Company for their financial support and permission to publish these findings. We would also like to thank T. Bill Waggoner of Mobay Chemical Company for his permission to publish data relevant to febantel and olaquindox.

Literature Cited

1. Klaassen, C. D. In *Toxicology: The Basic Science of Poisons*; Klaassen, C. D.; Amdur, M. O.; Doull, J., Eds; Macmillian Publishing Co.: New York, 1986, 3rd Edition, pp 11-32.
2. Albert, A. In *Selective Toxicology*; Chapman and Hall: London, 1985, 7th Edition, pp 3-20.
3. Hodgson, E.; Levi, P. E. In *A Textbook of Modern Toxicology*; Elsevier Science Publishing: New York, 1987; pp 1-22
4. Weber, N. E. *J. Environ. Pathol. Toxicol.* **1980**, *3*, 35-43.
5. Gallo-Torres, H. E. *J. Toxicol. Environ. Health* **1977**, *2*, 827-845.
6. Huber, W. G.; Becker, S. R.; Archer, B. P. *J. Environ. Pathol. Toxicol.* **1980**, *3*, 45-63.
7. Dorough, H. W.; Ballard, S. K. In *Biodegradation of Pesticides*; Matsumura, F.; Krishna-Murti; C. R., Eds.; Plenum Press: New York, 1982, pp 3-19.
8. Kovacs, M. F. *Residue Rev.* **1986**, *97*, 1-17.
9. Dorough, H. W. In *Pesticides: Minimizing the Risk*; Ragsdale, N. N.; Kuhr, R. J., Eds.; ACS Symposium Series No. 336, American Chemical Society: Washington, D. C., 1987; pp 106-114.
10. Khan, S. U.; Hamilton, H. A. *J. Agric. Food Chem.* **1980**, *28*, 126-132.
11. Dorough, H. W. *J. Environ. Pathol. Toxicol.* **1980**, *3*, 11-19.
12. Guengerich, F. P.; Liebler, D. C. *CRC Crit. Rev. Toxicol.* **1985**, *14*, 259-307.
13. Shimabukuro, R. H.; Lamoureux, G. L.; Frear, D. S. In *Biodegradation of Pesticides*; Matsumura, F.; Kirishna-Murti, C. R., Eds.; Plenum Press: New York, 1982, pp 21-65.
14. Coles, B. *Drug Metabol. Rev.* **1985**, *15*, 1307-1334.
15. Gillette, J. R.; Pohl, L. R. *J. Toxicol. Environ. Health* **1977**, *2*, 849-871.
16. Pohl, L. R.; Branchflower, R. V. *Method Enzymol.* **1981**, *77*, 43-50.

17. Delatour, P. In *Veterinary Pharmacology and Toxicology*; Ruckebusch, Y.; Toutain, P. L.; Koritz, G. D., Eds.; Proceedings 2nd European Associaiton for Veterinary Pharmacology and Toxicology; AVI Publishing Co.: Westpoint, CN, 1983, pp 659-670.
18. Khan, G. S. U. *J. Agric. Food Chem.* **1980**, *28*, 1096-1098.
19. Still, G. G.; Balba, H. M.; Mansager, E. R. *J. Agric. Food Chem.* **1981**, *29*, 739-746.
20. Schmitt, R. D. In *Pesticide Assessment Guidelines, Subdivision O: Residue Chemistry*; U. S. Environmental Protection Agency, Office of Pesticide and Toxic Substances: Washington, D. C., 1982.
21. *General Principles for Evaluating the Safety of Compounds Used in Food-Producing Animals*; U. S. Food and Drug Administration: Washington, D. C., 1983.
22. Montesissa, C.; Stracciari, J. M.; Fadini, L.; Beretta, C. *Xenobiotica* **1989**, 19, 97-100.
23. Delatour, P.; Daudon, M.; Garnier, F.; Benoit, E. *Ann. Rech. Vet.* **1982**, *13*, 163-170.
24. Armour, J. In *Pharmocological Basis of Large Animal Medicine*; Bogan, J. A.; Lees, P.; Yoxall, A. T., Eds.; Blackwell Scientific Publications: London, 1983, pp 174-199.
25. Friedman, P. A.; Platzer, E. G. *Biochim. Biophys. Acta* **1980**, *630*, 271-278.
26. Davis, J.; Libke, K. *Am. J. Vet. Res.* **1976**, *71*, 1257-1259.
27. Raynaud, P.; Branault, G.; Patterson, E. *Am J. Vet. Res.* **1981**, *42*, 51-53.
28. Bronsch, K.; Schneider, D. *Zeitschrift fur Tierphysiologie, Tiernahrung and Futtermittelkunde* **1976**, *36*, 211-221.
29. Beutin, P. E.; Kowalski, B. *Antimicrob. Agents Chemother.* **1981**, *20*, 336-343.
30. Tucker, M. J. *J. Natl. Can. Instit.* **1975**, *55*, 137-145.
31. Hermann, H. *Vet. Med. Rev.* **1978**, *1*, 93-100.
32. Castro, C. E.; Belser, N. O. *J. Agric. Food Chem.* **1981**, *29*, 1005-1008.
33. Castro, C. E.; Wade, R. S.; Besler, N. O. *J. Agric. Food Chem.* **1983**, *31*, 1184-1187.
34. Langebartels, C.; Harms, H. *Ecotoxicol Environ. Safety* **1985**, *10*, 268-279.
35. Green, J. W. In *Methods in Carbohydrate Chemistry*; Whistler, R. L.; Green, J. W.; BeMiller; J. N., Wolfrom, M. L., Eds.; Academic Press: New York, 1983, Vol. 3, pp 12-13.
36. Khan, S. U.; Kacew, S.; Molnar, S. J. *J. Agric. Food Chem.* **1985**, *33*, 712-717.
37. Wolf, F. J.; Alvaro, R.; Steffens, J. J.; Wolf, D. E.; Koniuszy, F. R.; Green, M. L.; Jacob, T. A. *J. Agric. Food Chem.* **1984**, *32*, 711-714.
38. Baer, J. E.; Jacob, T. A.; Wolf, F. J. *J. Toxicol. Environ. Health* **1977**, *2*, 895-903.
39. Marshall, T. C.; Dorough, H. W. *J. Agric. Food Chem.* **1977**, *25*, 1003-1008.

RECEIVED December 13, 1990

Chapter 18

Groundwater Contamination by Atrazine and Its Metabolites

Risk Assessment, Policy, and Legal Implications

D. A. Belluck[1], S. L. Benjamin[1], and T. Dawson[2]

[1]Minnesota Pollution Control Agency, 520 Lafayette Road, St. Paul, MN 55155
[2]Wisconsin Department of Justice, Madison, WI 53707

Atrazine can degrade in soil to metabolites that can enter groundwater. Principal metabolites are currently assumed to be at least of equal toxicity to atrazine, a possible human carcinogen. Recent groundwater sampling indicates that atrazine and its metabolites can occur alone or in combination. Total residues, atrazine plus its metabolites, can exceed atrazine regulatory limits even though atrazine does not. Atrazine metabolites themselves can exceed these limits. Routine monitoring of parent and metabolites will enable total resource degradation and potential human health risks to be appropriately evaluated. Options to regulate atrazine metabolites in groundwater include increasing the stringency of current regulatory limits for atrazine to account for the potential toxicity of the unmeasured metabolites.

Groundwater contamination by pesticides is of great and growing concern in the United States (1, 2). Widespread public concern exists that agricultural activities are contributing to the contamination of United States groundwater and surface water (3). For instance, rural Americans rank groundwater and drinking water contamination at the top of their environmental concerns (4). Approximately 70% of all pesticides used in the United States are applied to agricultural lands and more than $4 billion worth of agricultural pesticides are sold each year (5). An estimated 4 E+8 kg of pesticides were applied to agricultural lands in 1985. Less than 0.1% of applied pesticide actually reaches targeted pests. About 99.9% of applied pesticide, therefore, has the potential to move into other environmental compartments including groundwater (6). Although certain pesticides have been removed from the market because of environmental or human health risks, these actions have not slowed the growth of pesticide use in the United States (7).

Contamination can occur by normal agricultural activities or via spills, improper storage, backflushing during chemigation, and other accidents (8). A recent review (9) states that "many pesticides are

0097–6156/91/0459–0254$06.00/0
© 1991 American Chemical Society

leaching to groundwater from routine, nonpoint source use." Nonpoint source pollution of groundwater by agricultural chemicals is an increasing environmental problem (10). According to the Congressional Research Service (8), agricultural activities are the most pervasive contributors to nonpoint source pollution of groundwater. USDA states that the potential for groundwater contamination by agricultural chemicals is high in many major crop and livestock producing areas of the United States (3). Agricultural activities, irrigation and livestock usage account for approximately two-thirds of groundwater used daily in the United States (8).

Pesticide Contamination of Groundwater in the United States

Several reports have recently been published on the extent of contamination of the nation's groundwater. The U.S. Environmental Protection Agency's (U.S. EPA) "Pesticides In Ground Water Data Base, 1988 Interim Report" (11) found 74 pesticides in the groundwater of 38 states. A total of 46 pesticides detected in the groundwater of 26 states were attributed to normal agricultural use, while 32 pesticides in groundwater of 12 states were attributed to point sources or misuse. A 1986 paper from the U.S. EPA's Office of Pesticide Programs noted that at least 17 pesticides have been found in a total of 23 states as a result of agricultural practices (12). A 1989 review reported that a total of 39 pesticides have been detected in groundwater from 34 states or provinces (9). Another 1989 report found 67 pesticides in a total of 33 states (13). A recent special report in Chemical and Engineering News (14) stated that groundwater of more than half the states in the United States contains agricultural pesticides.

A recent State of Iowa groundwater monitoring study exemplifies the types of information being generated by states on the extent of groundwater contamination by pesticides and the magnitude of human exposure. Iowa's State-Wide Rural Well-Water Survey (SWRL) checked private drinking water supplies used by rural Iowans and found parent pesticides (atrazine, metribuzin, pendamethalin, metolachlor, cyanazine, alachlor, picloram, 2,4-D, dacthal, propachlor and trifluralin) as well as degradation products of atrazine (deethylatrazine and deisopropylatrazine), alachlor (hydroxy-alachlor) and carbofuran (3-hydroxy-carbofuran and 3-keto-carbofuran). State officials estimate that, based on SWRL findings, approximately 13.6% of private, rural drinking water wells (both shallow and deep) in the state are contaminated with one or more pesticides. Approximately 1.2% of the private rural drinking water wells, serving about 5,400 rural residents (about 0.7% of the rural population), may be contaminated with pesticide levels exceeding drinking water advisory levels and about 94,000 rural Iowans (about 13.1% of the rural population) are consuming drinking water containing one or more pesticides (15).

Despite debate in the literature concerning the absolute number of states reporting groundwater contamination by pesticides and the number of pesticides found per state, the number of wells with reports of pesticide contamination increase each year. "Pesticides are leaching through the soil and into groundwater far more commonly than the preconceptions of a decade ago would have predicted"(9).

Atrazine Use and Groundwater Contamination. Atrazine, a triazine herbicide, accounts for 12% of all pesticides used in the United States (9). Atrazine use in the U.S. is heaviest in the midwest with significant use in the eastern states and lesser amounts used throughout the rest of the country (16). It has been the most heavily used herbicide over the past 30 years in the United States for nonselective weed control on industrial or noncropped land and selective weed control in corn, sorghum, sugar cane, pineapple and certain other plants (17). Annual use is estimated at 75 to 90 million pounds (18).

Following field application, atrazine may eventually enter ground and surface waters (10). "Atrazine shares the characteristics of other triazine herbicides which make it a ground water contaminant: high leaching potential, high persistence in soils, slow hydrolysis, low vapor pressure, moderate solubility in water, and moderate adsorption to organic matter and clay," according to U.S. EPA (18). Others have described atrazine's propensity to leach to groundwater as high (19), slight (20), moderate (16) and as a leacher (21).

Certain metabolites of atrazine appear to be more water soluble than parent atrazine and could, therefore, be expected to have higher mobility to groundwater than exhibited by the parent compound. In fact, a recent study indicates that soil mobility profiles correspond well to soil-water partitioning coefficients and that desethylatrazine is more mobile than atrazine in both Plainfield sand and Honeywood silt loam (22).

In Quebec, Canada, soils are cold for a greater part of the year than in many areas of the United States where atrazine is used. The rapid movement of the water table from close to the surface in spring to around 3 meters in summer, reported in Quebec, could enhance leaching of atrazine and its phytotoxic dealkylated metabolites which have been shown to persist, in some cases, beyond the growing season (23).

Atrazine Groundwater Monitoring Data. Atrazine detections in groundwater are commonly 10 to 20 times more frequent than the next most frequently detected pesticide (9). Atrazine is the most commonly detected compound "in nearly all corn-belt areas" in the United States (9).

Atrazine has been found in the groundwater of approximately 25 states due to both point and nonpoint sources (24). U.S. EPA's 1988 Pesticides in Groundwater Data Base (11) reported atrazine in the groundwater of 13 states attributed to normal use (maximum concentration = 40 ppb, median = 0.5 ppb) and in 7 states from point sources.

Groundwater monitoring data compiled by U.S. EPA shows that approximately 4% of about 15,000 wells had atrazine contamination. The localized detection rate in areas of high atrazine use with vulnerable hydrogeologic conditions would probably be much higher (25). Oregon's State University Extension Service reports that about 14% of tested wells in 17 of 28 states contained atrazine (13). In Wisconsin, where an estimated 5.2 million pounds of atrazine are used each year, a random sample of 534 Grade A dairy wells found 71 wells with atrazine contamination. A recent investigation of Wisconsin pesticide mixing/loading facilities found atrazine in surface soil (mean = 639 ppm, median = 41 ppm, and range 0.9-5,900 ppm) and groundwater from water supply (maximum = 1,800 ppb) and monitoring wells (maximum = 23,900 ppb). Also found were decreasing atrazine concentrations with increasing soil depth (26). No atrazine metabolite data was reported from this study. Minnesota reports finding contamination by atrazine residues in 107 of 400 drinking water wells (27). Atrazine was the most commonly detected pesticide in Nebraska groundwater, being found in 13.5% of the samples (28). Ontario's 1986 Pesticide Monitoring Program found 50% of sampled wells contained atrazine and deethylatrazine (29). "If atrazine is eliminated from the statistics on frequency distribution, then detection of pesticides in water supplies becomes infrequent"(30).

Risk Assessment and Regulatory Status of Atrazine and its Metabolites.
U.S. EPA is reviewing data submitted in response to a 1983 Registration Standard and sub quent Data Call-Ins. Current information in the atrazine data base has led U.S. EPA to conclude that atrazine is a Group C (possible human) carcinogen based on the incidence of mammary tumors in female rats (18). U.S. EPA estimates that a drinking water exposure to 200 ppt of atrazine poses a one-in-a-million lifetime cancer risk and that, at the proposed MCL of 3 ppb, consumption of atrazine in drinking water poses a risk of about one-in-one-hundred-thousand (31). U.S. EPA's Office of Drinking Water has published a lifetime drinking water health advisory level of 3 ppb (17). USDA found atrazine mutagenic in 19 of 38 assays reviewed (32). Atrazine plant extracts have been shown to be mutagenic in microbial tests. A water soluble extract from maize plants treated with atrazine has been shown to be mutagenic in strain TA100 of Salmonella. While the causative agent for mutagenicity is unknown, the study authors suggest a link between an unidentified metabolite and the observed mutagenic activity (33).

Referring to the toxicity of atrazine metabolites, Ciba-Geigy, the manufacturer, has been quoted as stating that "the principal metabolites have toxicity which equals or exceeds that of the parent" (34) and that the chlorinated degradation products have "toxicological action similar to atrazine" (35). For dealing with the latter, the study author suggests using a method called "risk by analogy" whereby, in the case where there is inadequate data available to form health risk estimates, "the toxicology of deethylatrazine and deisopropylatrazine can be assumed to be similar to that of atrazine."

In response to groundwater monitoring data showing atrazine to be among the most commonly found pesticide groundwater contaminants, the federal government and individual states have taken steps to reduce the probability of groundwater contamination by this herbicide after normal field use. Atrazine label amendments are in the process of being implemented by U.S. EPA in an effort to reduce human exposure and groundwater contamination. Atrazine's maximum application rate is expected to be reduced to 3 pounds of active ingredient per acre per year for corn and sorghum. Application to non-cropland for industrial weed control will be limited to 10 pounds of active ingredient per acre per year (25).

A recent report (36) noted that several midwestern states in the United States are considering restrictions on atrazine usage due to groundwater contamination concerns. In Iowa, restrictions would cut atrazine usage to about 1.5 pounds of active ingredient per acre. According to Ciba-Geigy, this application rate is at the margin of efficacy. Iowa already restricts use of atrazine near waterways (37). Wisconsin's Department of Agriculture, Trade and Consumer Protection has proposed a plan that would ban most uses of atrazine in nine counties based on the propensity of the chemical to contaminate groundwater (37). Under this plan, farmers in other areas of the state would be restricted to using two pounds per acre or less, depending on the type of soil. Considering the availability of new herbicides used at significantly lower application rates, Felsot suggests that "perhaps it is time for agricultural scientists to recommend that use of atrazine be discontinued" (30).

Monitoring for Pesticide Metabolites in Groundwater

As the science of residue analysis for pesticides in water has matured over the last several decades, more emphasis has been placed on monitoring for metabolites, as well as parent compounds. Early investigations into groundwater contamination by pesticides analyzed for chlorinated hydrocarbon pesticides, their degradates (e.g. DDT and DDE, aldrin and dieldrin) and atrazine. Atrazine was found in all samples tested. In the late 1970s DBCP was found in California and aldicarb was found in New York. Findings of aldicarb and DBCP prompted additional monitoring during the 1980s with numerous nonpoint and point source contamination events being reported (9).

Current groundwater monitoring programs, including those in California (38), Iowa (39) and U.S. EPA's National Pesticide Survey (Boland, J.J., U.S. EPA, personal communication, 1989), have begun to stress monitoring for both parent pesticide and metabolites. Maine has reported ethylenethiourea (ETU), a metabolite and environmental degradation product of the dithiocarbamate fungicides, maneb and mancozeb, in the state's groundwater (40).

In Canada, Ontario's Ministry of the Environment (MOE) appears to routinely monitor for deethylatrazine in groundwater and drinking water (29). Nova Scotia has monitored for desethyl- and desisopropyl-atrazine in farm wells (41).

Atrazine Metabolic Breakdown Pathways. A recent review (42) on the degradation of atrazine noted that there is considerable interest in the fate of this herbicide in soil and water because of its persistence and migration to groundwater. Although considerable research has been performed to elucidate the environmental breakdown pathways of atrazine, understanding of this process is still incomplete. Microbial, as well as non-microbial, degradation appear to be important atrazine pathways in soil systems. Dealkylation appears to be the first step in microbial degradation and may be the slowest, or one of the slowest, steps in the biodegradation of atrazine. Atrazine can be degraded into several metabolites, some of which have equal or greater solubility than the parent compound. For extensive treatment of this subject area, the reader is referred to the work of Erickson and Lee (42) and papers referenced therein.

Groundwater Contamination by Atrazine Metabolites. A limited number of atrazine parent and metabolite monitoring studies are available in the open literature. One such study conducted by Ontario's Ministry of the Environment (MOE), looked at 37 domestic wells and five municipal supply wells for several pesticides, including atrazine, and their metabolites (29). The findings from this study illustrate the importance of monitoring for pesticide metabolites.

MOE used strict sample site selection criteria in an attempt to eliminate runoff contamination of wells. High risk wells were selected for sampling. Most private wells chosen were shallow (<15m) and situated in relatively permeable soils (sandy, sand loam). Several deeper private and municipal wells were included to represent deeper aquifer conditions. Wells surveyed were located within 60 meters of cultivated fields and down gradient from groundwater flow directions. MOE attempted to select only wells in good repair, where surface drainage was away from the well, and away from potential sources of contamination such as buildings where agricultural chemicals were stored or mixed (29).

Table I presents atrazine (ATZ) and deethylatrazine (DEA) concentrations, and ratios of the two chemicals for Ontario wells that had at least one deethylatrazine detection during the sampling year. For the column labelled Ratio ATZ/DEA, a ratio of more than 1 means there is more atrazine than deethylatrazine, 1 means there are equal amounts, and less than 1 means there is more deethylatrazine than atrazine in the groundwater sample. When both atrazine and deethylatrazine were present in a sample, ratios ranged from 3.8 to 0.1. Another column presents deethylatrazine as a percent of total residues (atrazine plus deethylatrazine). Percentages ranged from 0 to 100 indicating wide variations in deethylatrazine concentrations found in groundwater samples. Deethylatrazine concentrations were as high as 7.6 ppb indicating that the metabolite alone reached a concentration of potential public health concern.

Data in Table I shows that atrazine and its metabolite deethylatrazine occur alone and in combination in groundwater samples. It is clear from data in Table I that a given well can have differing ratios of atrazine and deethylatrazine depending on when sampling occurred.

If these findings were obtained for water in the United States, rather than in Canada, sampling for atrazine alone would not have, in most cases, detected concentrations above the proposed U.S. EPA MCL of 3 ppb. In several instances, however, these data would have indicated contamination exceeding the MCL, if the levels of parent and metabolites were summed. This analysis shows the need for routine monitoring of both parent and metabolite to obtain a clear picture of total atrazine residues in drinking water or groundwater.

In water samples collected from 91 wells on farms with mineral soils across Southern Ontario (43), atrazine residues ranged from 0.1 to 74 ppb. In half the reported cases, metabolite levels are greater than concentrations of the parent. In three of four cases metabolite concentrations exceed the current recommended U.S. EPA atrazine lifetime drinking water health advisory level and proposed drinking water standard (MCL) of 3.0 ppb (44).

U.S. EPA's National Pesticide Survey is currently underway to estimate pesticide and nitrate contamination of the nation's drinking water wells. The survey monitors for deethylatrazine at the relatively high minimum quantification limit of 4.4 ppb

Table I. Atrazine and deethylatrazine in Ontario, Canada drinking water derived from groundwater sources

Well I.D.	Atrazine (ATZ) ug/l	Deethyl- atrazine (DEA) ug/l	Ratio ATZ/ DEA	ATZ + DEA ug/l	DEA As % of Total Atrazine Residues
SW 306	ND	0.44	----	0.44	100%
SW 308	0.36	0.46	0.78	0.82	56%
SW 309	ND	0.54	----	0.54	100%
SW 309	2.40	ND	----	2.40	0%
SW 309	1.70	ND	----	1.70	0%
SW 309	2.90	0.84	3.5	3.74	22%
SW 309	1.10	0.66	1.7	1.76	38%
SW 309	0.34	0.94	0.36	1.28	73%
SW 309	3.00	1.20	2.5	4.20	29%
SW 309	ND	1.20	----	1.20	100%
SW 309	0.44	1.50	0.29	1.94	77%
SW 309	0.14	0.74	0.19	0.88	84%
SW 309	0.34	0.78	0.44	1.12	70%
SW 310	0.36	ND	----	0.36	0%
SW 310	0.55	ND	----	0.50	0%
SW 310	0.60	ND	----	0.60	0%
SW 310	0.38	ND	----	0.38	0%
SW 310	0.78	ND	----	0.78	0%
SW 310	0.84	0.44	1.9	1.28	34%
SW 310	1.20	0.83	1.4	2.03	41%
SW 310	1.40	0.96	1.5	2.36	41%
SW 310	1.60	0.94	1.7	2.54	37%
SW 310	4.20	1.10	3.8	5.30	21%
SW 310	1.20	0.94	1.3	2.14	44%
SW 310	3.00	0.80	3.8	3.80	21%
SW 310	1.20	0.90	1.3	2.10	43%
SW 310	0.74	0.33	2.2	1.07	31%
SW 310	0.66	0.46	1.4	1.12	41%
SW 313	0.20	0.24	0.8	0.44	55%
SW 313	0.28	0.50	0.6	0.78	64%
SW 313	0.33	0.63	0.5	0.96	66%
SW 313	0.26	TR	----	0.26	----
SW 313	0.40	0.60	0.7	1.00	60%
SW 316	0.23	0.21	1.1	0.44	48%
SW 320	0.90	1.60	0.6	2.50	64%
SW 320	0.79	ND	----	0.79	0%
SW 320	0.81	1.50	0.5	2.31	65%

Continued on next page

Table I continued.

Well I.D.	Atrazine (ATZ) ug/l	Deethyl- atrazine (DEA) ug/l	Ratio ATZ/ DEA	ATZ + DEA ug/l	DEA As % of Total Atrazine Residues
SW 320	1.10	ND	----	1.10	0%
SW 320	1.30	2.40	0.5	3.70	65%
SW 320	0.89	ND	----	0.89	0%
SW 320	2.90	7.60	0.4	10.50	72%
SW 320	1.10	2.50	0.4	3.60	69%
SW 320	0.61	1.30	0.5	1.91	68%
SW 320	0.70	1.80	0.4	2.50	72%
SW 320	0.76	1.50	0.5	2.26	66%
SW 320	1.60	3.10	0.5	4.70	66%
SW 320	0.94	1.60	0.6	2.54	63%
SW 320	1.30	2.20	0.6	3.50	63%
SW 320	1.00	1.90	0.5	2.90	66%
SW 320	0.84	1.40	0.6	2.24	63%
SW 320	2.00	2.30	0.9	4.30	53%
WCA 19	ND	0.26	----	0.26	100%
WCA 21	0.38	0.20	1.9	0.58	34%
WCA 21	0.41	0.46	0.9	0.87	53%
WCA 21	0.31	ND	----	0.31	0%
WCA 21	0.34	0.33	1.0	0.67	49%
WCA 22	0.24	0.11	2.2	0.35	31%
CP 886	0.55	0.35	1.6	0.90	39%
CP 886	0.65	ND	----	0.65	0%
CP 886	0.62	ND	----	0.62	0%
CP 886	0.36	3.90	0.1	4.26	92%
CP 886	0.39	2.10	0.2	2.49	84%
CP 886	0.48	ND	----	0.48	0%
CP 886	0.58	1.20	0.5	1.78	67%
CP 886	0.44	1.10	0.4	1.54	71%
CP 886	0.81	2.00	0.4	2.81	71%
CP 886	0.64	1.30	0.5	1.94	67%
CP 886	0.51	1.40	0.4	1.91	73%
CP 886	0.36	0.90	0.4	1.26	71%
CP 886	0.26	1.20	0.2	1.46	82%
CP 886	0.31	0.75	0.4	1.06	71%
CP 886	0.76	1.90	0.4	2.66	71%
CP 886	0.70	1.00	0.7	1.70	59%
CP 886	1.10	1.80	0.6	2.90	62%
CP 986	0.50	1.50	0.3	2.00	75%

SOURCE: Adapted from Ref. 29.
ND = No Detect
TR = Trace

(45) and minimum reporting limit of 2.2 ppb (44). No deethylatrazine has been reported by the survey as yet. This may be due, in part, to the relatively high detection and quantification limit used by U.S. EPA for deethylatrazine in groundwater.

A 1990 University of Wisconsin-Madison masters thesis, investigating the occurrence of atrazine metabolites in Wisconsin groundwater (35), detected deethylatrazine in all 32 study wells where atrazine was detected and in three wells where atrazine was below the method detection limit. Deisopropylatrazine was found in 11 wells with atrazine detects and in two wells where atrazine was below the detection limit. Concentrations of the two metabolites increased in groundwater samples with increasing atrazine levels. Summarizing her work, DeLuca stated that "deethylatrazine was consistently found in the wells studied and measurably contributed to total atrazine contamination. Deisopropylatrazine, although found less often than deethylatrazine, is present often enough and at high enough concentrations to be of importance to total atrazine contamination" (35).

Other Wisconsin researchers are measuring and comparing groundwater concentrations of atrazine and its metabolites deethylatrazine, deisopropylatrazine, diaminoatrazine, hydroxyatrazine, deethylhdroxyatrazine, deisopropylhydroxyatrazine, and diaminohydroxyatrazine. They have found groundwater samples containing concentrations of deethylatrazine exceeding the parent pesticide (46); containing metabolites when no parent was detected; less than the standard for the parent alone, but exceeding the Wisconsin groundwater enforcement standard for atrazine of 3.5 ppb when parent and metabolite levels were summed; above the atrazine enforcement standard for metabolites alone when the atrazine concentration was below the standard; and having metabolites at concentrations greater than the Wisconsin groundwater Preventive Action Limit of 0.35 ppb (46, 47). An explanation of the Wisconsin Groundwater Law is provided in a later section.

The above sections demonstrate that atrazine metabolites can be found in groundwater and can occur there at concentrations of environmental and human health concern. Drinking water and groundwater protection programs, designed to protect human health and the resource, face prodigious risk assessment and regulatory challenges if they are to respond to these metabolite residue findings in groundwater. Wisconsin's groundwater protection program will be used to illustrate the risk assessment, policy, and legal challenges facing a state considered to be at the forefront of groundwater protection.

Risk Assessment Implications of Atrazine Metabolites in Groundwater: Wisconsin's Groundwater Protection Program

Wisconsin's Groundwater Law (1983 Wisc. Act 410), requires the generation of risk assessment based groundwater standards for known or potential groundwater contaminants. Based on a noncarcinogen endpoint and an uncertainty factor that takes into account the chemical's possible carcinogenicity, a groundwater enforcement standard of 3.5 ppb was recommended and adopted for atrazine (Sec. NR 140.10, Wis. Adm. Code). Findings of atrazine groundwater contamination at, or above, the enforcement standard, following normal agricultural application, initiated a process that required the State of Wisconsin to determine whether the herbicide could be used without causing significant groundwater contamination.

Achieving Groundwater Protection through Standards: University of Wisconsin Concerns. The University of Wisconsin's Water Resources Management Program recently published a report entitled "Managing Pesticides In Groundwater, A Decision-Making Framework" (48) that raised the following concerns over the state's reliance on numeric groundwater standards for assessing the effects of regulatory policy and for evaluating the impact of atrazine and other chemicals on groundwater quality:
   *failing to account for interactive chemical effects;
   *failing to account for breakdown products; and
   *assuming that amounts below the standards will have no deleterious effects.
   Addressing these three concerns could result in a modified risk assessment approach and resultant numeric groundwater standard for atrazine. For example, researchers have yet to precisely identify whether the parent atrazine, one or more metabolites, or a combination of parent and metabolites cause the observed cancer in test animals. Lacking such information, a policy based approach will be needed to determine whether atrazine and its metabolites should be regulated in groundwater as one substance by summing parent and metabolites residue concentrations, or as several substances with the parent and each metabolite having separate groundwater standards. When atrazine and its metabolites are simultaneously found in a groundwater sample, it is probable that numeric standards will, by necessity, rely on a summation strategy, until a clear understanding of the mechanism by which atrazine causes its toxic action in animals, comprehensive health effects toxicology profiles and U.S. EPA validated analytical monitoring methods for atrazine metabolites at levels of environmental and human health relevance, become available.
   Concentrations of groundwater contaminants below a standard are currently assumed to pose no significant health risk. Without sufficient data to appropriately evaluate the potential carcinogenic and noncarcinogenic risks associated with a given groundwater or drinking water exposure to atrazine metabolites, however, this assumption may be incorrect. For example, it is possible that the observed toxic effects of atrazine are caused solely by one or more metabolites. Thus, groundwater containing non-measurable quantities of atrazine, but significant levels of one or more metabolites, could be of greater health concern than water having an equal amount of parent and no measurable metabolite concentrations. This is especially true when considering that certain metabolites are assumed to be of equal or greater toxicity than atrazine. Determining which, if any, of the atrazine metabolites pose significant human health or environmental degradation risks must await development of suitable data bases.

   University of Wisconsin Groundwater Metabolite Research. Based on University of Wisconsin groundwater research on atrazine and its metabolites, a mathematical model was constructed to predict atrazine breakdown products levels as a function of the concentration of atrazine in groundwater. This model adjusts the current atrazine parent enforcement standard of 3.5 ppb to 1.1 ppb to account for the possible presence and effects of metabolites that are not tested for when monitoring for the parent (35).

If this more stringent groundwater standard for atrazine is adopted, in lieu of routinely monitoring for its metabolites, there is little doubt that field application rates for atrazine would need to be dramatically modified to prevent groundwater contamination above the standard and that more areas of the state could become targets for banning use of the herbicide. If application rates were decreased, so that normal use of atrazine would not result in groundwater contamination above the suggested adjusted enforcement standard, the ability of the herbicide to be effective as a weed control agent might be reduced.

Legal and Policy Implications of Metabolites under the Wisconsin Groundwater Law

The Wisconsin Groundwater Law. The Wisconsin Groundwater Law created several statutory provisions, most intended to impose duties on regulatory agencies that already had groundwater protection authorities. The centerpiece of the Act was the creation of chapter 160 of the Wisconsin Statutes.

Chapter 160 requires the establishment of groundwater "enforcement standards" (ESs) and "preventive action limits" (PALs) for substances that may contaminate groundwater. The ESs are risk assessment based numbers set to protect the groundwater resource. The PALs are percentages of the ESs and are used, in part, as regulatory warning flags. In the case of carcinogens, the PAL is 10% of the ES; for substances of "public health concern", the PAL is 20% of the ES; and for substances of "public welfare concern", the PAL is 50% of the ES.

Under the law, regulatory agencies that have jurisdiction over substances or activities that contaminate groundwater generally must prohibit the activities that cause the attainment of the ES. The agencies are also given the duty to respond to contamination up to the ES, and to achieve compliance below the PAL level where "technically and economically feasible." (Sec. 160.23(1)(b), Wis. Stats) For a complete discussion of the Wisconsin Groundwater Law, see Belluck et al., _Bureau of National Affairs_ (in press).

Although the Wisconsin groundwater standards are based on public health risk assessment methodologies, they are not to be confused with drinking water standards or "groundwater quality standards". For example, the Wisconsin Department of Natural Resources (WDNR) has the authority under other statutory provisions to set groundwater quality standards based on other considerations besides minimum protection of public health. The WDNR has the ability, for example, to establish groundwater quality standards for areas of the state based on existing high quality of groundwater in furtherance of its authority to "protect, maintain, and enhance" water quality (Sec. 144.025, Wis. Stats.). Groundwater quality standards have not been established in Wisconsin.

The _authority_ to act, however, is not the _duty_ to do so. The enforcement standards set under the groundwater law are the duty to act. They are the points of groundwater contamination at which the legislature has commanded state agencies to wait no longer to take enforcement action. Under most authorizing statutes, agencies may take action. But, they also may do little or nothing. The groundwater law's enforcement "triggers" are the points of groundwater contamination where the Legislature has removed agency discretionary authority and, in effect, has defined when the agencies must take enforcement action. At these levels, the agencies must act, even though other statutes authorize the same agencies to take action before the standards are violated.

The groundwater law defines "substances", for which an ES and PAL must be established, as including any chemical "in its original form, or as a metabolite or a degradation product" (Sec. 160.02(8), Wis. Stats). The state continues to establish groundwater standards for substances known to have been detected in groundwater (Chapter NR 140, Wis. Adm. Code). Atrazine is listed as having an ES of 3.5 ppb, and a PAL of 0.35 ppb (Sec. NR 140.10, Wis. Adm. Code).

The Groundwater Law and Atrazine Metabolites. The WDNR and Wisconsin Department of Agriculture, Trade and Consumer Protection (WDATCP) conduct groundwater sampling programs that monitor for parent atrazine, but not its metabolites (LeMasters, G., Wisconsin Department of Agriculture, Trade and Consumer Protection, personal communication, 1990.). A 1990 University of Wisconsin thesis (35), however, addressed the regulatory problems associated with inadequate data for atrazine metabolites by suggesting that, when inadequate data is available to assess the risks associated with pesticide degradation products, toxicology data for substances with similar structures can be used to formulate a risk estimate. Termed "risk by analogy", this approach would suggest using atrazine's toxicology data base to predict the toxicological activity of its degradation products.

Equal Toxicity of Parent and Metabolites. According to DeLuca, in 1987 Darryl Summner of Ciba-Geigy stated that "[t]his agrees with the stance Ciba-Geigy has taken, which is to treat the chlorinated degradation products as having toxicological action similar to atrazine" (35). WDNR and WDATCP, with the concurrence of Ciba-Geigy, the major manufacturer of atrazine, have agreed to this additivity approach (summing parent plus metabolites) when evaluating total atrazine residues to determine if they exceed the ES of 3.5 ppb (LeMasters, G., Wisconsin Department of Agriculture, Trade and Consumer Protection, personal communication, 1990.).

The Wisconsin Department of Health and Social Services (WDHSS) recently apprised WDNR and WDATCP of its concern about the metabolite issue, urging the agencies to sample groundwater for the atrazine metabolites deethylated- and deisopropylated-atrazine (49). "Until suitable toxicity data becomes available," WDHSS noted "it seems appropriate to assume that these chemicals are at least as toxic as their parent." Furthermore, WDHSS "proposes that the existing groundwater enforcement standard of 3.5 ug/l be applied to total atrazine residues. Thus, detection of 2 ug/l atrazine and 1 ug/l of each of these dealkylated metabolites would be interpreted as a total atrazine concentration of 4 ug/l, and would exceed the current enforcement standard."

Concerns of the Agricultural Community. Wisconsin's agricultural community is concerned that, in order to deal with potential resource degradation and human health risks posed by unmeasured atrazine metabolites in groundwater, state agencies might adopt a more stringent atrazine enforcement standard. A second concern is that state agencies might implement a routine monitoring program for atrazine metabolites. Both concerns revolve around the fear that either option might result in a determination that atrazine contamination of Wisconsin's groundwater is a greater problem than currently believed. Should the state's regulatory agencies reach such a judgment, new or additional atrazine use restrictions could be introduced, reducing or eliminating atrazine use in some areas, and result in economic hardship for farmers using the herbicide.

In response to fears of economic impacts, Wisconsin state agencies have performed a technical and economic analysis to determine the feasibility of minimizing atrazine use by banning it. Under several scenarios, it was concluded that readily available, currently applied chemical and non-chemical alternatives to atrazine use are economically feasible. For example, a draft WDATCP report indicates that "reduction or elimination of atrazine in planned programs for control of broadleaf and grass weeds in continuous corn would only lead to cost increases on coarse soils where Bladex is not an option. For medium/fine soils, reduction or elimination of atrazine appears to have little if any impact on weed control costs. In both the reduced and non-atrazine strategies, comparable-cost alternatives are available." Furthermore, "continuous corn and corn/wheat/soybean rotations, non-atrazine options are available at comparable costs, particularly if cultivation is practical" (Wisconsin Department of Agriculture, Trade and Consumer Protection, Draft Analysis of the Technical and Economic Feasibility of Alternatives to Atrazine, 1990.).

Atrazine Use Limitations. Based on widespread detections of parent atrazine above the PAL of 0.35 ppb, WDATCP is currently proposing administrative regulations to limit use of the herbicide. The limits would be represented by cuts in field application rates, taking into consideration field conditions, such as soil types, groundwater table depth, and other related factors. WDATCP is not recommending a ban on atrazine, except in areas where the groundwater law compels it upon evidence of ES exceedence (proposed Wis. Adm. Code Chapter Ag 30, approved for hearing August 8, 1990). WDATCP is basing its proposed atrazine limits on parent compound groundwater monitoring data only, the majority of which are above the PAL, but below the ES. To date, no WDATCP or WDNR atrazine metabolite sampling data exists.

By routinely measuring only atrazine, WDATCP is most likely underestimating total groundwater contamination caused by a combination of atrazine and its metabolites. Thus, some areas that are currently judged to be marginal for atrazine contamination, based solely on parent compound monitoring data, may in fact be contaminated at, or above, the ES if metabolites had also been measured. Areas that have shown no evidence of parent compound contamination might have metabolites present. In such cases, areas once considered not impacted by atrazine groundwater contamination would need to be redesignated as atrazine impacted.

Considering such scenarios, if metabolite sampling data existed to demonstrate exceedence of the "combined" atrazine ES, it is strongly suspected that atrazine bans in many areas of the state would be compelled under the law. It is partly on this basis that environmental advocates are urging stronger measures, such as banning atrazine use where detection of atrazine parent are above PAL levels.

Groundwater Policy Implications of the Atrazine Experience. There are many significant policy implications arising from Wisconsin's atrazine groundwater contamination experience. It is not sufficient to only identify parent chemical compounds for groundwater monitoring programs. This is only a start. Identification of atrazine metabolites in groundwater samples must be part of routine atrazine monitoring programs (50). This can be accomplished only if responsible government agencies have available validated analytical methods that measure these substances at levels of environmental and human health concern. Once routine monitoring programs are in place, complete metabolite toxicology data bases will be needed to adequately evaluate their potential environmental and human health impacts. Where this data does not exist, or is not within the public domain, policies and regulations need to be developed to force its creation and/or public disclosure.

If It Is Not Monitored, It Will Not Be Detected. Non-detection of a substance is likely to result in its non-regulation. As the Wisconsin experience shows, groundwater sampling programs, as indispensable as they are, can seriously mislead us into believing that ignorance about substances in our water is bliss. Unless national and state drinking, surface, and ground water monitoring programs recognize the metabolite issue, and begin looking for parent plus metabolites in samples, the extent of contamination will remain unknown and will be under-reported.

Monitoring programs that do not look for the full spectrum of breakdown products produced from parent compounds run the real risk of underestimating the extent of potential harm to the environment and human health. Finding parent compounds at relatively low levels, or not at all, does not guarantee that their metabolites are not present. "Safety factors" and "uncertainty factors" used in the development of water protection standards, such as federal drinking water standards and state groundwater standards, may not provide an adequate or desirable level of human health protection from pesticide residues when considering the unknown extent and potential toxicity of unmeasured metabolites.

Clean-up of pesticide contaminated groundwater may not be complete without factoring in metabolite residues. Clean-up levels for contaminated sites are based on reducing single chemical concentrations below their regulatory limits or multiple contaminants below some risk threshold. The presence of undetected metabolites, however, may continue to expose humans to unacceptable health risks and the environment to unacceptable degradation when only parent pesticides are measured, regulated, and remediated.

Regulatory responses to underestimated residue levels and consequent underestimated risk may exacerbate, rather than remedy, remediation problems. For example, based on a calculation of negligible health risk to humans who may be exposed to contaminants, a site once thought to be remediated for parent pesticides may, in fact, need to be revisited and remediated again due to the presence of metabolites.

Legal and Regulatory Response to Technical Advances. The atrazine experience in Wisconsin demonstrates the potential for failure of legislatures and agencies to appreciate the importance of developments in the scientific arena that lend new and significant dimensions to health and environmental protection policies. While agencies should be given more than the mere authority to act, there must be room for adapting to new challenges. Striking this balance is a challenge in itself. One option under the Wisconsin law would have been for the regulatory agencies to myopically interpret the groundwater law as strictly directing them to set groundwater standards for parent compounds and their metabolites as separate "substances". As recognized by the agencies, however, this option fails to take into account the total risk presented by the use of chemicals that the law was clearly intended to regulate. Instead, Wisconsin agencies have appropriately interpreted and applied the law to take into account the total risk of atrazine parent and metabolites in groundwater.

This interpretation raises additional policy and regulatory challenges, however. One alternative to the approach apparently taken by the Wisconsin state agencies on atrazine would be to stake out a policy position that says, in the absence of actual metabolite monitoring data, the law will assume the presence of designated metabolites at the same (or predetermined) levels and the same (or predetermined) toxicities as the parent compound, unless demonstrated otherwise. Such a policy would provide a modicum of protection to the resource and put the burden, including costs, of the pollution's detection and risk assessment at least partially on those benefitting from the compound's use.

Another consideration is that the current approach implies the need to develop "enforcement standard formulas" that would take into account the total accumulated risk created by the presence of several unrelated parent compounds, their metabolites and formulation materials. Persons exposed to such an undefined chemical mixture in their water may take little consolation in the fact that no individual substance exceeds its health based standard, if the aggregate risk of the mixture can be above an acceptable risk level. For example, how should detection levels for several substances found together in groundwater be used in determining what regulatory responses are compelled by the groundwater law, such as when the PALs for several substances are exceeded, but not their ESs? This issue has not yet been tackled in Wisconsin, a state at the forefront of groundwater policy development.

Consideration of Total Risk. The total risk from exposure to groundwater contaminants is not usually taken into account in the setting of standards. For example, generic testing requirements for interactive effects toxicology and neurotoxicity are either lacking or nonexistent. While the potential for a "synergistic" or "potentiation" effect among chemical mixtures is recognized, government agencies are still debating how to handle chemical mixtures from a policy perspective.

The risks from unknown, un-investigated, or under-investigated toxicity endpoints may not be fully compensated for through the use of uncertainty factors. It is possible that these factors provide insufficient protection to exposed populations. Current standards, based on traditional toxicity endpoints, may provide policy makers, risk assessors, and risk managers with a false sense of security.

It is interesting to review the evolution of the atrazine groundwater metabolite debate in Wisconsin in terms of its public institutions. In the mid-1980s, Wisconsin's DHSS, in conjunction with the Department of Justice's Public Intervenor's Office, identified the potential environmental and human health threat posed by unidentified and unmeasured atrazine metabolites in the state's groundwater. These two agencies sought support from other state agencies, including certain segments of the university community. It was only WDNR that provided limited funding for a graduate student at the State Laboratory of Hygiene to investigate methodologies to monitor for certain atrazine metabolites and evaluate their significance.

Institutional resistance from numerous individuals and departments of the university was finally reduced to cooperation, in large part by the publication of a graduate student's thesis in 1990. Several university groups are presently vying to determine the extent of atrazine metabolite groundwater contamination and its health implications. Unfortunately, WDNR and WDATCP have yet to announce routine groundwater monitoring for atrazine metabolites in areas known or suspected to have atrazine usage. It is interesting that, in this case, the state health and justice departments lead the push for the previously described investigations, not the agencies directly charged with the work.

Thus, while Wisconsin appears to be on the leading edge of groundwater policy development, that edge marks the beginning of uncharted territory in terms of understanding and regulating environmental contaminants. There is plenty of opportunity for other states, local government units and the federal government to learn from the Wisconsin experience and, if need be, to outstrip Wisconsin, in order to establish policy that will meet the public's consistent demand for a clean water supply for now and the future.

Summary and Conclusions

Atrazine, one of the most widely used herbicides in the United States, has been found in the groundwater from point and nonpoint sources. After intentional or accidental introduction into the soil, atrazine can begin its migration to groundwater. While in the soil column, atrazine can degrade into several metabolites of equal or greater solubility than the parent herbicide. Depending on soil conditions, different concentrations of parent and metabolites can enter the groundwater. Atrazine parent and metabolite groundwater monitoring results presented in this chapter indicate that atrazine metabolites can occur alone or in combination with parent atrazine. Atrazine metabolite residues in groundwater can be of human health concern, based on the determination that certain metabolites should be considered of equal or greater toxicity than the parent, which is a U.S. EPA Group C (possible human) carcinogen.

Although the atrazine data base is relatively comprehensive, it is unknown how atrazine, its metabolites, or combination of atrazine and metabolites caused the reported mammary tumors in female rats. Lacking basic scientific information to determine the potential environmental and human health risks posed by atrazine metabolites, yet obligated to take regulatory action when metabolites are found in groundwater, government agencies will probably base their decisions on the assumption that certain metabolites are of equal or greater toxicity than the parent.

Actually being able to establish whether atrazine use, misuse, or spills have impacted groundwater depends on whether parent and metabolites have been looked for and the quality of the analytical methods employed. For example, programs that monitor for the parent, but not its metabolites, may underestimate total atrazine contamination in a groundwater sample. Based on such sample results, well owners might be told their groundwater has not been impacted by atrazine, when in reality measurement of metabolites would indicate the groundwater was, indeed, contaminated, perhaps at levels of environmental or human health concern.

Inadequate analytical methods can also result in reports of no atrazine contamination when atrazine and/or its metabolites are present below the ability of an analytical technique to detect. For example, if atrazine metabolites are present in water samples in the 0.1 to 1.0 ppb range, analytical methods that report only concentrations in excess of 2 ppb would show no evidence of metabolites in a sample. In this instance, although metabolites are in the water, they are reported as not present, due to limitations of the analytical method. The number of metabolites to be routinely monitored will be determined by the definition of what constitutes significant resource degradation or human health risks.

Failure to either routinely monitor for atrazine metabolites, or create more stringent groundwater regulatory limits, could result in under-protection of the resource or human health. Two possible mechanisms have been explored in this chapter to regulate atrazine metabolites in groundwater. The first mechanism is routine monitoring for parent and metabolites. This would provide agencies the ability to evaluate total resource degradation and potential human health risks associated with parent and metabolite groundwater residues. To implement this strategy the summation approach could be adopted -- comparing the sum of the parent and its metabolites against the existing atrazine regulatory limit -- or a regulatory limit could be set for the parent and each metabolite. This type of program, however, could be resource and time intensive.

The second mechanism would be to establish more stringent regulatory limits for parent atrazine to take into account the potential presence of unmeasured metabolites. This has the advantage of being relatively inexpensive and simple to implement, since no additional analytical capability is needed because no metabolites would be measured in groundwater samples. On the other hand, this approach would require government agencies to make assumptions on the toxicity, frequency of occurrence, and concentrations of metabolites found in groundwater.

Government agencies will need to decide how to factor atrazine metabolites into the development and implementation of groundwater and

drinking water protection plans. Will agencies begin routine
monitoring for some or all atrazine metabolites and set regulatory
limits for each? Or will agencies find this too great of an economic
and risk assessment burden and choose, instead, to promulgate more
stringent regulatory limits for the parent, basing metabolite
regulation on a series of assumptions? Atrazine is but one
substance that degrades into metabolites not typically monitored for
in groundwater. For other substances, the technical, policy,
regulatory and legal implications of coping with atrazine metabolites
should serve as a useful model to guide risk assessors, risk managers
and other professionals who will formulate monitoring programs, numeric
standards or other mechanisms for managing groundwater contamination
by the myriad additional substances that could pose analogous
environmental degradation or potential human health impacts.

Institutional and technological barriers have retarded movement
toward adequate monitoring for metabolites. The challenge to policy
makers, however, resides in the need to encourage and accept advances
in analytical technologies and to apply these technologies within a
policy framework that meets the environmental and human health needs
identified in this chapter and by other researchers. Responsible
federal agencies, as well as some state agencies, already have the
authority to take such actions.

Disclaimer: The views expressed and conclusions reached in this chapter
are solely those of the authors and do not necessarily reflect or
represent the views and policies of the Minnesota Pollution Control
Agency.

Literature Cited

1. Stara, J.F.; Patterson, J.; Dourson, M.L.   In Evaluation of
   Pesticides in Ground Water; Garner, W.J.; Honeycutt, R.C.; Nigg,
   H.N., Eds.; ACS Symposium Series 315; American Chemical Society:
   Washington, D.C., 1986; 445-461.
2. Szmedra, P.; Osteen, C. Agricultural Outlook. 1988, 26-27.
3. USDA Research Plan For Water Quality; USDA/ARS: Washington, D.C.,
   1989.
4. Ground Water Age. 1990, 24(7), 10.
5. Ground Water Safety Act of 1988. Calender No. 1054, Report 100-
   583; 100th Congress, 2d Session; U.S. Senate: Washington, D.C.,
   1988.
6. Younos, T.M.; Weigmann, D.L. J. Water Poll. Control Fed. 1988, 60,
   1199-1205.
7. Osteen, C.D.; Szmedra, P.I. Agricultural Pesticide Use Trends and
   Policy Issues: Agricultural Economic Report Number 622; USDA/ERS:
   Washington, D.C., 1989.
8. Feliciano, D.V. Agricultural Effects on Groundwater Quality;
   Congressional Research Service, The Library of Congress:
   Washington, D.C., 1986.
9. Hallberg, G.R. Agr. Ecosyst. Environ. 1989, 26, 299-367.
10. Rostad, C.E.; Pereira, W.E.; Leiker, T.J. Biomed. Environ. Mass
    Spec. 1989, 18, 820-827.
11. Williams, W.M.; Holden, P.W.; Parsons, D.W.; Lorber, M.N.
    Pesticides In Ground Water Data Base. 1988 Interim Report; U.S.

Environmental Protection Agency Office of Pesticide Programs: Washington, D.C., 1988.

12. Cohen, S.Z.; Eiden, C.; Lorber, M.N. In Evaluation of Pesticides In Groundwater; Garner, W.J., Honeycutt, R.C.; Nigg, H.N.,Eds.; ACS Symposium Series 315; American Chemical Society: Washington, D.C., 1986; 170-196.

13. Parsons, D.W.; Witt, J.M. Pesticides In Groundwater In The United States Of America. A Report of a 1988 Survey of State Lead Agencies. EM 8406; Oregon State University Extension Service: 1989.

14. Hileman, B. Chem. Engin. News. 1990, 68, 26-40.

15. Iowa State-Wide Rural Well-Water Survey. Summary of Results: Pesticide Detections; Iowa Department of Natural Resources: 1990.

16. Barrett, M.R.; Williams, W.M. In Pesticides in Terrestrial and Aquatic Environments. Proceedings of a National Research Conference; Weigmann, D.L., Ed.; Virginia Water Resources Research Center: 1989; 39-61.

17. U.S. Environmental Protection Agency. Drinking Water Health Advisory: Pesticides; Lewis Publishers: Chelsea, MI, 1989.

18. U.S. EPA Office of Pesticide Programs. Environmental Fact Sheet, Atrazine Label Amendments; January 23, 1990.

19. Nielson, E.G.; Lee, L.K. The Magnitude and Costs of Groundwater Contamination From Agricultural Chemicals: A National Perspective; USDA Economic Research Services, Agricultural Economic Report No. 576; 1987.

20. Brach, J. Agriculture and Water Quality. Best Management Practices for Minnesota; 1989.

21. Council For Agricultural Science and Technology. Agriculture and Groundwater Quality; Report No. 103; 1985.

22. Bowman, B.T. Environ. Toxicol. Chem. 1990, 9, 453-461.

23. Prasher, S.; Masse, L. Groundwater Pollution From Agricultural Pesticides; 1990.

24. United States Environmental Protection Agency. Office of Drinking Water. Drinking Water Regulations and Health Advisories; 1990.

25. U.S. Environmental Protection Agency. EPA Restricts Pesticide Atrazine; 1990.

26. Habecker, M.A. Environmental Contamination at Wisconsin Pesticide Mixing/Loading Facilities: Case Study, Investigation and Remedial Evaluation; Wisconsin Department of Agriculture, Trade and Consumer Protection, Agricultural Resource Management Division: 1989.

27. LeMasters, G.; Doyle, D.J. Grade A Dairy Farm Well Water Quality Survey; Wisconsin Department of Agriculture, Trade and Consumer Protection and Wisconsin Agricultural Statistics Service: 1989.

28. Spalding, R.F.; Burbach, M.E.; Exner, M.E. Groundwater Monitoring Review. 1989, 9, 126-133.

29. Pesticides In Ontario Drinking Water - 1986; Ontario Ministry of the Environment: 1987, Toronto, Canada.

30. Felsot, A. In Proc. 16th Ann. IL Crop Protection Workshop; 1990; 32-42.

31. Andreasen, J. Memo to U.S. EPA Atrazine Team; August 4, 1988.

32. USDA Forest Service. Human Health Risk Assessment for the Use of Pesticides in USDA Forest Service Nurseries; FS-412; 1987.

33. Means, J.C.; Plewa, M.J.; Gentile, J.M. Mutation Research. 1988, 197, 325-336.

34. Anderson, H. <u>Public Hearing Testimony on Proposed Ag 30 Relating to the Use of Atrazine;</u> Wisconsin Division of Health: October 8, 1990.
35. DeLuca, D.B. <u>Analytical Determination of Atrazine, Alachlor and Their Selected Degradation Products in Contaminated Groundwater: Implications for Wisconsin Groundwater Standards;</u> Masters of Science Thesis in Land Resources, Institute for Environmental Studies, University of Wisconsin-Madison: Madison, WI, 1990.
36. <u>Chemicalweek.</u> January 3-10, 1990, 9.
37. <u>Wisconsin State Journal.</u> July 19, 1990, 1a-2a.
38. <u>Sampling For Pesticide Residues In California Well Water, 1988 Update, Well Inventory Data Base;</u> California Department of Food and Agriculture: 1988.
39. <u>Iowa State-Wide Rural Well-Water Survey. Summary of Results: Atrazine Detections;</u> Iowa Department of Natural Resources: 1990.
40. Frakes, R. <u>Northeast Regional Environmental Public Health Center.</u> 1987, <u>1,</u> 4.
41. Nova Scotia Departments of the Environment, Agriculture and Marketing, and Health and Fitness. <u>Nova Scotia Farm Well Water Quality Assurance Study, Phase I - Final Report;</u> 1990.
42. Erickson, L.E.; Lee, K.H. <u>Crit. Rev. Environ. Control.</u> 1989, <u>19,</u> 1-14.
43. Frank, R.; Ripley, B.D; Braun, H.E.; Clegg, B.S.; Johnston, R.; O'Neill, T.J. <u>Arch. Environ. Contam. Toxicol.</u> 1987, <u>16,</u> 1-8.
44. U.S. EPA. <u>Release of National Pesticide Survey Report;</u> Memo From Jeanne Briskin, Director, National Pesticide Survey to State Health, Environmental, Drinking Water and Agriculture Directors; October 29, 1990.
45. Munch, D.J.; Graves, R.L.; Maxey, R.A.; Engel, T.M. <u>Environ. Sci. Technol.</u> 1990, <u>24(10),</u> 1446-1451.
46. Chesters, G.; Simsiman, G.V.; Levy, J.; Read, H.W.; Gustafson, D.P.; Fathulla, R.N.; Matt, F.; Purabi, D.; Xiang, X. <u>Fourth Quarterly Report To Wisconsin Department of Agriculture, Trade and Consumer Protection and Wisconsin Department of Natural Resources;</u> 1990.
47. Chesters, G.; Simsiman, G.V.; Levy, J.; Purabi, D.; Xiang, X. <u>Third Quarterly Report To Wisconsin Department of Agriculture, Trade and Consumer Protection and Wisconsin Department of Natural Resources;</u> 1990.
48. <u>Managing Pesticides In Groundwater, A Decision-Making Framework;</u> Water Resources Management Program, Institute for Environmental Studies, University of Wisconsin-Madison: Madison, WI, 1988.
49. Knobeloch, L. <u>Memo to N. Neher, WDATCP, and K. Kessler, WDNR;</u> April 23, 1990.
50. Belluck, D.A.; Benjamin, S.L.; Dawson, T. <u>J. Pest. Reform.</u> 1990, <u>9,</u> 28-31.

RECEIVED December 11, 1990

Chapter 19

Pesticide Degradation Products in the Atmosphere

Jack R. Plimmer and W. E. Johnson

Environmental Chemistry Laboratory, Natural Resources Institute, U.S. Department of Agriculture, Beltsville, MD 20705

A large fraction of the pesticides that are manufactured enters the environment. Losses to the air are well documented and depend on vapor pressure of the active ingredient, type of formulation, application technique, climate, etc. Pesticide transformations in the air are driven primarily by solar irradiation. Atmospheric transport and deposition to water and terrestrial surfaces occurs. In some cases pesticide transformation products may be more toxic than the parent compound, however, the amounts detected and their dissipation pathways in the atmosphere suggest that although this may not be a major issue, some transformation products may be of concern.

The lack of hard data makes assessment of the significance of pesticide metabolites and transformation products in the atmosphere extremely difficult. The situation is made more complex because atmospheric processes give rise to a complex mixture of degradation products and it may be difficult to distinguish whether organic air pollutants are derived from pesticides or from other sources.

Better information relating to the amounts of pesticides and other synthetic chemicals entering the environment is essential. There are a number of programs at national and regional levels to survey the quantities used and the patterns of use. A considerable fraction of applied pesticides is lost to the environment and it is difficult to arrive at a quantitative estimate of the amount lost to the atmosphere and subsequently deposited at the earth's surface. However, even greater difficulties arise in determining or predicting the biological effects of such residues at any but the most unrealistically high concentrations.

Our knowledge of the impact of chemicals on ecosystems is limited by test systems in which isolated chemicals are evaluated against test organisms in a laboratory environment. Most studies must be conducted by specialists, and coordination of experimental approaches and integration of results may be poor. Modelers who try to integrate data and provide useful predictive systems may be frustrated by the lack of high quality data and disagreements over the importance of specific factors.

The need for multidisciplinary studies was emphasized by Moore and Ramamoorthy (1): "Active and strong input from individuals with diverse training will provide a broad understanding of the long-term consequences of specific environmental problems." To illustrate their contention that a restricted approach to study of the perturbations affecting an organism or ecosystem will be ineffective or lead to conflicting recommendations, they give the example of 2,3,7,8-tetrachlorodibenzo-p-dioxin (TCDD) in water. Although this is not a pesticide metabolite, TCDD has received great attention as a trace impurity in some chlorinated phenoxy compounds used in pest control. For example, coho salmon exposed to TCDD for a short period showed no measurable effects (food consumption, weight gain, survival) over a 60-day post exposure period (2). Further studies showed that growth and survival of salmon decreased over a 114-day exposure to higher concentrations of TCDD and the body burden increased with level and duration of exposure. They also contrast the results of other approaches to the determination of the biological effects of TCDD and point out that it would be unnecessary to undertake some of these investigations if TCDD is rapidly photodecomposed in the environment. In the opinion of these authors, a false evaluation of the hazards of TCDD could have resulted from "undisciplined investigations."

This issue is important in assessment of the environmental effects of specific chemicals. Data are usually scarce, particularly important physicochemical data that determine the likelihood that a particular chemical will reside in a particular environmental compartment and its persistence at a site. In addition, processes for obtaining data on toxicology and biological effects are usually costly and time-consuming. Transport of pesticides in the atmosphere is important as a major source of loss from application sites, but without adequate data on wet and dry deposition and rates of reaction in the atmosphere it is impossible to predict the amounts of pesticides that will be deposited from the atmosphere or their effects.

Transformation Processes and Products

An important criterion for design or registration of new synthetic organic pesticides is the readiness by which they are transformed to innocuous products by environmental biological and chemical transformation processes. A second criterion is their potential for transport from the site of application by leaching to ground water, runoff to surface water, or volatilization. Under ideal conditions, the most extensively used compounds will represent those which present negligible environmental hazard, and this will also apply to their principal metabolites which receive the same regulatory scrutiny as the parent compound. Murphy (3) has commented on the importance of the atmospheric route by which toxic organics may ultimately contaminate water bodies, and it must also be borne in mind that organics may also be lost to the atmosphere by processes at the water surface.

The loss of pesticides to the atmosphere is governed initially by volatilization or drift during application. Residues may subsequently volatilize from soil or foliar surfaces. Metabolites or photoproducts may also enter the atmosphere if they are sufficiently volatile. Ultimately, most of these materials will be deposited on the earth's surface and it would be anticipated that rainfall, snow, and other forms of wet or dry deposition would contain not only the parent pesticides, but also photochemical or chemical transformation products.

Photolytic Processes in the Atmosphere

Estimation of atmospheric lifetimes of organic species requires knowledge of their rates of deposition and their rates of reaction with the major radical species present in the atmosphere. Radiation of wavelengths 290-400 nm comprises about 4% of the solar spectrum. The more energetic region (290-320 nm) is sometimes known as the UV-B region. Many pesticides absorb light in this region and these may undergo direct photolysis in sunlight. Additionally, a variety of indirect processes can occur and there is good evidence of the importance of attack by photochemically-generated oxidants. Organic compounds in the atmosphere may be rapidly transformed or degraded by the combined action of free radicals, for example nitrophenols may be formed by the action of hydroxyl and NO_x radicals (4).

Oxidants in the atmosphere include hydroxyl radicals, ozone, singlet oxygen, nitric oxide, hydroperoxy radicals, and other radical species. Therefore airborne organic pollutants may be substantially degraded through a series of radical chain reactions.

Hydroxyl radicals are probably the most important of the oxidant species responsible for the oxidative transformation of organic compounds that enter the atmosphere because these have been shown to dominate the transformation of a variety of organic compounds in air. When rates of transformation and atmospheric residence time of a number of organic compounds were measured, it was found that atmospheric residence times were generally less than 4 days (longer for dichlorobenzene and PCBs)(5).

Because they are most important in the early stages of oxidation of volatile organic compounds, information on the rate constants for the rate of attack of hydroxyl radicals on pesticides is desirable for estimation of atmospheric residence times. Kinetic and mechanistic data for the reactions of photolytically-generated OH, and NO_3 radicals and ozone with organophosphorus compounds were obtained in a study of vapor-phase reactions conducted in a 6400 L. Teflon environmental chamber (6). Rate constants for reaction with the three radical species were measured and combined to calculate atmospheric lifetimes for a series of organophosphorus compounds. Values of 0.8 hr to 2.1 days were reported for reaction with OH radicals, corresponding values for NO_3 radicals were 2.9 hr to >48 days and for ozone >42 days to > 275 days. In product studies, results of experimental observations were in accordance with the expected conversion of the thiophosphoryl to the phosphoryl bond (7). These findings are consistent with the results of field experiments.

In a series of field experiments to measure the pesticide content of fogs in California and Maryland, pesticides and transformation products were detected and their concentrations were measured in fog water and in ambient air. Methyl parathion, malathion, parathion, chlorpyrifos, methidathion, and diazinon were found together with the corresponding oxons. These pesticides were extensively used on agricultural crops in the area near Monterey, CA where the fog was collected. The ratios of concentrations of pesticides and transformation products in the fog water to those in ambient air were often much higher than would have been predicted by the Henry's Law relationships and, in some cases, the concentration of the oxons was greater than that of the parent organophosphorus compounds. In one instance, the concentration of malaoxon was 10 times that of malathion and the concentration of chlorpyrifos oxon was 20 times that of chlorpyrifos (8).

Measurements of methyl parathion and methyl paraoxon at sites in the Sacramento Valley, CA showed thion/oxon ratios in air that varied, averaging about 10:1 at one site during May, although ratios fell to 2:1 on a few days (9).

The authors consider thion-to-oxon conversion to occur during atmospheric transport from agricultural to nonagricultural areas. The conversion of thion to oxon in dilute aqueous hydrogen peroxide is too slow (10) to account for concentrations of oxon observed within the fog droplets. Ratios of oxon to thion also increased after sunup indicating that photochemical conversion might be important, and it was suggested that thion to oxon most probably takes place in the vapor phase, followed by partitioning into the aqueous phase of the droplet.

There is little direct information concerning the reactions of pesticides adsorbed on particulates present in air or pesticides in wetfall. For example, trifluralin is rapidly lost by volatilization if it is not incorporated in soil (11) and its half-life in the atmosphere is estimated at 20 minutes (12). The vapor-phase photolysis products include mono- and di-dealkylated trifluralin, a benzimidazole formed by a cyclization process involving one N-propyl group and a nitro group, and the corresponding dealkylated benzimidazole. Free radical oxidation by atmospheric oxygen is considered responsible for the dealkylation reactions.

Trifluralin and its photoproducts were detected in the air above both surface and soil-incorporated fields (13) and it is suggested that these products probably arose primarily from photolysis of trifluralin on the soil surface followed by volatilization. However, the contribution of vapor phase photolysis probably increases with air-residence time of trifluralin.

There are other examples of vapor-phase reactions of pesticides: aldrin is converted to photoaldrin and dieldrin; photoaldrin or dieldrin gave photodieldrin (14); fenitrothion gives 3-methyl-4-nitrophenol and other products; and 2,4-D esters give 2,4-dichloroanisole. These reactions were conducted in a specially designed photoreactor. Although it has been suggested that products observed may be ascribed to wall reactions, it may be very difficult to distinguish between homogeneous and heterogeneous reactions in the environmental situation.

Photochemical transformations in the atmosphere may be facilitated by adsorption on solid surfaces and trifluralin or parathion adsorbed on dust readily photodecompose. Changes in wavelength at which light is absorbed occur when compounds are adsorbed on solids. Adsorption of pesticides on silica caused bathochromic shifts or hypsochromic shifts in the ultraviolet absorption spectrum depending on the hydrogen bonding interaction. It may also improve contact with oxygen or facilitate contact with sensitizers (15,16).

For other reasons photochemical reactions in the laboratory photoreactor may differ from those in the environment. For example, molinate (S-ethyl N,N-hexamethylenethiocarbamate), a rice herbicide, is photodegraded in the field but did not degrade when exposed to UV light in the laboratory. It is likely that the difference in reaction rates is due to the higher ambient ozone concentrations outdoors (0.1-0.3 μl/L compared with 0.01 μl/L in the photoreactor)(17,18). Ozone addition also increases the rate of photolysis of parathion and trifluralin and its addition appears to be desirable if the course of the photoreaction in the laboratory is to mimic closely that in the field.

Measurements of Environmental Concentrations

Global monitoring studies show that high molecular weight organic compounds are transported in the environment and many of these are chlorinated compounds associated with pest control application. Hexachlorocyclohexanes (HCHs) and chlorinated benzenes are among the most abundant. These conclusions are substantiated by findings in a number of studies from different regions of the world (19). However, many organochlorine insecticides are composed of mixtures of isomers or related compounds. Their composition may vary with the method of manufacture and, in some parts of the world, there may be very little information concerning the actual composition of the technical product. Technical chlordane is an example. Another example is the insecticide, lindane, which is the pure γ isomer of 1,2,3,4,5,6-hexachlorocyclohexane (HCH) but technical HCH has been used extensively. Technical HCH contains α-(55-60%), ß-(5-14%), γ-(10-18%), δ-(6-10%), and ϵ-(3-4%) isomers (20). Its use was phased out in Japan in 1970 although its manufacture for export continued and its use was banned in China in 1983. Thus, the composition of atmospheric samples may be extremely difficult to interpret because photochemical interconversions may further complicate the picture.

An eleven year study of long-range transport of pesticides showed that α-HCH was the most abundant species present among the organochlorine insecticides (21). Levels of HCH and DDT isomers were measured in water and atmospheric samples collected in survey cruises of the Pacific and Indian Oceans. Highest concentrations of these species were in the mid-Northern Hemisphere. The source of the HCH's is technical HCH and lindane. This led the authors to conclude that the low ratio of α- to γ-HCH in the Southern Hemisphere may reflect greater use of lindane.

Although technical HCH may be the source of the α-HCH, it was suggested that some caution is required in explaining atmospheric ratios. The atmospheric ratio of α- to γ- was greater than 5 in many locations and increased from south to north. Temporally, total HCH concentrations declined over the eleven-year period indicating reductions in use as prohibitions imposed by several governments became effective. However, an increase in the ratio of α- to γ- in the Southern Hemisphere may indicate an change in use pattern from lindane to technical BHC over the period of the observations. Kurtz and Atlas (22) also found high atmospheric concentrations of α-HCH in both Northern and Southern Hemispheres.

The ratios of α- to γ-HCH measured in a 1986 Pacific survey cruise originating and ending in Hawaii were found to be about 10-15 (23). The two isomers accounted for over 50% of the total high molecular weight organochlorines measured in the north Pacific atmosphere. It is suggested that possible causes for the difference in isomer ratio between that used in agriculture (less than 1 to 6: α- to γ-HCH) may be: photoisomerization (24,25), enhanced washout and removal of γ-HCH due to differences in Henry's Law constants and volatilization of already altered and/or degraded HCHs to treated soils. There is a need for more information on the removal of HCHs from the atmosphere and on the atmospheric stability of the α and γ isomers (26).

A report from the Netherlands summarizes available data for that country (27) and concludes that in addition to photodecomposition, dry deposition is an important way by which organic chemicals are removed from the atmosphere. Yearly deposition of organic compounds in the Netherlands is estimated at about 89,000 tons. Some information concerning pesticide concentrations in air is included (Table I). The ratios of α-HCH to lindane reported are lower than those found in the Pacific area.

Methyl bromide, 1,2-dibromoethane, dieldrin, aldrin and endrin were also among the pesticides measured. Heptachlor is used as a pesticide, but heptachlor epoxide is probably a photoproduct. Of interest are the group of compounds related to p,p'-DDT. Their concentrations are similar but the ratios observed do not reflect those in technical DDT. Formation of DDE and DDD by photolysis in solution has been explained by formation of a radical by loss of Cl˙ from the CCl₃ group (28) but little is known of interconversion mechanisms in the atmosphere.

Table I. Pesticide concentrations in air (pg/m³)

Compound	Mean	Maximun
α-HCH	1250	1200
Lindane	360	3400
o,p'-DDT	25	130
p,p'-DDT	57	540
o,p'-DDE	25	140
p,p'-DDE	17	90
o,p'-DDD	17	120
p,p'-DDD	7	44
Heptachlor	37	190
Heptachlor epoxide	32	360
α-Endosulfan	168	1130

The isomeric content of atmospheric samples of chlordane is variable and the ratio of trans- (γ) to cis- (α) declined from May to July in samples collected over the Pacific (21). Atmospheric samples of chlordane generally contain a lower content of the trans-isomer than the technical mixture.

Specific Examples

Parathion. Parathion and methyl parathion are examples of insecticidal chemicals that are used in large quantities in the United States. Their metabolism and pathways of environmental degradation are well established (29). An important intermediary metabolite in plant and other tissues is paraoxon, a powerful inhibitor of cholinesterase.

Irradiation of parathion in the vapor phase in laboratory experiments resulted in the rapid formation of paraoxon. The end product was 4-nitrophenol and diethylphosphoric and diethylthiophosphoric acids were detected (12). Experiments in an orchard treated with parathion led to the conclusion that parathion was converted rapidly to paraoxon in air, that the reaction took place largely in the vapor-phase and was promoted by sunlight. Detection of parathion, paraoxon, and 4-nitrophenol together in fog indicates that mechanisms exist by which residues may reenter crop or forest environments.

Conversion of parathion to paraoxon occurs on plant surfaces where it may accumulate in dry weather. By its persistence, it may present a hazard for workers many weeks after application. Recoveries of radioactivity after ^{14}C-labelled parathion was applied to cotton plants in environmental growth chambers, greenhouse, controlled-exposure field, and open field (30) were 11.2% to 15.4% (58% to 68% unchanged parathion) suggesting that volatility losses may be considerable.

Methyl parathion has similar properties to parathion but is somewhat less toxic to mammals and does not persist so long as a foliar residue or in soils.

Methyl parathion, parathion, and paraoxon have been measured in air and fog. At Stoneville, Mississippi, analyses of air sampled in 1972, 1973, and 1974 showed the presence of p,p'-DDT, o,p-DDT, and methyl parathion during a 156 week sampling period (31). Pesticide residues could be detected by the sampler in both vapor and particulate phases. Other organophosphorus pesticides were detected only during the growing season.

4-Nitrophenol has been determined in rain and snow (32). It has also been analyzed in fog (33) where it occurs together with parathion and paraoxon (34). Ultraviolet irradiation of phosphorus esters in the presence of water or moisture causes hydrolysis and, as previously stated, parathion gives 4-nitrophenol as a photolysis product. It is a stable end-product of prolonged irradiation of parathion and paraoxon was detected soon after irradiation commenced (28). In the environment, it may arise from other environmental sources besides pesticide use. Together with other nitrophenols, it may be formed by atmospheric oxidation of aromatic hydrocarbons and subsequent nitration and it has been suggested that such products may be among the factors influencing forest decline (35).

U.S. manufacture of 4-nitrophenol amounted to about 19,000 metric tons annually (36) and it is listed as a priority pollutant by the USEPA. According to an EPA report, 27% of its use is for parathion manufacture and 13% is used for the synthesis of dye components (37). Residues of nitrophenol in effluent from one chemical manufacturing plant were reported as 9300 $\mu g/L^{-1}$ (38).

Malathion. Although malathion has been used extensively as an insecticide, attempts to monitor large-scale application programs have shown that residues were predominantly those of the parent compound and that malaoxon, a primary product of photolysis, could be found only in extremely small amounts.

In a Californian program to eradicate the Mediterranean fruit fly, malathion was applied at 2.4 ounces of 91% malathion per acre (plus 9.6 ounces Staley's fluid protein bait) weekly for 6 weeks. During this period, about 250 square miles were intensively monitored for malathion and malaoxon. About 76% of the malathion could be accounted for by sampling deposited material. Air concentration was measured using XAD-2 resin in high and low volume samplers. Malathion concentrations never exceeded 1 ug/m^3 and malaoxon concentrations were generally one to two orders of magnitude lower, never exceeding 0.1 $\mu g/m^3$.   In water, the average concentration of malathion was 3.5 ppb over the 6-week period (high value 32 ppb).   In swimming pools, malathion concentrations in water averaged 0.9 ppb (high value 23 ppb). Chemical additives to swimming pools are strong oxidants and the value for malaoxon was 8 ppb, however, degradation to non-detectable levels occurred within hours.

In natural waters, levels showed greater variation with a maximum of 703 ppb malathion (average 25 ppb) and an average malaoxon concentration of 2 ppb. Extensive sampling showed that malathion was rapidly degraded and dissipated and there was no accumulation of residues (39).

Conclusions

The powerful reactivity of oxidants and free radicals generated by solar irradiation is responsible for extensive degradation of many organic compounds in the atmosphere. Persistent organochlorine compounds are an exception, although some of these may undergo transformations. An important need for assessments of environmental impact is representative data on the atmospheric half-lives of compounds that are likely to enter the atmosphere by reason of their use pattern or their physical properties.

Pesticide metabolites and transformation products entering the environment generally represent a fractional increment of the environmental burden due to the parent compound. Most environmental studies have been concerned with the parent compound and the detection of metabolites in air samples from most environmental sources generally lies beyond the capabilities of routine analytical methods, with the exception of some persistent organochlorine compounds.

A knowledge of chemical and physical properties may be adequate for an assessment of the environmental fate of many metabolites. Although in some cases they may be more toxic than the parent compound, the amounts detected and the variety of pathways by which they may be dissipated, the accumulated evidence suggests that to focus on this issue would be to divert resources from other, more pressing problems.

In few cases are we able to appraise the effects of the parent compound after transport in air to land or water surfaces and a major task is to assess potential loading and biological consequences. Consequently, the question of transformation products often becomes secondary. That is not to underestimate the importance of the issue, but to emphasize that it must be recognized that we need better predictors of future environmental damage or deterioration than sample analysis on a compound by compound basis.

Literature Cited

1. *Organic Chemicals in Natural Waters*; Moore, J.W.; Ramamoorthy, S., Eds.; Springer-Verlag: New York, 1984; pp 1-3.

2. Miller, R.A.; Norris, L.A.; Loper, B.R. *Trans. Am. Fish. Soc.* **1979**, *108*, 401-407.

3. Murphy, T.J. In *Toxic Contamination in Large Lakes, Sources, Fates, and controls of Toxic Contaminants*; Schmidtke, W., Ed.; Lewis Publishers, Inc.: Chelsea, MI, 1988, Vol. 3; pp 83-96.

4. Nishioka, M.G.; Howard, C.C.; Contos, D.A.; Ball, L.M.; Lewtas, J. *Environ. Sci. Tech.* **1988**, 22, 908-914.

5. Cupitt, L.T. *Fate of Toxic and Hazardous Materials in the Air Environment*; USEPA Report EPA-600/53080-084; Athens, GA, 1980, pp 1-26.

6. Winer, A.M.; Atkinson, R. In *Long Range Transport of Pesticides;* Kurtz, D., Ed.; Lewis Publisher, Inc.: Chelsea, MI, 1990, pp 115-126.

7. Atkinson, R.; Aschmann, S.M.; Arey, J.; McElroy, P.A.; Winer, A.M. *Environ. Sci. Technol.* **1989**, 23, 243-244.

8. Schomburg, C.J.; Glotfelty, D.E.; Seiber, J.N. *Environ. Sci. Technol.* in press.

9. Seiber, J.N.; McChesney, M.M.; Woodrow, J.E. *Environ. Toxicol. Chem.* **1989**, 8, 577-588.

10. Draper, W.M.; Crosby, D.G. *J. Agric. Food Chem.* **1984**, 32, 231-237.

11. Taylor, A.W. *APCA Journal.* **1978**, 28, 922-927.

12. Woodrow, J.E.; Crosby, D.G.; Mast, T.; Moilanen, K.W.; Seiber, J.N. *J. Agric. Food Chem.* **1978**, 26, 1312-1326.

13. Soderquist, C.J.; Crosby, D.G.; Moilanen, K.W.; Seiber, J.N.; Woodrow, J.E. *J. Agric. Food Chem.* **1975**, 23, 304-309.

14. Crosby, D.G.; Moilanen, K.W. *Arch. Environ. Contam. Toxicol.* **1974**, 2, 62-74.

15. Plimmer, J.R. In *Fate of Pesticides in the Environment;* Tahori, A.S., Ed., Gordon and Breach: New York, NY, 1972, pp 47-76.

16. Gaeb, S.; Nitz, S; Parlar, H.; Korte, F. *Chemosphere* **1975**, *4*, 251-256.

17. Moilanen, K.W.; Crosby, D.G.; Seiber, J.N. 4th Int. Congr. Pestic. Chem. IUPAC, Zurich, V623, 1978.

18. Moilanen, K.W.; Crosby, D.G.; Seiber, J.N. 173rd. National Meeting of the American Chemical Society, Abstract Pesticide Division, paper no. 02, 1977.

19. *Long Range Transport of Pesticides;* Kurtz, D., Ed.; Lewis Publishers, Inc.: New York, NY, 1990.

20. Brooks, G.T. *Chlorinated Insecticides, Vol.1 Technology and Application;* CRC Press: Cleveland, OH, 1974; p 189.

21. Tatsukawa, R.; Yamaguchi, Y.; Kawano, M.; Kannan, N.; Tanabe, S. In *Long Range Transport of Pesticides;* Kurtz, D., Ed.; Lewis Publishers, Inc.: Chelsea, MI, 1990, pp 127-141.

22. Kurtz, D.; Atlas, E.L. In *Long Range Transport of Pesticides;* Kurtz, D. Ed.; Lewis Publishers, Inc.: Chelsea, MI, 1990, pp 143-160.

23. Atlas, E.L.; Schauffler, S. In *Long Range Transport of Pesticides;* Kurtz, D., Ed.; Lewis Publishers, Inc.: Chelsea, MI, 1990, pp 161-183.

24. Oehme, M.; Ottar, B. *Geophys. Res. Lett.* **1984**, 11, 1133-1136.

25. Pacyna, J.M.; Oehme, M. *Atmos. Environ.* **1988**, 22, 243-257.

26. Bidleman, T.F.; Patton, G.W.; Hinckley, D.A.; Walla, M.D.; Cotham, W.E. In *Long Range Transport of Pesticides;* Kurtz, D., Ed.; Lewis Publishers, Inc.: Chelsea, MI, 1990, pp 347-372.

27. Guicherit R., Schulting, F.L. 1985. The Science of the Total Environment, 43:193-219.
28. Plimmer, J.R.,Klingebiel U.I., Hummer, B.E. 1970. Science 167,67-69.
29. Eto, M. *Organophosphorus Pesticides Organic and Biological Chemistry*; CRC Press: Cleveland, OH, 1974, pp 241.
30. Joiner, R.H.; Baetcke, K.P. *J. Agric. Food Chem.* **1973**, *21*, 391-396.
31. Arthur, R.D.; Cain, J.D.; Barrentine, B.F. *Bull. Environ. Contam. Toxicol.,* **1976**, 15, 129-139.
32. Alber, M.; Böhm, H.B.; Brodesser, J.; Feltes, J.; Levsen, K.; Schöler, H.F. *Fresenius Z. Anal. Chem.* **1989**, 334, 540-545.
33. Venkat, J.A.; Plimmer, J.R.; Glotfelty, D.E. 198 Nat. Meet. American Chemical Society, Agrochemicals Division, Sept. 1989, paper 98.
34. Glotfelty, D.E.; Seiber, J.N.; Liljeddahl, L.A. *Nature.* **1987**, 325, 602-605.
35. Rippen, G.; Zietz, E.; Frank, R.; Knacker, T.; Klöpffer, W. *Environ. Technol. Letters.* **1987**, 8, 475-482.
36. *Organic Chemicals in Natural Waters*; Moore, J.W.; Ramamoorthy, S., Eds.; Springer-Verlag: New York, NY, 1984; pp. 143.
37. Marke, R.A.; Fentiman, A.F.; Steadman, T.R.; Meyer, R.A. *Potentially Toxic and Hazardous Substances in the Industrial Chemicals and Organic Dyes and Pigment Industries.* EPA 600/2-80-056; NTIS: Springfield, VA, 1980.
38. Lopez-Avila, V.; Hites, R.A. *Environ. Sci. Technol.* **1980**, 14, 1382-1390.
39. Oshima, R.J.; Neher, L.A.; Mischke, T.M.; Weaver, D.J.; Leifson, O.S. *A characterization of sequential aerial malathion applications in the Santa Clara Valley of California, 1981*; California Dept. of Food and Agriculture: Sacramento, CA, 1982.

RECEIVED December 6, 1990

Chapter 20

Pesticide Transformation Products Research

A Future Perspective

L. Somasundaram and Joel R. Coats

Pesticide Toxicology Laboratory, Department of Entomology, Iowa State University, Ames, IA 50011

Pesticide research has largely been restricted to the synthesis, effects, and fate of parent molecules. The identification of primary, secondary, and subsequently formed transformation products, and the toxicological evaluation of those products of potential environmental concern are crucial to create a much needed data-base on pesticide transformation products. The establishment of new regulatory policies will largely depend upon scientific information generated in the coming years.

Research on synthetic pesticides has been conducted for the past 50 years. Most research efforts, however, have focused on the parent compound per se. Except for studies concerning metabolism or the identification of products, little attention has been paid to the fate and significance of the products formed from pesticide transformation (1-4). The slow progress in the field of pesticide transformation products can be attributed to three factors: [1] low concentrations at which such products are formed in the environment, [2] assumption that pesticides are generally mineralized to insignificant products, and [3] lack of suitable analytical techniques.

Analytical Techniques

Many transformation products are more water soluble than their respective parent compounds. Difficulty in extracting, separating, and analyzing polar transformation products present in the environmental matrices is one of the many constraints on transformation products research. Volatile metabolites also present unique challenges to sample processing and analysis. Because pesticide metabolites are usually present at trace levels, isolating sufficient quantities of metabolites for identification can be a formidable problem. In addition to the extraction difficulties,

0097–6156/91/0459–0285$06.00/0

analytical techniques commonly used to detect parent compounds are not always adequate to detect the transformation products. Liquid and gas chromatography, mass spectrometry, and radiotracers are used to qualitatively and quantitatively detect trace levels of degradation products in complex matrices. New technologies that are being employed include supercritical fluid extraction and chromatography, fluorescence, electrochemical and photodiode array detection, immunoassays (radio and enzyme-linked), biosensors, headspace analysis, and solid-phase nuclear magnetic resonance. As even more sensitive, efficient, and economical techniques of analysis are developed, our understanding of the significance of pesticide transformation products will be enhanced.

Radiotracers. Although the first radiolabeled pesticides were synthesized in 1952 (5), the use of radioassays in plant, soil, and animal metabolism studies has become widespread only since the 1970's. Before the use of radiotracers, the environmental fate of most pesticides was not adequately understood, and the concept of bound residues did not exist. The high cost and difficulty involved in synthesis have resulted in few degradation products being available in radiolabeled form. But labeled degradation products of commonly used pesticides such as atrazine and 2,4-D have been instrumental in detailed investigations of these compounds. Because certain pesticide degradation products are of significance in crop protection and environmental contamination, the synthesis of radiolabeled degradation products in addition to that of the parent compound is emerging as a requisite for the development of new pesticides.

Secondary Degradation Products

The degradation of a pesticide is not completed with the formation of primary degradation products. For several pesticides, the primary degradation products are unstable and are converted to secondary and tertiary degradation products. Relatively few studies have investigated the significance of the latter. The enhanced biodegradation of isofenphos is attributed to the nutritional substrate value of its secondary metabolite, salicylic acid, rather than to that of isopropyl salicylate, its primary metabolite, which does not induce enhanced biodegradation of the parent compound (3). Although the enhanced biodegradation of carbofuran has been well documented, the mechanisms involved are not completly understood. It has been suggested that the secondary metabolite, methylamine, is the possible substrate involved in the enhanced biodegradation process (3). Feusulfothion, an organophosphorus insecticide, is hydrolyzed by <u>Klebsiella pneumoniae</u> to 4-methylsulfinyl phenol, which is reduced to a secondary metabolite, 4-methylthiophenol. Because of the inhibitory effect on <u>K. pneumoniae</u>, this secondary metabolite is not biodegradable (6). Some secondary metabolites, including aldicarb sulfone, have been detected in surface-water and groundwater samples (7).

Interactions

Intensive agriculture requires the use of several agrochemicals (pesticides, fertilizers, growth hormones). In crop protection, herbicides, insecticides, and fungicides are applied to control a wide range of pests. The degradation of these chemicals often results in the formation of many breakdown products. The presence of several degradation products from a number of agrochemicals could result in interactions of these products either among themselves or with the parent compounds. The interactions could result in additive, synergistic, antagonistic, or potentiation effects. The implications of these interactions in crop protection and environmental contamination could have an important bearing on the use of the agrochemicals.

Regulatory Policies

Regulatory policies on transformation products are less than adequate, primarily because of the lack of scientific data supporting the development of new policies. The U.S. Environmental Protection Agency's requirements for the registration and reregistration of pesticides include identification of degradation products formed at levels > 10% of the applied pesticide. A mere identification is not sufficient, and to understand the significance of these products, more environmental fate and toxicological studies are required. In Canada, plans are in the offing to include in the Pest Control Product Act required toxicity tests for degradation products formed at concentrations exceeding 10% of the applied pesticide in laboratory or field studies (8).

An area of major concern with reference to regulatory policies and degradation products is groundwater contamination (9). In recent years, several degradation products have been detected in groundwater (10). The cumulative concentration of the degradation products may be higher than that of the parent compound (11). The environmental behavior and toxicological effects of some degradation products may differ from those of their parent compounds. The current enforcement standards, however, are only for the parent compounds. Because of the lack of toxiciology data, the toxicological effects of atrazine degradation products are assumed to be similar to those of atrazine itself (9). Atrazine is only the first instance, and in the future the same concerns may arise for other pesticides. A greater understanding of the environmental chemistry and environmental toxicology of pesticide degradation products will contribute to the informed regulatory policies of the future. The cost involved in conducting these studies will also be an important factor in deciding the extent to which transformation products will be investigated.

Future Research Needs

Research on the bioactive products of propesticides, on the role of degradation products in influencing the fate of parent compounds, and on phytotoxic and pesticidal potentials will promote understanding of the role of degradation products in crop

protection. From the environmental perspective, a vast majority of
the pesticides are transformed to products of less toxicological
concern. Toxicological studies (including carcinogenicity and
mutagenicity potentials) and environmental fate studies (movement
and sorption characteristics, persistence in the surface soil and
subsoils, and identification and bioavailability of bound residues)
need to be conducted for transformation products of potential
environmental and/or public health concern. The influence of the
interactive effects of degradation products with other chemicals
present in the environmental matrices also needs to be elucidated.
The scientific information generated will be useful in developing
health advisory levels and regulatory policies for transformation
products. If this information is generated before the introduction
of new pesticides, pesticides possessing transformation products of
potential environmental significance could be identified in the
early stages of development.

Acknowledgments

The authors thank the United States Department of Agriculture's
Management System Evaluation Area Program and North Central Region
Pesticide Impact Assessment Program, and Leopold Center for
Sustainable Agriculture for funding their research on pesticide
degradation products. Journal Paper No. J-14311 of the Iowa
Agriculture and Home Economics Experiment Station, Ames, Iowa.
Project No. 2306.

Literature Cited

1. Baarschers, W.H.; Bharath, A.I.; Hazenberg, M.; Todd, J.E. *Can. J. Bot.* **1980**, *58*, 426-431.
2. Smelt, J.H.; Dekker, A.; Leistra, M.; Houx, N.W.H. *Pestic. Sci.* **1983**, *14*, 173-181.
3. Somasundaram, L.; Coats, J.R.; Racke, K.D. *J. Environ. Sci. Health* **1989**, *B24*, 457-478.
4. Hiraoka, Y.; Tanaka, J.; Okuda, H. *Bull. Environ. Contam. Toxicol.* **1990**, 44, 210-215.
5. Harthoorn, P.A.; Gillam, M.G.; Wright, A.N. In *Progress in Pesticide Biochemistry and Toxicology*, Volume 4, Hutcson, D.H.; Roberts, T.R.; Eds.; John and Wiley and Sons: New York, 1985, pp 262-355.
6. Mac Rae, I.C.; Cameron, A.J. *Appl. Environ. Microbiol.* **1985**, *49*, 236-237.
7. Cohen, S.Z.; Creeger, S.M.; Carsel, R.F.; Enfield, C.G. In *Treatment and Disposal of Pesticide Wastes*; Krueger, R.F.; Seiber, J.N.; Eds.; Am. Chem. Soc.: Washington, D.C., 1984, pp 287-325.
8. Day, K. **1991**, chapter in this volume.
9. Deluca, D.B.; M.S. Thesis in Land Resources, Institute for Environmental Studies, University of Wisconsin, Madison, Wisconsin, 1990.
10. Somasundaram, L.; Coats, J.R. chapter 1 in this volume.
11. Muir, D.C.; Baker, B.E. *J. Agric. Food Chem.* **1976**, 24, 122-125.

RECEIVED December 13, 1990

INDEXES

Author Index

Affiliation Index

Subject Index

Hydroxychlor, fungitoxicity, 213
Hydroxyl radicals, oxidative
    transformation of organic
    compounds, 276
2-Hydroxypyridine, degradation in soil, 100

I

Inducers, degradation products,
    162,167t,168
Inorganic redox agents, effect on
    pesticide degradation, 11–22
Insecticidal activity of insecticide
    degradation products
    cholinesterase inhibition and toxicity
        of organophosphorus insecticides,
        173,174t,175
    discovery, 173
    products in plants, 175,176t,177–178
    products in soil, 178,179–180t,181
Insecticide(s)
    pesticidal activity on fungi, 183–184
    toxicity of transformation products to
        aquatic biota, 218–230
Insecticide degradation products,
    insecticidal activity, 173–181
Intensive agriculture
    importance in providing food to global
        population, 2
    role of pesticides, 2
Isofenphos, secondary degradation product
    formation, 286
Isomerizations, rearrangement reaction,
    21–22

K

Kitazin P
    oxidation, 66
    structure, 65f

L

Laterite soil, effect on degradation
    product formation, 44,45f,46
Lead arsenate sprays, phytotoxicity to
    apples, 173

Leptophos, production of
    desbromoleptophos, 26–27
Light, effect on pesticide degradation, 11
Lindane, environmental concentration,
    278,280t
Lipophilicity, importance as molecular
    descriptor, 155
Liver, bound residues of febantel and
    olaquindox, 246,247t,248
Livestock, toxicological significance of
    bound residues, 242–252
Long-term pesticide usage, potential
    hazards, 194

M

Malaoxon, predicted toxicity, 156,158f
Malathion
    degradation product formation, 71–72
    groundwater contamination, 71
    oxidation, 64f,66
    predicted metabolites, 151,154f
    predicted toxicity, 156,157f
    structure, 65f
    toxicity of transformation products to
        aquatic biota, 223
Mercaptans
    oxidation, 64f,66
    structure, 65f
METABOLEXPERT
    function, 150
    use in metabolism prediction, 150–155
Metabolic activities of soil
    microorganisms
    examples, 209–210
    nitrogen transformations, 210–211
    respiration, 210
    substrate utilization, 211,212t
    toxicity, 211,213
Metabolic pathways, classifications, 149
Metabolism of organophosphorus
    insecticides
    active detoxification, 38
    aliesterases, effect, 37–38
    desulfuration reaction, effect, 35,37
    glutathione transferases, effect, 39
    4-nitrophenol production, effect, 39
    phosphorothionate parathion, 35,36f
    prior exposure to chemicals, effect, 39

Production: Kurt Schaub
Indexing: Deborah H. Steiner
Acquisition: A. Maureen R. Rouhi

Books printed and bound by Maple Press, York, PA

Paper meets minimum requirements of American National Standard for Information Sciences—Permanence of Paper for Printed Library Materials, ANSI Z39.48–1984 ∞